ENCYCLOPEDIA OF STATISTICAL SCIENCES

SUPPLEMENT VOLUME

ENCYCLOPEDIA OF STATISTICAL SCIENCES

SUPPLEMENT VOLUME

A WILEY-INTERSCIENCE PUBLICATION

John Wiley & Sons

NEW YORK · **CHICHESTER** · **BRISBANE** · **TORONTO** · **SINGAPORE**

Library of Congress Cataloging-in-Publication Data:

Encyclopedia of statistical sciences.

   "A Wiley-Interscience publication."
   Includes supplement volume.
   Includes bibliographies.
   Contents: v. 1. A to Circular probable error—
v. 3. Faà di Bruno's formula to Hypothesis testing—
[etc.]—v. 7. Plackett family of distributions to
Regression, wrong.
   1. Mathematical statistics—Dictionaries.
2. Statistics—Dictionaries. I. Kotz, Samuel.
II. Johnson, Norman Lloyd. III. Read, Campbell B.

QA276.14.E5 1982   519.5′03′21   81-10353
ISBN 0-471-81274-9 (Supplement vol.)

Printed in the United States of America

10 9 8 7 6 5 4 3 2 1

In memory of Boris M. Kotz (1903–1988)

CONTRIBUTORS

A. C. Atkinson, *Imperial College, London, England.* Optimum Design of Experiments

R. A. Bailey, *Rothamsted Experimental Station, Harpenden, Herts, England.* Cross-validation

D. L. Banks, *Carnegie-Mellon University, Pittsburgh, Pennsylvania.* Bootstrapping-II

R. E. Barlow, *University of California, Berkeley, California.* Influence Diagrams

Th. Bezembinder, *Katholieke Universiteit, Nijmegen, The Netherlands.* Ostrogorski Paradox

R. J. Bhansali, *University of Liverpool, Liverpool, England.* Wiener-Kolmogorov Prediction Theory

J. V. Bilenas, *Long Island University, Greenvale, New York.* Regression, Iterative

P. J. Bjerve, *Oslo, Norway.* Kiaer, Anders Nicolai

W. Böge, *Universität Heidelberg, Heidelberg, Federal Republic of Germany.* Learn-Merge Invariance

J. Burbea, *University of Pittsburgh, Pittsburgh, Pennsylvania.* Rao Distance

R. W. Butler, *University of Michigan, Ann Arbor, Michigan.* Retrodiction

E. Carlstein, *University of North Carolina, Chapel Hill, North Carolina.* Central Statistic

R. J. Carroll, *Texas A & M University, College Station, Texas.* Redescending M-Estimators

P. S. Cleary, *Institute of Statisticians, London, England.* Institute of Statisticians, The

M. L. Cohen, *University of Maryland, College Park, Maryland.* Synthetic Estimation

R. K. Conway, *Economic Research Service, U.S. Department of Agriculture, Washington, D.C.* MARL Estimator

N. A. C. Cressie, *Iowa State University, Ames, Iowa.* Cressie-Read Statistic

M. H. DeGroot, *Carnegie-Mellon University, Pittsburgh, Pennsylvania.* Refinement

A. S. Dennis, *Bureau of Reclamation, U.S. Department of Interior, Boulder, Colorado.* Weather Modification-II

A. F. Desmond, *University of Waterloo, Waterloo, Ontario, Canada.* Estimating Equations, Theory of

N. R. Draper, *University of Wisconsin, Madison, Wisconsin.* Rotatability Index

M. L. Eaton, *University of Minnesota, Minneapolis, Minnesota.* FKG Inequality

E. S. Edgington, *University of Calgary, Calgary, Alberta, Canada.* Stochastically Closed Reference Sets

R. C. Elandt-Johnson, *University of North Carolina, Chapel Hill, North Carolina.* Rates

R. E. Fay, *U.S. Bureau of the Census, Washington, D.C.* Rao and Scott (Type) Tests

V. V. Fedorov, *Institute for System Studies, Moscow, USSR.* Optimum Design of Experiments

E. D. Feigelson, *Pennsylvania State University, University Park, Pennsylvania.* Astronomy, Statistics in

S. E. Fienberg, *Carnegie-Mellon University, Pittsburgh, Pennsylvania.* Undercount in the U.S. Decennial Census

N. I. Fisher, *Commonwealth Scientific and Industrial Organization, North Ryde, New South Wales, Australia.* Girdle (Equatorial) Data/Distributions

J. A. Flueck, *Environmental Sciences Group, National Oceanic and Atmospheric Administration, Boulder, Colorado.* Weather Modification-II

K. R. Gabriel, *University of Rochester, Rochester, New York.* Weather Modification-II

J. Gani, *University of California, Santa Barbara, California.* Literature and Statistics-II

W. Gibson, *Queen's College, City University, New York.* Regression, Iterative

I. Guttman, *University of Toronto, Toronto, Canada.* Rotatability Index

J. Haigh, *University of Sussex, Falmer, Brighton, Sussex, England.* Daniell-Kolmogorov Theorem, The

P. Hall, *Australian National University, Canberra, ACT, Australia.* Spitzer-Rosén Theorem

E. F. Harding, *University of Cambridge, Cambridge, England.* Statistical Modelling

S. A. Harding, *Rothamsted Experimental Station, Harpenden, Herts, England.* Cross-Validation

L. V. Hedges, *University of Chicago, Chicago, Illinois.* Meta-Analysis

W. G. S. Hines, *University of Guelph, Guelph, Ontario, Canada.* Eberhardt Statistic; kth Nearest Point—kth Nearest Neighbor Sampling; Nearest Point-Nearest Neighbor Sampling

N. T. Jazairi, *York University, North York, Ontario, Canada.* Stock Market Price Indexes

G. G. Judge, *University of Illinois, Champaign, Illinois.* Stochastic Regression Models

N. Keiding, *Statistical Research Unit, Copenhagen, Denmark.* Woodroofe's Inversion Formula

A. W. Kemp, *University of Bradford, Bradford, England.* Even-Point Estimation

C. D. Kemp, *University of Bradford, Bradford, England.* Even-Point Estimation

A. C. Kimber, *University of Surrey, Guildford, Surrey, England.* Eulerian Numbers

J. B. Knight, *Institute of Economics and Statistics, Oxford, England.* Oxford Bulletin of Economics and Statistics

H. S. Konijn, *Tel Aviv University, Tel Aviv, Israel.* DeWitt, Johan

W. H. Kruskal, *University of Chicago, Chicago, Illinois.* Kiaer, Anders Nicolai

E. L. Lehmann, *University of California, Berkeley, California.* Group Families

D. V. Lindley, *Minehead, Somerset, England.* De Finetti, Bruno

J. Matatko, *York University, North York, Ontario, Canada.* Stock Market Price Indexes

I. Mellin, *University of Helsinki, Helsinki, Finland.* Linear Model Selection, Criteria and Tests

P. W. Mielke, Jr., *Colorado State University, Fort Collins, Colorado.* Moment Approximation Procedures; Weather Modification-II

R. C. Mittelhammer, *Washington State University, Pullman, Washington.* MARL Estimator

J. Möcks, *Universität Heidelberg, Heidelberg, Federal Republic of Germany.* Learn-Merge Invariance

E. Ott, *University of Maryland, College Park, Maryland.* Chaos

A. N. Pettitt, *University of Queensland, St. Lucia, Queensland, Australia.* Change-Point Problem

B. M. Pötscher, *Technische Universität Wien, Vienna, Austria.* Generic Uniform Laws of Large Numbers

R. F. Potthoff, *Burlington Industries, Burlington, North Carolina.* Run Lengths, Tests of

A. Prat, *Universitat Politécnica de Catalunya, Barcelona, Spain.* Qüestiió

I. R. Prucha, *University of Maryland, College Park, Maryland.* Generic Uniform Laws of Large Numbers

D. Quade, *University of North Carolina, Chapel Hill, North Carolina.* Partial Correlation

T. R. C. Read, *Iowa State University, Ames, Iowa.* Cressie-Read Statistic

J. **Rissanen,** *International Business Machines Corporation, San José, California.* Stochastic Complexity

P. K. **Sen,** *University of North Carolina, Chapel Hill, North Carolina.* Affluence and Poverty Indexes; Antiranks

G. L. **Smith,** *Rothamsted Experimental Station, Harpenden, Herts, England.* Cross-Validation

R. L. **Smith,** *University of Surrey, Guildford, Surrey, England.* Gringorten's Formula

W. L. **Smith,** *University of North Carolina, Chapel Hill, North Carolina.* Logistic Processes

M. **Stone,** *University College, London, England.* Approgression

T. **Teräsvirta,** *Research Institute of the Finnish Economy, Helsinki, Finland.* Linear Model Selection, Criteria and Tests

M. L. **Tiku,** *McMaster University, Hamilton, Ontario, Canada.* Modified Maximum Likelihood Estimation

D. M. **Titterington,** *University of Glasgow, Glasgow, Scotland.* Compositional Data; Logistic-normal Distribution; Self-consistent Estimators

R. L. **Trader,** *University of Maryland, College Park, Maryland.* Bayes, Thomas; Edgeworth, Francis Ysidro

G. **Wahba,** *University of Wisconsin, Madison, Wisconsin.* Spline Functions

T. **Wansbeek,** *University of Groningen, Groningen, The Netherlands.* Permutation Matrix-II

G. H. **Weiss,** *National Institutes of Health, Bethesda, Maryland.* Simulated Annealing

L. **Weiss,** *Cornell University, Ithaca, New York.* Efron-Morris Estimators

D. G. **Wellington,** *Office of Policy, Planning and Evaluation, Environmental Protection Agency, Washington, D.C.* Rai and Van Ryzin Dose-Response Model

INTRODUCTORY NOTE

This volume constitutes a completion of, rather than a supplement to, the nine volumes of the *Encyclopedia of Statistical Sciences*. There are, indeed, a few entries representing repair of serious omissions, for which the Editors must accept responsibility, but the bulk of the material is composed of (a) entries received too late to be included at their proper places in the alphabet and (b) additional cross-references intended to improve the usefulness of the *Encyclopedia*.

ERRATA (VOLUMES 1–9)

Volume	Page	Column	Line	Correction
1	v	2		Delete D. R. Cox item
	129	2	2f	Replace ARUMA with ARIMA
	174	2	2 of A9	$1 < j$ should be $i < j$
	182	1	1f of entry BALDUCCI HYPOTHESIS	Replace "increases" with "decreases"
	282	1	Above references	Add: The following earlier uses of the term "black box" have come to our attention: McArthur, D. S. and Heigh, J. J. (1957) *Trans. Amer. Soc. Qual. Control*, 1–18; Good, I. J. (1954) *J. Inst. Actu.*, **80**, 19–20 (Contribution to discussion of a paper by R. D. Clarke); Good, I. J. "Subjective probability as the measure of a nonmeasurable set," *Logic, Methodology and Philosophy of Science: Proc. 1960 Congress*, Stanford University Press, Stanford, CA (pp. 319–329) (Published 1962) Reprinted in *Good Thinking*, 73–82 (1983)
	305	2	4f	Biship should be Bishop
	306	1	2f	$(y - 1)/\lambda$ should be $(y^\lambda - 1)/\lambda$
			Fig. 2 legend	Delete Oiunts; delete 1 after extreme value
2	100	2	6–8	Delete entry COMPUTER ASSISTED TELEPHONE SURVEYS
	132	2	1	Replace second greenhouse with first greenhouse
			2–4	These lines are legend to Table 9 and not text
	189	2	1 of equation (10)	First line of right hand side should be $u_p + \frac{1}{6}(u_p^2 - 1)k_3$
	229	2	1 of CUMULANTS entry	$(it)j$ should be $(it)^j$
	276	1	Ref. [2]	483 should be 493
			Ref. [3]	*Ann.* should be *Appl.*
3	215	2	2f–1f	Should read $\dfrac{1}{B(m_1, m_2)} x^{m_1 - 1}(1 - x)^{m_2 - 1}$ $0 \leqslant x \leqslant 1, m_1 > 0, m_2 > 0$
	396	2	1	Replace p/q with q/p
			2	Replace p/q^2 with q/p^2

Volume	Page	Column	Line	Correction
	604	2	First display	Insert $\frac{1}{2}$ at beginning of right hand side
			5f	$r_p(\theta, \theta') = 1$ should be $r_p(\theta, \theta') = 0$
4	425	2	8f	b_2 should be β_2; m_2; m_4 should be μ_2; μ_4 (twice)
			6f	"arithmetic mean" should be "expected value"
	426	1	2	b_2 should be β_2
			6	g_2 should be b_2
			7	Replace by "m_4/m_2^2 where m_2, m_4 are the second and fourth moments, respectively, about the arithmetic mean"
			12	g_2 should be b_2
5	103	2	2f	Replace (4) with (8)
	104	2	17f	Replace "model (9) which includes (10)" with "model (13) which includes (14)"
	105	1	5	Replace "model (10)" with "model (14)"
			6	Replace "to include (11)" with "to include (15)"
			8	Replace "model (10)" with "model (14)"
			12	Replace (11) with (15)
7	v	1	6f–5f	Affiliation for S. Blumenthal should be *Ohio State University, Columbus, Ohio*
	57	1	8f	Replace $\varnothing$ with ϕ
8	ix	1	16–18	Delete Potthoff item
	306	1	5	Delete $=$ so that equation reads $y_i = \{\exp(x_i - \theta) + 1\}^{-1}$
	329	2	Entry SECULAR TREND	Change to "Now often replaced by trend*. Secular usually implies long-term (relative to period of observation). (TIME SERIES; TREND)"
	378	2	16	Change y_i to Y_i
	381	1	8f	Change "insufficient" to "is sufficient"
	382	1	8	Insert a colon after "states"
		1	10	Change $\theta = \theta$ to $\theta = \theta_1$
	383	1	21	Change $\epsilon\theta$ to $\epsilon\Omega$
		1	11f	Change u_n to U_n
	384	1	21f	Change h_0 to H_0
		1	22f	Change h_1 to H_1
		2	16	Change h_0 to H_0
	385	1	4	Delete a
	386	2	Ref. [24]	Change 53 to 553
	739	1	2	Delete entry STATISTICS, OPTIMIZATION IN
	747	1	7	Add OPTIMIZATION, STATISTICS IN
9	53		Running Head	SUBEXPOTENTIAL should be SUBEXPONENTIAL

Volume	Page	Column	Line	Correction
	107	1	eq. 7	The discussion of the complexity of eq. 7 was only preliminary. A much more thorough discussion of the complexity of apparently numerological formulae and assertions appears in items C333, C334, and C335 of *J. Statist. Comput. & Simul.* (1989). A relationship between surprise indexes and *P*-values is also pointed out in C332.
	236	2	1	Replace Gauss with Camp–Meidell
	280	1	2 below eq. 2	Insert "the" after "of"
	289	1	Ref. [5]	Delete *Ser.*
	323	2	Ref. [4]	Delete *Ser.*
	347	1	5	t_{tg} should be T_{tg}
			12	t_{tg} should be T_{tg}
	348	1	1	First sentence should read "An additional application for the two-sample case concerns censoring."
	364	1	1 of text	Insert "distribution-free" before "test"
		2	3	Insert "confidence interval" after "graphical"
		1	1f	"$\max(n_1, n_2) \leqslant 20$" should be "$|n_1 - n_2| \leqslant 20$"
	489	1	10	$\otimes$ should be $\odot$
		2	8f	Replace $\sum_{i=1}^{m} a_i = 0$ with $\sum_{i=1}^{r} a_i = 0$
	490	1	10f	Replace $\sum_{n(\mathbf{h})}$ with $\sum_{N(\mathbf{h})}$
			7f	Delete $\pm$
		2	2	Replace $\tilde{\gamma}$ with $\bar{\gamma}$
	493	1	Ref. [13]	(1984) should be (1884)

—

ENCYCLOPEDIA OF STATISTICAL SCIENCES

SUPPLEMENT VOLUME

A

AFFLUENCE AND POVERTY INDEXES

Affluence of a society or community is generally quantified by the proportion of its rich (or affluent) people and the concentration of their wealth [or (real) income distribution*]. Poverty is usually defined as the extent to which individuals in a society or community fall below a minimal acceptable standard of living, so that it can be quantified in terms of the proportion of the poor people and their income inequity. It seems that for a proper description of affluence and poverty, there is a need to consider the following:

(a) Assessment of real income of individuals or families in terms of a single quantitative criterion which can be incorporated in the definition of the indexes to follow.

(b) Arbitration of a physically meaningful and interpretable *affluence line* (i.e., minimal level of wealth or real income) leading to an unambiguous identification of the rich or affluent people.

(c) Arbitration of a similar *poverty line* relating to an appropriate minimal acceptable standard of living and leading to a proper definition of the poor.

(d) Some measure of concentration or inequality of wealth (or real income) among the affluent.

(e) Similar measures of income inequality among the poor.

Granted proper guidelines for (a), an affluence index based solely on (b) or (d) may not depict the real picture on affluence, and a more meaningful index incorporates both (b) and (d) in this formulation. Likewise, granted (a), a poverty index* based on both (c) and (e) provides a more realistic picture of poverty than a poverty index based on either one of them. In this context, we may remark that for (a), (b), and (c), mostly, socioeconomic and other monetary utility functions dominate the formulation, whereas for both (d) and (e), there is ample room for statistical considerations. The arbitration of affluence or poverty lines may be quite different in a socialistic or capitalistic society; even for the same society, it may vary considerably over time. However, once these guidelines are agreed upon, the basic statistical issues are to land on suitable measures of income inequality (among the poor or the rich) and to incorporate them fruitfully in the formulation of the appropriate indexes. In this context, we may refer to Kakwani [5] and Sen [6] for some useful discussions.

INCOME INEQUALITY MEASURES is a good source of references on the various measures of income inequalities and their rationality. However, in the specific context of affluence or poverty, there is a greater need for a more detailed statistical treatment, and this is intended here.

With reference to the basic formulation of real income [see (a)], we denote by $F(x)$, $x \geq 0$, the income distribution (for a society or community). Also, with reference to (b) and (c), we conceive of a poverty line ω and an affluence line θ (such that $0 < \omega < \theta < \infty$), so that $\alpha = F(\omega)$ represents the proportion of the poor, and $\rho = 1 - F(\theta) = \bar{F}(\theta)$ represents the proportion of the rich.

1

Note that the income distribution of the poor (the truncated income distribution at ω) is given by

$$F_\alpha(x) = \alpha^{-1}F(x), \quad \text{for} \quad x \in [0, \omega]$$
$$= 1, \qquad \text{for} \quad x \geq \omega. \quad (1)$$

Therefore, the *average income of the poor* is given by

$$\mu_\alpha = \int_0^\omega y \, dF_\alpha(y) = \alpha^{-1}\left(\int_0^\omega y \, dF(y)\right). \quad (2)$$

The *income gap ratio of the poor* is then equal to

$$\beta = 1 - \omega^{-1}\mu_\alpha = 1 - (\omega\alpha)^{-1}\int_0^\omega y \, dF(y). \quad (3)$$

For an arbitrary (income) distribution F (defined on $[0, \infty)$), the *Gini coefficient* (of income inequality) is given by

$$G(F) = (E|Y_1 - Y_2|)/(E(Y_1) + E(Y_2)), \quad (4)$$

where Y_1 and Y_2 are two independent random variables, each having the distribution F. $G(F)$ is closely related to the Lorenz curve* for the income distribution F. For the income distribution F_α in (1), we denote the corresponding Gini coefficient by $G_\alpha = G(F_\alpha)$. Then, typically, an index of poverty (π) is based on the triplet (α, β, G_α). If we let $\pi = \alpha$, then the index fails to take into account the income gap ratio as well as the income inequality G_α in the formulation of the poverty picture, and hence may not be very realistic. On the other hand, an index solely based on G_α (or the income gap ratio) may not properly take into account the relative proportion of the poor, and hence may cease to be very sensitive. Based on a set of axioms, Sen [7] suggested the simple poverty index

$$\pi_A = \alpha\beta = \text{income gap ratio adjusted}$$

$$\text{proportion of the poor.} \quad (5)$$

Since this index does not explicitly take into account the Gini coefficient G_α, a refined index of poverty has also been considered by

Sen [7]:

$$\pi_S = \alpha\{\beta + (1 - \beta)G_\alpha\}. \quad (6)$$

Takayama [10] proposed an alternative poverty index based on a somewhat different set of axioms. His index is given by

$$\pi_T = G_\omega^c, \quad (7)$$

the Gini coefficient for a special censored distribution F_ω^c, defined as

$$F_\omega^c(x) = F(x), \quad \text{for } x < \omega,$$
$$F_\omega^c(x) = 1, \qquad \text{for } x \geq \omega. \quad (8)$$

It may be noted that G_ω^c depends on α and G_α as well as β. In fact (Sen [8]),

$$G_\omega^c = \alpha G_\alpha + (1 - \alpha)(1 - \alpha\beta)^{-1}\{\beta - G_\alpha\},$$
$$G_\alpha \leq \beta, \quad (9)$$

$$\alpha G_\alpha \leq G_\omega^c \leq \alpha\beta,$$

$$\text{for all censored distributions.} \quad (10)$$

This also suggests that $\pi_T^* = \alpha G_\alpha$ is a plausible poverty index, and we have the following ordering of the poverty indexes:

$$0 \leq \pi_T^* \leq \pi_T \leq \pi_A \leq \pi_S \leq \alpha\beta(2 - \beta) \leq \alpha, \quad (11)$$

for all $0 \leq \alpha$, $\beta \leq 1$ (i.e., all income distributions). Although each of π_S and π_T is justified on the grounds of certain plausible axioms (see refs. 7 and 10), from a statistical point of view, generally, for smaller values of α and β, π_T is somewhat more conservative, whereas π_S is more anticonservative than they should be ideally. This differential picture is mainly due to the two forms of the Gini coefficients G_α and G_ω^c which (though related to the common income distribution of the poor) behave rather differently with the variation of α, β and the inequality of incomes among the poor. For this reason, Sen [8] initiated a more intensive study of the behavior of the Gini coefficient under various patterns of income inequality, and this led to consideration of a more robust version of π_S, namely,

$$\pi^* = \alpha(\beta^{1 - G_\alpha}). \quad (12)$$

There is a simple interpretation for π^* and

its affinity to π_S. In (6), we may rewrite

$$\pi_S = \alpha G_\alpha + \pi_A(1 - G_\alpha)$$

= weighted arithmetic mean

of the two bounds α and π_A

with the relative weights G_α and $1 - G_\alpha$.

(13)

If we replace the weighted arithmetic mean* in (13) by the weighted geometric mean* with the same relative weights, then the resulting measure is π^*, given by (12). Since, by (11), α is a crude upper bound for the poverty indexes under consideration and the geometric mean is smaller than or equal to the arithmetic mean, it is clear that π^* is less sensitive than π_S; this picture has been studied thoroughly by Sen [8]. Incorporating α^* in the spectrum in (11), we obtain

$$0 \leqslant \pi_T^* \leqslant \pi_T \leqslant \pi_A \leqslant \pi^* \leqslant \pi_S$$

$$\leqslant \alpha\beta(2 - \beta) \leqslant \alpha, \quad \text{for all } (\alpha, \beta). \quad (14)$$

Based on an alternative normative approach, Blackorby and Donaldson [1] proposed a poverty index of the form

$$\pi_{BD} = f(\alpha, \beta^*),$$

$$f(x, y) \text{ defined on } [0, 1]^2, \quad (15)$$

where β^*, the *representative income gap ratio* is a distribution-adjusted parameter, and suitable side conditions are imposed on $f(\cdot)$. All the indices in (14) are special cases of (15) where $f(x, y) = xy$; $\beta^* = \beta$ for π_A, $\beta + (1 - \beta)G_\alpha$ for π_S and β^{1-G_α} for π^*. This suggests that one may generally consider a poverty index of the form (15) with $\beta^* = \beta^{1-G_\alpha}$, which would then satisfy both the normative and robustness aspects.

Recall that for a set affluence line θ, $\rho = 1 - F(\theta) = \bar{F}(\theta)$ represents the proportion of affluent people. Thus, from the statistical point of view, for a society or community an affluence picture has to be drawn from the upper tail of the income distribution F, where the cutoff point θ is to be determined by various socioeconomic utility functions*. Generally, there are difficulties in determining this cutoff point precisely as well as in measuring accurately the real income of rich people, particularly the excessively rich ones. Such inadequate or erroneous mensuration may generally lead to an imprecise picture relating to the conventional mean income, income gap ratio, the Gini coefficient and other measures of the income distribution of the affluent. This drawback may be alleviated by the use of some alternative measures which are more robust. The harmonic mean*, harmonic income gap ratio, and harmonic Gini coefficient of the income distribution of the rich play important roles in this context.

For the affluent, the income distribution (F_ρ^*) is the truncated version of F with left truncation at θ, i.e.,

$$F_\rho^*(y) = [F(y) - F(\theta)]/\rho, \quad \text{for } y \geqslant \theta,$$

$$F_\rho^*(y) = 0, \quad \text{for } y < \theta.$$

(16)

The average income of the affluent is equal to

$$\mu_\rho^* = \int_0^\infty y \, dF_\rho^*(y) = \left\{ \int_\theta^\infty y \, dF(y) \right\} \Big/ \rho.$$

(17)

By definition, μ_ρ^* is greater than or equal to θ, so that unlike in (3), β^*, the income gap ratio of the rich, has to be defined in a different manner. One possibility is to set

$$\beta^* = \beta_1^* = 1 - \theta/\mu_\rho^*$$

$$= 1 - \theta\rho \left(\int_\theta^\infty y \, dF(y) \right)^{-1}. \quad (18)$$

Similarly, parallel to (4), the Gini coefficient of the income distribution F_ρ^* may be defined as

$$G(F_\rho^*) = G_\rho^* = \frac{E[|Y_1^* - Y_2^*|]}{E[Y_1^* + Y_2^*]}, \quad (19)$$

where Y_1^* and Y_2^* are two independent random variables, each having the distribution F_ρ^*. A left-censored version of F may also be defined as

$$F_\rho^{*c}(y) = 0, \quad \text{for } y < \theta,$$

$$F_\rho^{*c}(y) = F(y), \quad \text{for } y \geqslant \theta,$$

(20)

and the Gini coefficient G_ρ^{*c} for this distribution may then be defined in the usual manner. Once these coefficients are defined, an affluence index may then be defined as in (5), (6), (7), or (12). Note that often wealth in a form other than income needs to be transferred into an income form for the formulation of real income, and in addition, there are other difficulties for reliable mensuration of high incomes. Thus there is generally sufficient scope for gross errors or outliers* in the formulation of F_ρ^*, and as a result, the measures in (17), (18), (19), and (20) are all vulnerable to gross errors or outliers. A better picture can be drawn from the economic utility theory point of view. The marginal value of income (money) may not generally remain constant; it is usually a decreasing function of personal income. Hence it may be quite natural to introduce a utility function $u(t_1, t_2)$ which is nonnegative and nonincreasing in each of t_1, t_2, and then to consider a utility oriented Gini coefficient as

$$G_\rho^{u*} = \frac{E[u(Y_1^*, Y_2^*)|Y_1^* - Y_2^*|]}{E[u(Y_1^*, Y_2^*)\{Y_1^* + Y_2^*\}]}, \quad (21)$$

where Y_1^*, Y_2^* are defined as in (19). Note that a flat utility function leads to the usual coefficient in (19), whereas for a harmonic utility function [i.e., $u(t_1, t_2) = (t_1 t_2)^{-1}$], (21) reduces to the harmonic Gini coefficient

$$G_\rho^{H*} = \frac{E[|(Y_1^* - Y_2^*)/Y_1^* Y_2^*|]}{E[(Y_1^* + Y_2^*)/Y_1^* Y_2^*]}. \quad (22)$$

Side by side, we may introduce the harmonic mean income of the rich as

$$\mu_\rho^{H*} = \left\{\int_\theta^\infty y^{-1} dF_\rho^*(y)\right\}^{-1}$$

$$= \rho \bigg/ \left\{\int_\theta^\infty y^{-1} dF(y)\right\}. \quad (23)$$

As such, a second measure of the income gap ratio among the rich can be posed as

$$\beta_2^* = 1 - \theta/\mu_\rho^{H*}$$

$$= 1 - (\theta/\rho)\left\{\int_\theta^\infty y^{-1} dF(y)\right\}. \quad (24)$$

It is known [9] that $\beta_2^* \leqslant \beta_1^*$. The measures in (22), (23), and (24) are all more robust

than their counterparts in (17), (18), and (19). Keeping the analogy with the poverty indexes and the above robustness picture in mind, Sen [9] considered the following affluence indexes:

$$\kappa_A = \rho\beta_2^*$$

$$\kappa_S = \rho\{\beta_2^* + (1 - \beta_2^*)G_\rho^{H*}\},$$

and

$$\kappa^* = \rho\{(\beta_2^*)^{1 - G_\rho^{H*}}\}. \quad (25)$$

All these indexes may be characterized as in (15) with $f(x, y) = x \cdot y$, and for β^* one needs to take β_2^* (for κ_A), $\beta_2^* + (1 - \beta_2^*)G_\rho^{H*}$ (for κ_s), or $(\beta_2^*)^{1 - G_\rho^{H*}}$ (for κ^*). Finally, as an alternative to the harmonic coefficient in (22), one may consider the Gastwirth coefficient [3] defined by

$$G_\rho^{0*} = E[|Y_1^* - Y_2^*|/(Y_1^* + Y_2^*)]. \quad (26)$$

Unlike the Gini coefficients, the Gastwirth coefficient may not be directly obtainable from the associated Lorenz curve. However, it has one nice property in that the coefficient based on the actual incomes and the harmonic incomes are the same. This particularly reveals the robustness property of G_ρ^{0*}, not shared by (19). For the indexes in (25), one may as well use G_ρ^{0*} instead of G_ρ^{H*}, and the relationship of such measures has been studied in detail in Sen [9]. This study provides some intuitive reasons for prescribing κ^* as an appropriate index of affluence.

We conclude this article with the remark that interdistribution inequality measures have also been considered by various workers to study the relative degree of affluence of one population with respect to others; see INCOME INEQUALITY MEASURES, Dagum [2] and Gastwirth [3], among others. Comparing appropriate poverty (or affluence) indexes instead of the related income distributions may provide a more realistic picture with added emphasis on the specific directions of divergence of the population distributions.

References

[1]   Blackorby, C. and Donaldson, D. (1980). *Econometrica*, **48**, 1053–1060.

[2]   Dagum, C. (1980). *Econometrica*, **48**, 1791–1803.

[3] Gastwirth, J. L. (1975). *Proc. Int. Statist. Inst.*, Vienna **1**, 368–372.

[4] Gastwirth, J. L. (1975). *J. Econometrics*, **3**, 61–70.

[5] Kakwani, N. C. (1980). *Income Inequality and Poverty: Methods of Estimation and Policy Applications*. Oxford University Press, New York.

[6] Sen, A. K. (1973). *On Economic Inequality.* Oxford University Press, London.

[7] Sen, A. K. (1976). *Econometrica*, **44**, 219–232.

[8] Sen, P. K. (1986). *J. Amer. Statist. Ass.*, **81**, 1050–1057.

[9] Sen, P. K. (1988). *Math. Soc. Sci.*, **5**, 65–76.

[10] Takayama, N. (1979). *Econometrica*, **47**, 747–760.

(DIVISIA INDICES
ECONOMETRICS
GINI COEFFICIENT
INCOME DISTRIBUTION MODELS
INCOME INEQUALITY MEASURES
INDEX NUMBERS
LORENZ CURVE
TRUNCATION
UTILITY THEORY)

P. K. SEN

AITCHISON DISTRIBUTIONS

These form a class of multivariate distributions with density functions:

$$f_X(x|\alpha, \beta)$$

$$\propto \left[\prod_{i=1}^{p} x_i^{\alpha_i - 1} \right]$$

$$\times \exp\left[-\frac{1}{2} \sum_{i<j}^{p} \beta_{ij} (\log x_i - \log x_j)^2 \right],$$

$$0 < x_i, \qquad \sum_{i=1}^{p} x_i = 1,$$

$$\alpha_i \geqslant 0, \qquad \beta \text{ nonnegative definite.}$$

Although there are p variables $x_1, \ldots, x_p$, the distribution is confined to the $(p-1)$-dimensional simplex: $0 \leqslant x_i, \Sigma_{i=1}^{p} x_i = 1$.

If $\beta = 0$ we have a *Dirichlet distribution**; if $\alpha_i = 0$ for all i we have a *multivariate logistic-normal distribution*.

Methods of estimating the parameters (α and β) are described by Aitchison [1, pp. 310–313].

Reference

[1] Aitchison, J. (1986). *Statistical Analysis of Compositional Data*. Chapman and Hall, London and New York.

(COMPOSITIONAL DATA
DIRICHLET DISTRIBUTION
FREQUENCY SURFACES, SYSTEMS OF
LOGISTIC-NORMAL DISTRIBUTION)

ALEATORY VARIABLE

An obsolete term for random variable*.

ANTIRANKS

Nonparametric tests and estimates are generally based on certain statistics which depend on the sample observations $X_1, \ldots, X_n$ (real valued) only through their *ranks** $R_1, \ldots, R_n$, where

$$R_i = \text{number of indices } r$$
$$(1 \leqslant r \leqslant n): X_r \leqslant X_i, \qquad (1)$$

for $i = 1, \ldots, n$. If $X_{n:1} \leqslant \cdots \leqslant X_{n:n}$ stand for the sample order statistics*, then we have

$$X_i = X_{n:R_i}, \qquad i = 1, \ldots, n. \qquad (2)$$

Adjustment for ties can be made by dividing equally the total rank of the tied observations among themselves. Thus if $X_{n:k} < X_{n:k+1} = \cdots = X_{n:k+q} < X_{n:k+q+1}$, for some k, $0 \leqslant k \leqslant n-1$, and $q \geqslant 1$ (where $X_{n:0} = -\infty$ and $X_{n:n+1} = +\infty$), then for the q tied observations (with the common value $X_{n:k+1}$), we have the *midrank* $k + (q+1)/2$.

Let us now look at (2) from an opposite angle: For which index (S_k), is X_{S_k} equal to $X_{n:k}$? This leads us to define the *antiranks* $S_1, \ldots, S_n$ by

$$X_{n:i} = X_{S_i}, \quad \text{for } i = 1, \ldots, n. \qquad (3)$$

Note the inverse operations in (2) and (3) as depicted below:

$$X_1, \ldots, X_i, \ldots, X_{S_i}, \ldots, X_n$$

$$X_{n:1}, \ldots, X_{n:i}, \ldots, X_{n:R_i}, \ldots, X_{n:n}, \tag{4}$$

so that, we have

$$R_{S_i} = S_{R_i} = i, \quad \text{for } i = 1, \ldots, n, \tag{5}$$

and this justifies the terminology: Antiranks.

Under the null hypothesis that $X_1, \ldots, X_n$ are exchangeable random variables, $\mathbf{R} = (R_1, \ldots, R_n)$, the vector of ranks, takes on each permutation of $(1, \ldots, n)$ with the common probability $(n!)^{-1}$. By virtue of (5), we obtain that under the same null hypothesis, $\mathbf{S} = (S_1, \ldots, S_n)$, the vector of antiranks, has the same (discrete) uniform permutation distribution. In general, for the case of ties neglected, (5) can be used to obtain the distribution of $\mathbf{S}$ from that of $\mathbf{R}$ (or vice versa), although when the X_i are not exchangeable, this distribution may become quite cumbrous. Under the null hypothesis of exchangeability*, for suitable functions of $\mathbf{R}$ (i.e. rank- statistics), *permutational central limit theorems** provide asymptotic solutions, and by virtue of (5), these remain applicable to antirank statistics as well.

For mathematical manipulations, often $\mathbf{S}$ may have some advantage over $\mathbf{R}$. To illustrate this point, consider a typical *linear rank statistic** (T_n) of the form $\sum_{i=1}^n c_i a_n(R_i)$, where the c_i are given constants and $a_n(1), \ldots, a_n(n)$ are suitable scores. By (5), we have

$$T_n = \sum_{i=1}^n c_{S_i} a_n(i), \tag{6}$$

and this particular form is more amenable to censoring schemes (*see* PROGRESSIVE CENSORING). If we have a type II censoring (at the kth failure), then the censored version of T_n in (6) is given by

$$T_{nk} = \sum_{i=1}^k \left(c_{S_i} - \bar{c}_n \right) \left[a_n(i) - a_n^*(k) \right],$$

$$k \geq 0, \tag{7}$$

where

$$\bar{c}_n = n^{-1} \sum_{i=1}^n c_i,$$

$$a_n^*(k) = (n-k)^{-1} \sum_{j=k+1}^n a_{n(j)}, \quad k < n$$

and

$$a_n^*(n) = 0.$$

For the classical *Kolmogorov–Smirnov tests**, these antiranks may be used to express the statistics in neat forms and to study their distributions in simpler manners; we may refer to Hájek and Šidák [1] for a nice account of these.

In life-testing* problems and clinical trials*, for rank procedures in a time-sequential setup, the antiranks play a vital role; for some details, see Sen [2, 3].

References

[1] Hájek, J. and Šidák, Z. (1967). *Theory of Rank Tests*. Academic, New York.

[2] Sen, P. K. (1981). *Sequential Nonparametrics*. Wiley, New York.

[3] Sen, P. K. (1985). *Theory and Application of Sequential Nonparametrics*. SIAM, Philadelphia.

(EDF STATISTICS
KOLMOGOROV–SMIRNOV TESTS
LIMIT THEOREMS, CENTRAL
LINEAR RANK STATISTICS
ORDER STATISTICS
PERMUTATIONAL CENTRAL LIMIT
   THEOREMS
PROGRESSIVE CENSORING
RANKS
TIME-SEQUENTIAL TESTS)

P. K. SEN

APPROGRESSION

The use of regression functions (usually linear) to approximate the truth, for simplicity and predictive efficiency. Some *optimality* results are available.

The term seems to be due to H. Bunke in

Bunke, H. (1973). Approximation of regression functions. *Math. Operat. Statist.*, **4**, 314–325.

A forerunner of the concept is the idea of an *inadequate* regression model in

Box, G. E. P. and Draper, N. R. (1959). A basis for the selection of a response surface design. *J. Amer. Statist. Ass.*, **54**, 622–654.

The paper:

Bandemer, H. and Näther, W. (1978). On adequateness of regression setups. *Biom. J.*, **20**, 123–132

helps one to follow some of the treatment in Section 2.7 of

Bunke, H. and Bunke, O., eds. (1986). *Statistical Inference in Linear Models.* Wiley, New York,

which is exceedingly general and notationally opaque. The Bunke and Bunke account has about 12 relevant references. More recent work is to be found in

Zwanzig, S. (1980). The choice of appropriate models in non-linear regression. *Math. Operat. Statist. Ser. Statist.*, **11**, 23–47

and in

Bunke, H and Schmidt, W. H. (1980). Asymptotic results on non-linear approximation of regression functions and weighted least-squares. *Math. Operat. Statist. Ser. Statist.*, **11**, 3–22,

which pursue the concept into nonlinearity. A nice overview is now available in

Linhart, H. and Zucchini, W. (1986) *Model Selection*. Wiley, New York.

(LINEAR MODEL SELECTION, CRITERIA AND TESTS
REGRESSION (various entries)
RESPONSE SURFACES)

M. STONE

ARTIFICIAL INTELLIGENCE *See* STATISTICS AND ARTIFICIAL INTELLIGENCE

ASTRONOMY, STATISTICS IN

Perhaps more than other physical sciences, astronomy is frequently statistical in nature. The objects under study are inaccessible to direct manipulation in the laboratory. The astronomer is restricted to observing a few external characteristics of objects populating the Universe, and inferring from these data their properties and underlying physics. From the seventeenth through nineteenth centuries, European astronomers were engaged in the application of Newtonian theory to the motions of bodies in the solar system. This led to discussions of the statistical treatment of scientific data, and played a critical role in the development of statistical theory. The twentieth century has seen remarkable success in the applications of electromagnetism and quantum mechanics* to heavenly bodies, leading to a deep understanding of the nature and evolution of stars, and some progress in understanding galaxies and various interstellar and intergalactic gaseous media. Statistical theory has played a less important role in these advances of modern astrophysics. However, the last few years have seen some reemergence of interest in statistical methodology to deal with some challenging data analysis problems. Some examples of these contemporary issues are presented.

EARLY HISTORY

Celestial mechanics in the eighteenth century, in which Newton's law of gravity was found to explain even the subtlest motions of heavenly bodies, required the derivation of a few interesting quantities from numerous inaccurate observations. As described in detail by Stigler [40], this required advances in the understanding of statistical inference* and error distributions. Mayer, in his 1750 study of lunar librations, suggested a procedure of reconciling a system of 27 inconsistent linear equations in three unknowns by solving the equations in groups. Laplace*, in a 1787 analysis of the influence of Jupiter's gravity on Saturn's motion, suggested a more unified approach that led to Legendre's invention of the least-squares method in an 1805 study of cometary orbits. Shortly thereafter, in an 1809 monograph on the mathematics of planetary orbits, Gauss* first presented the normal (or Gaussian) distribution of errors in overdetermined systems of equations using a form of Bayes theorem*, though the actual derivation was flawed.

Many other individuals also contributed substantially to both astronomy and statistics [40, 14]. Galileo gave an early discussion of observational errors concerning the distance to the supernova of 1572. Halley, famous for his early contributions to celestial mechanics, laid important foundations to mathematical demography and actuarial science. Bessel, codiscoverer of stellar parallaxes and the binary companion of Sirius, introduce the notion of probable error*. Quetelet*, founder of the Belgian Royal Observatory, led the application of probability theory to the social sciences. Airy, a Royal astronomer, is known both for his text on statistical errors and his study of telescope optics. *See also* LAWS OF ERROR.

STATISTICAL ASTRONOMY

As comprehensive all-sky catalogs of star positions were compiled and astronomical photography permitted faint stars to be located, a field known as "statistical astronomy" rose to importance in the first half of this century. It is concerned with various collective properties of stars including their luminosity and mass distribution functions, their spatial distribution in the Milky Way,

their distances from us and the related problem of light absorption in the interstellar medium, and their motions with respect to the Sun and to the center of the galaxy. The principal results of these studies are discussed in the monumental 1953 monograph of Trumpler and Weaver [41]. Prephotographic statistical discussions are reviewed in ref. 37 and more recent findings can be found in refs. 38 and 12.

From these studies we have learned that the Sun resides about 25 thousand light-years off-center in a disk of a differentially rotating spiral galaxy, with stars of increasing ages occupying the galaxy's spiral arms, smooth disk, and halo. By comparing the galactic mass inferred from star counts with that inferred from their rotational velocities around the galactic center, the existence of a "dark matter" component in the outer regions of the galaxy is inferred. Dynamical studies of other galaxies confirm that the mass in visible stars and gas is dwarfed by the dark matter in galaxy halos; yet astronomers do not know whether the matter is in the form of planets, elementary particles, black holes, or some more exotic form.

We give two modern examples of studies in galactic astronomy. Murray [25] derives the joint densities of the observed parallaxes, proper motions, and brightnesses for 6125 stars, and computes the luminosity function, scale height in the galactic disk, streaming motions, and outliers with high velocities. A maximum likelihood technique is used; the principal limitation is that parametric (e.g., Gaussian luminosity functions and observational errors, exponential scale heights for stars in the disk) forms are assumed throughout. Caldwell and Ostriker [8] seek fits of a three-component model of the mass distribution of the galaxy to 14 observationally derived quantities constraining the size, density, and motions in the galaxy. A nonlinear least-squares minimization algorithm is used to find the minimum chi-squared* solution.

Perhaps the central problem in statistical astronomy is the derivation of the *intrinsic* luminosity distribution function of a class of stars from a survey of the stars with the greatest *apparent* brightnesses (usually called a magnitude- or flux-limited survey). The observed population contains an excess of high luminosity stars, which can be seen out to large distances, and a deficit of low luminosity stars, which are bright enough to appear in the sample only if they lie close to us. Intrinsic or experimental errors scatter faint stars preferentially into flux-limited samples. These and related problems are called the "Malmquist effects," after the Swedish astronomer who derived a correction in 1920 for the bias for Gaussian luminosity distributions.

Interest in luminosity functions reemerged during the last decade with attempts to understand the phenomenon of quasar evolution [42]. The density of quasars, which are galaxies possessing extremely violent activity in their nuclei probably due to an accreting massive black hole, per unit volume was thousands of times higher when the Universe was young than it is today, indicating evolution of the shape and/or amplitude of the luminosity function. In 1971, Lynden-Bell [21] derived a remarkably simple nonparametric maximum likelihood* procedure for extrapolating from a flux-limited data set (assumed to be randomly truncated) of quasar counts to obtain a complete luminosity function. The method was largely unused by astronomers, and unknown to statisticians until its reexamination in 1985 by Woodroofe [43]. A similar method was independently developed by Nicoll and Segal [30] in support of Segal's proposed chronometric cosmology. A more common approach to the quasar evolution problem is to fit the data to parametric evolution formulas using least-squares* or maximum likelihood criteria; see, for example, Marshall [23].

STATISTICAL ANALYSIS OF MODERN ASTRONOMICAL DATA

The modern observational astronomer is typically schooled only in elementary techniques such as the chi-squared test* and least-squares regression, using computer software such as that given by Bevington [6].

Some nonparametric methods such as Kolmogorov–Smirnov* and rank correlation* tests have come into frequent use, and computer codes distributed by Press et al. [32] are likely to bring other methods into the astronomer's repertoire. Unfortunately, very few astronomers are familiar with the major statistical software* packages such as SAS or BMDP.

Interest in more sophisticated and specialized statistical techniques of data analysis has emerged during the last decade. Murtagh and Heck [26] have written a monograph on multivariate techniques for astronomers, with a thorough bibliography of astronomical applications. The proceedings of a 1983 conference devoted specifically to the subject of statistical methods in astronomy is available [39]. An informal international Working Group for Modern Astronomical Methodology has formed, with a newsletter and announcements of workshops published in ref. 7. Articles in the statistical literature include those of Kendall [19], Scott [36], and Narlikar [28]. Following are brief descriptions of a few topics of current interest.

Censored data. A common experiment in astronomy entails the observations of a preselected sample of objects at a new spectral band, but some of the objects are not detected. For example, only about 10% of optically selected quasars are detected with the most sensitive radio telescopes and about 50% with the most sensitive satellite-borne X-ray telescopes, unless unacceptably long exposure times are devoted to the experiment. The data thus suffer type I left-censoring in apparent brightness, and a quasi-random censoring in intrinsic luminosities because the quasars lie at different distances. The studies seek to measure the mean radio or X-ray luminosities of the sample, differences in the luminosity functions between subsamples, correlations and linear regressions between luminosities in the various spectral bands, and so forth.

Until recently astronomers were unaware that survival analysis* statistical methods used by biostatisticians and others were available that provide many solutions to these questions. Avni, an astrophysicist, independently derived the "redistribute-to-the-right" formulation of the Kaplan–Meier product-limit estimator* [3], and a maximum likelihood linear regression* assuming normally distributed residuals [2]. Schmitt [35] has developed a linear regression procedure, based on the two-dimensional Kaplan–Meier estimator and bootstrap error analysis, which can be applied to data censored in both the dependent and independent variables. Schmitt, Isobe, and Feigelson have brought previously known survival analysis methods into use for astronomical applications [35, 10, 18, 17]. The principal difficulties in adapting standard survival analysis to astronomical situations are that the censoring levels are usually not known precisely, and the censoring patterns in flux-limited data are usually not completely random. These prob-lems have yet to be addressed in detail.

Spatial analysis of galaxy clustering. Our Milky Way is but one of 10^{11} detectable galaxies, each containing up to $\sim 10^{11}$ luminous stars and a greater but uncertain amount of nonluminous matter. The galaxies appear to be rushing away from each other in a universal expansion that started 10–20 billion years ago. Correct modeling of the spatial and velocity distribution of the visible galaxies is critical for understanding the history of the Universe, and for determining whether or not the Universe will cease expanding and recollapse.

Galaxies are not distributed randomly but are strongly clustered on many spatial scales. Neyman and Scott [29, 36] were among the first to study this during the 1950s, using a hierarchy of clusters distributed as a uniform

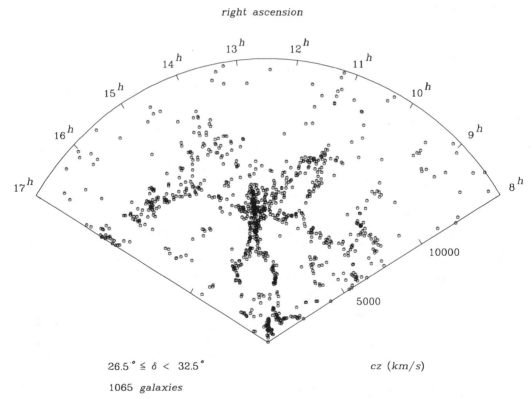

Figure 1 A "slice of the Universe" showing the spatial distribution of bright galaxies in a portion of the sky [9]. It shows the strongly anisotropic clustering pattern of "filaments" surrounding "voids." Courtesy of M. Geller, Center for Astrophysics.

Poisson point process*. Other statistical analyses followed including the nearest-neighbor* distance distribution, the multiplicity function, and, most extensively, the two- and three-point correlation functions [31]. The power law two-point correlation function found for galaxies could be explained as the result of simple gravitational interactions of matter in an initially homogeneous Universe with an appropriate spectrum of density fluctuations.

By the middle 1980s, however, surveys of thousands of galaxy recessional velocities had been completed, revealing an unexpectedly anisotropic clustering pattern, as illustrated in Fig. 1 [9]. The galaxies seem to be concentrated along the edges of giant shells. The effect is sometimes referred to as "filaments" or "sheets" surrounding "voids," with a "spongelike" topology. The largest structures may exceed one-tenth the size of the observable Universe. In addition, large regions of the Universe may be moving in bulk with respect to other regions [22]. None of these phenomena can be easily explained by simple gravitational effects in an initially homogeneous Universe. Statistical modeling of these data has just begun. Suggested methods include minimal spanning trees* [5], random walks* [20], ridge-finding algorithms [24], measures of topological curvature [13], and a quadrupole elongation statistic [11]. The study of Lynden-Bell et al. [22] is also of methodological interest, illustrating how sophisticated maximum likelihood models of galaxy location and motions have become.

Analysis of periodic time series. Stellar systems often exhibit periodic behavior, from vibrations or rotations of single stars to orbits of two or more bodies in mutual gravitational orbits. Time scales run from milliseconds to hundreds of years, and the data can involve any portion of the electromagnetic spectrum. For example, time series of X-ray emission from binary star systems where one companion of an accreting neutron star or black hole are quite interesting. Some systems show one or more strict periodicities, others random shot noise or white noise, others occasional sudden bursts of x-rays,

and yet others combinations of $1/f$ noise and quasi-periodic oscillations. Certain of these behaviors are understood as rotational or orbital effects, whereas others are still mysterious.

A time-series* problem that has attracted recent methodological interest is the difficulty of establishing the existence of binary star orbits from optical photometric or velocity time series, which often consist of sparse, unevenly spaced, and noisy data. The issue is important because as many as half of the "stars" one sees at night may in fact be wide binaries. Classical Fourier analysis*, which assumes evenly spaced data points, is not an adequate method, and a variety of alternatives have been proposed. One is the periodogram*, which can be equivalent to least-squares fitting of sine curves to the data [34, 16]. Several nonparametric methods have also been evaluated [15]. The principal difficulty with all methods is usually not finding the highest peak in the spectrum, but in evaluating its statistical significance and eliminating false alarms. Both analytical and bootstrap methods for measuring confidence limits have been suggested, but no consensus has emerged on the best treatment of the problem.

Image restoration techniques. Many classes of objects in the sky have complicated morphologies, and much effort is devoted to imaging them accurately. This entails compensation for the distortions caused by imperfect optics of the telescope or turbulence in the Earth's atmosphere, and nonlinearities in detectors such as photographic plates. Perhaps the greatest need for sophisticated image restoration is in "aperture synthesis" interferometry, a technique developed in radio astronomy that combines the signals from separated individual radio antennas to produce a single high-resolution image [33, 27]. The data consist of the two-dimensional Fourier transform of the brightness distribution in the sky, but the coverage in the Fourier plane is incomplete. Simple Fourier transformation thus gives an image contaminated by strong side lobes. Two restoration methods are commonly used: The "CLEAN" algorithm, which gives a least-squares fit to a superposition of many point sources; and the maximum entropy method*, which gives the most probable nonnegative smooth image consistent with the data. Both require prior knowledge of the coverage in the Fourier plane. The resulting CLEANed map can then be used as an improved model of the sky distribution to "self-calibrate" unpredicted instrumental or atmospheric disturbances. After many iterations, which can take hundreds of CPU-hours on large computers, images with extraordinarily high fidelity have been achieved with dynamic range (brightest spot in the image divided by the root mean square noise level) of order $10^5 : 1$. Maximum entropy image enhancement techniques are sometimes also used in optical and X-ray astronomy as well as radio interferometry.

Statistics of very few events. Important branches of astronomy have emerged in the last few decades from experimental physics that involve the detection of small numbers of discrete events. These include cosmic ray, X-ray, gamma-ray, and neutrino astronomy. With the proliferation of photon-counting detectors like charged-coupled devices, such data are becoming increasingly common in optical astronomy as well. The statistical procedures for interpreting these data are traditionally based on the Poisson distribution, though use of nonparametric and maximum likelihood techniques is also appearing.

A premier example of this problem was the detection of neutrinos from the supernova SN1987A, initiated by the gravitational collapse of a star in a satellite galaxy of our Milky Way. The detection represented the first astronomical object other than the Sun ever detected in neutrinos, and provides a unique opportunity to test a host of models of high-energy and nuclear astrophysics. The data consist of 11 events in 13 seconds in the Kamiokande-II underground detector, and 8 events in 6 seconds in the IMB detector. The timing of the events constrain the mass of the neutrino, which could be the most massive component of the Universe [4, 1]. However, different treatments of the same data give considerably different neutrino mass limits.

SUMMARY

Many aspects of modern astronomy are statistical in character, and demand sophisticated statistical methodology. In light of this situation, and the rich history of the interaction between astronomy and statistics through the nineteenth century, it is surprising that the two communities have been so isolated from each other in recent decades. Astronomers dealing with censored data, for example, were unaware of the relevant progress in biostatistics* and industrial reliability applications until recently. Most are not familiar with multivariate techniques extensively used in the social sciences. Conversely, statisticians were unaware of Lynden-Bell's maximum likelihood density estimation technique for a truncated distribution. There is little communication between astronomers analyzing the spatial distribution of galaxies (Fig. 1) and statisticians involved with point spatial processes* arising in other fields. Improved interactions between the two fields is clearly needed. It would give astronomers more effective techniques for understanding the Universe, and statisticians challenging and important problems to address.

References

[1] Arnett, W. D. and Rosner, J. L. (1987). *Phys. Rev. Lett.*, **58**, 1906.

[2] Avni, Y. and Tananbaum, H. (1986). *Astrophys. J.*, **305**, 57.

[3] Avni, Y., Soltan, A., Tananbaum, H., and Zamorani, G. (1980). *Astrophys. J.*, **235**, 694.

[4] Bahcall, J. N. and Glashow, S. L. (1987). *Nature*, **326**, 476.

[5] Barrow, J. D., Bhavsar, S. P., and Sonoda, D. H. (1985). *Monthly Notices R. Astron. Soc.*, **216**, 17.

[6] Bevington, P. R. (1969). *Data Reduction and Error Analysis for the Physical Sciences.* McGraw-Hill, New York.

[7] *Bulletin d'Information du Centre de Données Stellaires*, Observatoire de Strasbourg, 11 rue de l'Université, 67000 Strasbourg, France.

[8] Caldwell, J. A. and Ostriker, J. P. (1981). *Astrophys. J.*, **251**, 61.

[9] de Lapparent, V., Geller, M. J., and Huchra, J P. (1986). *Astrophys. J. Lett.*, **302**, L1.

[10] Feigelson, E. D. and Nelson, P. I. (1985). *Astrophys. J.*, **293**, 192.

[11] Fry, J. N. (1986). *Astrophys. J.*, **306**, 366.

[12] Gilmore, G. and Carswell, B., eds. (1987). *The Galaxy*. Reidel, Dordrecht, Netherlands.

[13] Gott, J. R., Melott, A. L., and Dikinson, M. (1986). *Astrophys. J.*, **306**, 341.

[14] Hald, A. (1986). *Int. Statist. Rev.*, **54**, 211–220.

[15] Heck, A., Manfroid, J., and Mersch, G. (1985). *Astron. Astrophys. Suppl.*, **59**, 63.

[16] Horne, J. H. and Baliunas, S. L. (1986). *Astrophys. J.*, **302**, 757.

[17] Isobe, T. and Feigelson, E. D. (1987). ASURV: Astronomy Survival Analysis, software package.

[18] Isobe, T., Feigelson, E. D., and Nelson, P. I. (1986). *Astrophys J.*, **306**, 490.

[19] Kendall, D. G. (1985). In *Celebration of Statistics*, A. C. Atkinson and S. E. Fienberg, eds. Springer Verlag, Berlin.

[20] Kuhn, J. R. and Uson, J. M. (1982). *Astrophys. J. Lett.*, **263**, L47.

[21] Lynden-Bell, D. (1971). *Monthly Notices R. Astron. Soc.*, **136**, 219.

[22] Lynden-Bell, D., Faber, S. M., Burstein, D., Davies, R. L., Dressler, A., Terlevich, R. J., and Wegner, G. (1988). *Astrophys. J.*, **326**, 19.

[23] Marshall, H. L. (1987). *Astron. J.*, **94**, 620.

[24] Moody, J. E., Turner, E. L., and Gott, J. R. (1983). *Astrophys. J.*, **273**, 16.

[25] Murray, C. A. (1986). *Monthly Notices R. Astron. Soc.*, **223**, 649.

[26] Murtagh, F. and Heck, A. (1987). Multivariate Data Analysis, *Astrophysics and Space Science Library*, **131**. Reidel, Dordrecht, Netherlands.

[27] Narayan, R. and Nityananda, R. (1986). *Ann. Rev. Astron. Astrophys.*, **24**, 127.

[28] Narlikar, J. V. (1982). *Sankhyā B*, **42**, 125.

[29] Neyman, J. and Scott, E. L. (1952). *Astrophys J.*, **116**, 144.

[30] Nicoll, J. F. and Segal, I. E. (1983). *Astron. Astrophys.*, **118**, 180.

[31] Peebles, P. J. E. (1980). *The Large-Scale Structure of the Universe*. Princeton University Press, Princeton, NJ.

[32] Press, W. H., Flannery, B. P., Teukolsky, S. A., and Vetterling, W. T. (1986). *Numerical Recipes*: *The Art of Scientific Computing*. Cambridge University Press, London.

[33] Roberts, J. A., ed (1984). *Indirect Imaging*. Cambridge University Press, London.

[34] Scargle, J. D. (1982). *Astrophys. J.*, **263**, 835.

[35] Schmitt, J. H. (1985). *Astrophys. J.*, **293**, 178.

[36] Scott, E. L. (1976). In *On the History of Statistics and Probability*, D. B. Owen, ed. Dekker, New York.

[37] Sheynin, O. B. (1984). *Archive for History of Exact Sciences*, **29**, 151–199.

[38] *Stars and Stellar Systems (1962–75)*. University of Chicago Press, Chicago. 9 volumes.

[39] *Statistical Methods in Astronomy* (1983). SP-201, European Space Agency, ESA Scientific and Technical Publications, c/o ESTEC, Noordwijk, Netherlands.

[40] Stigler, S. M. (1986). *The History of Statistics: The Measurement of Uncertainty Before 1900*. Harvard University Press, Cambridge, MA.

[41] Trumpler, R. J. and Weaver, H F. (1953). *Statistical Astronomy*. University of California Press, Berkeley, CA.

[42] Weedman, D. (1986). *Quasar Astronomy*. Reidel, Dordrecht, Netherlands.

[43] Woodroofe, M. (1985). *Ann. Statist.*, **13**, 163.

Bibliography

Di Gesù, V., Scarsi, L., Crane, P., Friedman, J. H., and Levaldi, S., eds. (1985; 1986). *Data Analysis in Astronomy I*; *II*. Plenum, New York. (Proceedings of International Workshops held at Erice, Sicily, Italy in 1984 and 1985.)

Maistrov, L. E. (1974). *Probability Theory: A Historical Sketch*, translated by S. Kotz. Academic, New York.

EDITORIAL NOTE

S. R. Searle [*Commun. Statist. A*, **17**, 935–968 (1988)] notes that the earliest appearances of variance components* are in two books on astronomy, namely:

Airy, G. B. (1861). *On the Algebraical and Numerical Theory of Errors of Observations and the Combination of Observations*. Macmillan, London.

and

Chauvenet, W. (1863). *A Manual of Spherical and Practical Astronomy.* Lippincott, Philadelphia.

Searle also draws attention to the fact that the development of the least-squares* estimation of parameters in linear models was presented in books on astronomy, namely:

Gauss, K. F. (1809). *Theoria Motus Corporum Coelestium in Sectionibus Conics Solem Ambientium.* Perthes and Besser, Hamburg.

and

Legendre, A. M. (1806). *Nouvelles Methodes pour Determination des Orbites des Comètes.* Courcier, Paris.

(CHAUVENET'S CRITERION
DISCRIMINANT ANALYSIS
GAUSS, CARL FRIEDRICH
HIERARCHICAL CLUSTER ANALYSIS
LEAST SQUARES
LINEAR REGRESSION
MULTIVARIATE ANALYSIS
NEWCOMB, SIMON
STATISTICS, AN OVERVIEW)

ERIC D. FEIGELSON

B

BAYES, THOMAS

Born: 1701, in London, England.
Died: April 7, 1761, in Tunbridge Wells, England.
Contributed to: statistical inference, probability.

Relatively little is known about Thomas Bayes, for whom the Bayesian school of inference and decision is named. He was a Nonconformist minister, a Fellow of the Royal Society, and according to the certificate proposing him for election to that society, was "well skilled in Geometry and all parts of Mathematical and Philosophical Learning" (see ref. 6, p. 357).

The Nonconformist faith played a major role in the scientific ideas of the eighteenth century, a time when religion and philosophy of science were inextricably linked. The growth of Nonconformism was influenced by the rise of Natural Philosophy, which encouraged a scientific examination of the works of the Deity so that men could come to understand His character. The Nonconformists included Deists and Arians (forerunners of modern Unitarians) among them, all sharing a rejection of the Trinity and a skepticism about the divinity of Christ. The Royal Society is known to have had a strong antitrinitarian component [6, p. 356].

Thomas was born to Joshua and Ann Bayes in 1702 in London. Joshua was the first Nonconformist minister to be publicly ordained (1694) after the Act of Uniformity was passed. He became minister of the Presbyterian meeting house at Leather Lane in 1723; Thomas served as his assistant.

Thomas was educated privately, as was the custom with Nonconformists at that time. After serving as an assistant to his father, he went to Tunbridge Wells as minister of a dissenting Presbyterian congregation. It is not known precisely when he began that post, but he was there by 1731 when his religious tract, *Divine Benevolence, or an attempt to prove that the Principal End of the Divine Providence and Government is the Happiness of his Creatures*, was printed by John Noon at the White Hart in Cheapside, London [1].

Bayes was elected a Fellow of the Royal Society in 1742. The signers of his certificate for election included officers of the Royal Society, John Eames, Mathematician and personal friend of Newton, and Dr. Isaac Watts, Nonconformist minister and well-known composer of hymns. Unfortunately, there is little indication of published work by Bayes meriting his election to the Royal Society. In fact, there is no concrete evidence of any published paper, mathematical or theological by Bayes from his election in 1742 until his death in 1761 [6, p. 358]. There is however, an anonymous 1736 tract, published by John Noon, which has been ascribed to Bayes, entitled *An Introduction to the Doctrine of Fluxions, and a Defence of the Mathematicians against the objections of the Author of the Analyst, in so far as they are designed to affect the general method of reasoning.* This work was a defense of the logical foundations of Newtonian calculus against attacks made by Bishop Berkeley, the noted philosopher. Augustus DeMorgan, in an 1860 published query about Bayes, noted the practice of Bayes' contemporaries of writing in the author's name on an anonymous tract, and DeMorgan (and the British Museum) accepted the ascription.

DeMorgan gives some sense of Bayes' early importance by writing, "In the last century there were three Unitarian divines, each of whom has established himself firmly among the foremost promoters of a branch of science. Of Dr. Price and Dr. Priestley... there is no occasion to speak: their results are well known, and their biographies are sufficiently accessible. The third is Thomas Bayes..." [5]. De Morgan recognized Bayes as one of the major leaders in the development of the mathematical theory of probability, claiming that he was "of the calibre of DeMoivre and Laplace in his power over the subject," and

noting in addition that Laplace* was highly indebted to Bayes, although Laplace made only slight mention of him [5].

Bayes is also known to have contributed a paper in 1761, published in the *Philosophical Transactions* in 1763 (see ref. 2), on semiconvergent asymptotic series. Although a relatively minor piece, it may have been the first published work to deal with semiconvergence [6, p. 358].

Bayes was succeeded in the ministry at Tunbridge Wells in 1752, but continued to live there until his death on April 17, 1761. He was buried in the Bayes family vault in Bunhill Fields, a Nonconformist burial ground. The bulk of his considerable estate was divided among his family, but small bequests were made to friends, including a £200 bequest to Richard Price.

Bayes' greatest legacy, however, was intellectual. His famous work, "An Essay towards Solving a Problem in the Doctrine of Chance," was communicated by Richard Price to the Royal Society in a letter dated November 10, 1763, more than two years after Bayes' death, and it was read at the December 23 meeting of the society. Although Bayes' original introduction was apparently lost, Price claims in his own introduction (see ref. 3) that Bayes' purpose was "to find out a method by which we might judge concerning the probability that an event has to happen in given circumstances, upon supposition that we know nothing concerning it but that under the same circumstances it has happened a certain number of times and failed a certain other number of times." Price wrote that the solution of this problem was necessary to lay a foundation for reasoning about the past, forecasting the future, and understanding the importance of inductive reasoning. Price also believed that the existence of the Deity could be deduced from the statistical regularity governing the recurrency of events.

Bayes' work begins with a brief discussion of the elementary laws of chance. It is the second portion of the paper in which Bayes actually addresses the problem described by Price, which may be restated as: Find $\Pr[a < \theta < b | X = x]$, where $X \sim$ Binomial(n, θ). His solution relies on clever geometric arguments (using quadrature* of curves instead of the more modern integral calculus), based on a physical analog of the binomial model—relative placements of balls thrown upon a flat and levelled table. Bayes' findings can be summarized as follows [7]:

$$\Pr[a < \theta < b, X = x]$$
$$= \int_a^b \binom{n}{x} \theta^x (1 - \theta)^{n-x} d\theta,$$

$$\Pr[X = x] = \int_0^1 \binom{n}{x} \theta^x (1 - \theta)^{n-x} d\theta,$$

and finally,

$$\Pr[a < \theta < b | X = x]$$
$$= \frac{\int_a^b \binom{n}{x} \theta^x (1 - \theta)^{n-x} d\theta}{\int_0^1 \binom{n}{x} \theta^x (1 - \theta)^{n-x} d\theta}.$$

The uniform prior distribution* for the unknown θ followed naturally by construction in Bayes' analogy. According to Price, Bayes originally obtained his results by assuming that θ was uniformly distributed, but later decided that the assumption may not always be tenable and resorted to the example of the table. Bayes added a *Scholium* to his paper in which he argued for an a priori uniform distribution for unknown probabilities.

The use of a uniform prior distribution to represent relative lack of information (sometimes called the *Principle of Insufficient Reason*) has long been controversial, partly because such a distribution is not invariant under monotone transformation of the unknown parameter. Many statisticians and philosophers have interpreted the *Scholium* in terms of the principle of insufficient reason, but more recently, Stigler [7] has argued that Bayes was describing a uniform predictive distribution for the unknown X, and *not* a prior distribution for the parameter θ. Such an interpretation seems plausible, particularly due to its emphasis on probabilities for ultimately observable quantities and due to Bayes' view of probability as an expectation

(see also ref. 4 for a view of probability as expectation). The assumption of a uniform predictive distribution is more restrictive than that of a uniform prior distribution; it does not follow, in general, that the implied distribution of θ would be uniform. One advantage of the uniform predictive distribution is that the problem of invariance no longer arises: If $\Pr[X = x]$ is constant, then so is $\Pr[f(X) = f(x)]$, for monotone functions f.

Thomas Bayes did not extend his results beyond the binomial model, but his views of probability and inductive inference have been widely adopted and applied to a variety of problems in statistical inference and decision theory.

References

[1] Barnard, G. A. (1958). *Biometrika*, **45**, 293–295. (A biographical note that introduces a "reprinting" of Bayes' famous paper.)

[2] Bayes, T. (1763). *Philos. Trans. R. Soc.*, London, **53**, 269–271.

[3] Bayes, T. (1763). *Philos. Trans. R. Soc.*, London, **53**, 370–418. (Also reprinted in *Biometrika*, **45**, 296–315. The reprinted version has been edited and modern notation has been introduced.)

[4] De Finetti, B. (1974–1975). *Theory of Probability*, Vols. 1 and 2. Wiley, New York.

[5] DeMorgan, A. (1860). *Notes and Queries*, January 7, 9–10.

[6] Pearson, K. (1978). *The History of Statistics in the 17th and 18th Centuries*, E. S. Pearson, ed. Macmillan, New York, pp. 355–370.

[7] Stigler, S. M. (1982). *J. R. Statist. Soc. A*, **145**, 250–258. (Argues that uniform distribution is intended for the predictive or marginal distribution of the observable events; discusses reasons behind common "misinterpretations" of Bayes' *Scholium*.)

Bibliography

Holland, J. D. (1962). *J. R. Statist. Soc. A*, **125**, 451–461. (Contains many references to primary sources concerning Bayes' life.)

Laplace, P. S. (1951). *A Philosophical Essay on Probabilities*. Dover, New York. (Laplace's indebtedness to Bayes is apparent in his General Principles.)

Stigler, S. M. (1983). *Amer. Statist.*, **37**, 290–296. (Evidence is presented that Nicholas Saunderson, an eighteenth century mathematician, may have first discovered the result attributed to Bayes.)

Todhunter, I. (1865). *A History of the Mathematical Theory of Probability*. Chelsea, New York (reprint 1965).

(BAYESIAN INFERENCE
LAPLACE, P. S.
LOGIC OF STATISTICAL REASONING
PRIOR DISTRIBUTIONS)

R. L. TRADER

BOOTSTRAPPING—II

Bootstrapping is a strategy for estimating standard errors and setting confidence intervals for parameters when the form of the underlying distribution is unknown. It is closely related to the jackknife*, the infinitesimal jackknife, and the nonparametric delta method; however, the bootstrap is far more computationally intensive. In fact, the bootstrap is a prominent example of a new class of nonparametric statistical procedures that substitute intensive computation for mathematical analysis. This is particularly valuable when the parameter of interest is a complicated functional of the true distribution.

Efron [11] first recognized the feasibility of the bootstrap method, and many variants have been subsequently proposed. The key idea is that the relationship between the true cumulative distribution function (CDF) F and the sample is similar to the relationship between the empirical cumulative distribution function (ECDF) $\hat{F}_n$ and a secondary sample drawn from it. Usually one cannot draw multiple samples from F, but modern computers allow one to draw large numbers of samples from $\hat{F}_n$. So one uses the primary sample to form an estimate $\hat{F}_n$ of F, and then calculates the sampling distribution of the parameter estimate under $\hat{F}_n$. This calculation is done by drawing many secondary samples and finding the estimate, or a function of the estimate, for each. (The name "bootstrap" derives from the reflexive nature of the secondary samples.) If $\hat{F}_n$ is a

good approximation of F, then H_n, the sampling distribution of the estimate under $\hat{F}_n$, is generally a good approximation to the sampling distribution for the estimate under F. H_n is commonly called the "bootstrap distribution" of the parameter.

VARIATIONS ON THE BOOTSTRAP

The bootstrap strategy is not uniquely defined, which has led to a variety of implementations. Efron [11, 12, 14–16] has proposed the percentile bootstrap, the bias-corrected percentile bootstrap, the pivoted bootstrap, the convolved bootstrap, the exponentially tilted bootstrap, and, most recently, a bootstrap that sets confidence regions that are asymptotically second-order correct (in a sense to be explained shortly). Rubin [30] has proposed a Bayesian bootstrap; Beran [3] puts forward a prepivoted bootstrap that automatically makes higher-order corrections; Loh [27] examines a calibrated bootstrap. Many other approaches have also been studied.

The simplest bootstrap is the *percentile bootstrap*, which is closely related to the ordinary jackknife. Instead of systematically deleting observations and recalculating the estimate, one randomly deletes and duplicates observations; as follows.

1. Draw an independent, identically distributed (i.i.d.) sample $X_1, \ldots, X_n$ from an unknown CDF F.

2. Form the empirical cumulative distribution function (ECDF) $\hat{F}_n$ and calculate the point estimate of the parameter of interest as $\hat{\theta} = \theta(\hat{F}_n)$.

3. Draw a large number B of i.i.d. samples of size n from $\hat{F}_n$; form the B corresponding ECDFs $\hat{G}_{n1}, \ldots, \hat{G}_{nB}$.

4. For each secondary sample, calculate $\theta_j = \theta(\hat{G}_{nj})$. Denote the ECDF of $\theta_1, \ldots, \theta_B$ by $\hat{H}_B$. This stands proxy for the sampling distribution of $\theta(F)$ in all subsequent inference. The standard deviation of $\hat{H}_B$ is an estimate of

the standard error of $\theta(\hat{F}_n)$, and $[\hat{H}_B^{-1}(\alpha), \hat{H}_B^{-1}(1 - \alpha)]$ is an approximate $(1 - 2\alpha)$ central confidence interval on $\theta(F)$.

As $B \to \infty$, the Glivenko–Cantelli theorem* ensures that $\hat{H}_B$ converges to H_n. In most applications, Efron and Tibshirani [19] suggest that reasonable agreement is obtained for $B = 1000$ when setting confidence intervals, or $B = 100$ when estimating the standard error. Hall [24] provides a more detailed discussion.

The percentile bootstrap can be improved in many ways. One modification of it is the convolved bootstrap, which smooths the ECDF $\hat{F}_n$ and sometimes $\hat{G}_{nj}$ by convolution with a suitable kernel. For many parameters and CDFs, this gives a slight improvement over the percentile bootstrap confidence intervals. Smoothing by convolution inflates the variance of the secondary samples over that obtained from sampling from F, and Silverman and Young [32] show it is often desirable to correct for this.

A more significant modification of the bootstrap strategy is bias correction; this version is usually called the "BC bootstrap." For many estimators $P[\theta_j \leqslant \hat{\theta}] \neq 0.5$; this type of median bias can often be eliminated or reduced through a suitable adjustment. Let Φ^{-1} denote the inverse function of the standard normal, let z_α satisfy $\alpha = \Phi(z_\alpha)$, and set $z_0 = \Phi^{-1}(\hat{H}_B(\hat{\theta}))$. Then the bias-corrected $(1 - 2\alpha)$ central confidence interval is

$$\left[\hat{H}_B^{-1}(\Phi(2z_0 - z_\alpha)), \hat{H}_B^{-1}(\Phi(2z_0 + z_\alpha)) \right].$$

Notice that when $\hat{H}_B$ is centered at $\hat{\theta}$, corresponding to median-unbiased estimates* from the secondary samples, then $z_0 = 0$ and no adjustment is made.

When exact confidence intervals exist for univariate parameters, the form of the endpoint is typically

$$\hat{\theta} = \hat{\sigma}\left(z_\alpha + \frac{A_{n,\alpha}}{\sqrt{n}} + \frac{B_{n,\alpha}}{n} + \cdots \right),$$

where $\hat{\sigma}$ is an estimate of the standard error of $\hat{\theta}$. Efron [14] shows that as $B \to \infty$, the BC confidence interval on θ becomes exactly second-order correct (i.e., gets the $A_{n,\alpha}$ term right) provided that there exists a monotonic transformation $g(\cdot)$ with $\hat{\phi} = g(\hat{\theta})$, $\phi = g(\theta)$ such that

$$(\hat{\phi} - \phi)/\tau \sim N(-z_0, 1),$$

where τ is the constant standard error of the transformed estimate and N is the normal distribution. Schenker [31] points out that such transformations do not exist for many typical applications.

A generalization of the BC bootstrap yields the bias-corrected percentile acceleration interval (BC$_a$ interval). Suppose there exists a monotonically increasing transformation $g(\cdot)$, a bias constant z_0, and an "acceleration constant" a such that for $\hat{\phi} = g(\hat{\theta})$, $\phi = g(\theta)$, one has

$$(\hat{\phi} - \phi)/\tau \sim N(-z_0\sigma_\phi, \sigma_\phi^2),$$

where τ is any constant and $\sigma_\phi = 1 + a\phi$. (This reduces to the BC bootstrap when $a = 0$.) For this case, Efron [16] shows that as $B \to \infty$, a central $(1 - 2a)$ confidence interval* for $\theta(F)$ of the form

$$\left[\hat{H}_B^{-1}(\Phi(z(\alpha))), \hat{H}_B^{-1}(\Phi(z(1 - \alpha))) \right]$$

is exactly second-order correct, where

$$z(\alpha) = z_0 + \frac{z_0 + z_\alpha}{1 - a(z_0 + z_\alpha)}$$

and similarly for $z(1 - \alpha)$.

In applying the BC or BC$_a$ bootstrap, one need not know the form of the transformation or the constants; these can be estimated from the data, as described in ref. 16. One can also smooth these bootstraps by convolution, or use histospline smoothing, or otherwise fine-tune the method.

A different approach to improving the accuracy of bootstrap confidence intervals depends on the use of a pivot. Bickel [5] shows that the BC$_a$ bootstrap confidence interval may be viewed as using a particular extract pivot, but approximate pivots are also available. These may be more attractive from the standpoint of nonparametric analysis.

The most popular pivot bootstrap uses studentization*. Let $\hat{\sigma}(\hat{\theta})$ and $\hat{\sigma}(\theta_j)$ be estimates of the standard error based on the original sample and the jth secondary sample, respectively. Then the studentized pivot bootstrap uses $(\theta_j - \hat{\theta})/\hat{\sigma}(\theta_j)$ as an approximation to the distribution of $(\hat{\theta} - \theta(F))/\hat{\sigma}(\hat{\theta})$; this is often more accurate than the percentile bootstrap's use of the distribution of θ_j to approximate the distribution of $\hat{\theta}$. Confidence intervals are formed in the usual way, by inverting the pivot.

Several alternative bootstraps involve repeated applications of the bootstrap idea. For example, Beran [3] suggests a prepivoting technique that is based upon a sequence of nested bootstraps. It can be iterated until, in the limit, the asymptotically optimal pivotal function is obtained. Loh [27] suggests an iterated bootstrap in which the error in confidence interval coverage for the first bootstrap is estimated by a second round of bootstrapping. Then one recalibrates the first interval accordingly. Hall [23] also discusses a bootstrap that uses iteration to obtain arbitrarily high-order asymptotic corrections for the endpoints of confidence intervals.

A somewhat different approach is the Bayesian bootstrap. This produces an estimate of the posterior distribution* of $\theta(F)$. In the univariate case, Rubin [30] proposes the following:

1. Draw an i.i.d. sample from F.

2. Drawn i.i.d. random vectors $\mathbf{V}_1, \ldots, \mathbf{V}_B$ from a Dirichlet distribution* with all n parameters equal to unity.

3. For each random vector $\mathbf{V}_j$, assign the weight of the ith component V_{ij} to observation $X_{(i)}$; the weights sum to unity, so this defines a random CDF F_j^*.

4. Calculate $\theta_j^* = \theta(F_j^*)$ for all $j = 1, \ldots, B$. Use the ECDF of these as a proxy for the posterior distribution of the parameter $\theta(F)$.

This description uses a noninformative Dirichlet prior, but one can choose other Dirichlet distributions* that put weight on order statistics in regions where there is prior belief that probability mass is likely to concentrate. Other Bayesian formulations are possible; for example, Laird and Louis [25] have recently suggested a method for setting empirical Bayes* confidence intervals based on bootstrap samples.

The bootstrap can be applied in parametric inference, although its natural formulation is nonparametric. Assume $F \in \{F_\xi: \xi \in \Xi\}$, where Ξ is finite dimensional. Then one can carry out the bootstrap algorithm exactly as before, but instead of resampling* from $\hat{F}_n$, one resamples from $F_{\hat{\xi}}$, where $\hat{\xi}$ is perhaps the maximum likelihood estimator of ξ. Efron [15] extends this parametric bootstrap to the problem of setting confidence intervals in the presence of nuisance parameters*.

As a final point, there are a variety of techniques designed to reduce the computational burden of bootstrap analysis. These are potentially most important for simulation studies and the iterated bootstraps. One technique is balanced resampling, first suggested by Davison et al. [8] and substantially extended by Wynn and Ogbonmwan [36]. Rather than randomly resampling from the ECDF of the original sample, one picks secondary samples so that the numbers of times each primary observation is resampled are equal, the numbers of times each pair of primary observations appears together in a resample are equal, etc. The aim is to obtain a set of resamples that are representative of the space of all possible resamples. DiCiccio and Tibshirani [9] offer a different approach to the problem of reducing computation; they give an approximation to the BC_a confidence interval that agrees with the endpoints to $O_p(n^{-1})$ but uses only $n + 2$ evaluations of the estimator. Alternative approaches include saddlepoint approximation* of the bootstrap distribution, importance sampling*, and exponential tilting. *See also* RESAMPLING PROCEDURES and SAMPLE REUSE.

ASYMPTOTICS

Many sorts of asymptotic results are available for various implementations of the bootstrap. Precise statements of the theorems are usually quite technical, so this summary will be qualitative. The interested reader should examine the references cited. However, before proceeding, it must be emphasized that the asymptotic behavior of particular bootstraps is often quite uninformative with respect to small sample (about $n \leqslant 30$) properties. Schenker [31], Wu [35], and Loh and Wu [28] make this point quite forcefully, and it is ratified by other simulation studies scattered throughout the literature.

The first significant asymptotic examinations of bootstrap properties were made by Singh [33] and Bickel and Freedman [6]. For inference on the mean, Singh establishes very general conditions under which the percentile bootstrap approximation to the distribution of $\sqrt{n}\,(\mu(F) - \mu(\hat{F}_n))$ is consistent and obtains precise bounds on the discrepancy through the Berry–Esséen theorem. Similar results are obtained for the studentized pivot bootstrap distribution for the mean and the percentile bootstrap distribution for the quantile. Bickel and Freedman give conditions for consistency of the percentile bootstrap distribution estimate of the sampling distribution of certain U-statistic* estimators, and show that percentile bootstrapping of the empirical and quantile processes* is very generally consistent. They also provide some illuminating examples in which the bootstrap distribution is not consistent.

Usually, bootstrap distributions are consistent estimators; sometimes slight modifications are required to achieve this. Ghosh et al. [22] give an example in which heavy tailweight causes the bootstrap distribution for the median to be inconsistent (the asymptotic variance is infinite); however, they also describe a small adjustment to prevent such misbehavior. Parr [29] gives a simple proof that the bootstrap distribution is consistent for the asymptotic distribution of statistical functionals* that are Frechét differentiable* with respect to the sup norm.

A different sort of asymptotic analysis has been undertaken by Lo [26]. He shows that almost surely, after suitable scaling and centering, the percentile bootstrap distribution, Rubin's Bayesian bootstrap distribution and the posterior of Ferguson's Dirichlet process* with very general shape parameter can all be simultaneously approximated by a Kiefer process* in absolute deviation with rate

$$O\left(n^{-1/4}(\log \log n)^{1/4}(\log n)^{1/2}\right).$$

Similar results are obtained for smoothed versions of these distributions. It follows that for a general class of estimators, the different procedures converge almost surely to the same limiting conditional distribution at the same rate.

Asymptotically, there is strong evidence for preferring the studentized pivot form of the bootstrap over the percentile bootstrap. Abramovitch and Singh [1] show that for estimators that are asymptotically normal, the studentized pivot form of the bootstrap improves the asymptotic error in the coverage probabilities from $O(n^{-1/2})$ to $O(n^{-1})$. Additional improvement can be obtained by making higher-order Edgeworth* corrections to the pivot; this line of reasoning then leads to the iterated bootstraps proposed by Hall [23] and Beran [3].

Regarding the asymptotic advantage of smoothing the original ECDF by convolution, Silverman and Young [32] show that this can increase the mean squared error (MSE) of the bootstrap distribution for certain linear or approximately linear functionals. Write the linear functional as $\theta(F) \doteq \int \psi(t)\, dF(t)$; then the basic result is that smoothing asymptotically reduces MSE if and only if $\psi(X)$ and $\ddot{\psi}(X)$ are negatively correlated. This result extends to the case of multivariate observations, where it depends upon the covariance matrix of F. A similar theorem holds if one smooths the ECDF by convolution and then deflates the variance before resampling.

APPLICATIONS

The bootstrap has been applied to an enormous variety of problems. The consensus is that it offers a generally reliable analysis for problems which would otherwise demand an unrealistic degree of modeling. This overview cannot describe all of the noteworthy applications, but will attempt to indicate the diversity of application.

One of the benchmark examinations of bootstrap performance is the estimation of a confidence interval for the correlation between the average grade-point average for entering classes of first-year law students, and their average LSAT scores, based on data from 14 U.S. law schools. Many different versions of the bootstrap have been tested against this problem, and results are reported in Efron [14], Rubin [30], Efron and Diaconis [17], Efron and Gong [18], Efron and Tibshirani [19], and Lo [26]. For this problem, the generally good performance of the percentile bootstrap and the percentile bootstrap smoothed by convolution are supported by a body of simulation evidence.

The performance of bootstrap method in regression has also been widely studied. Freedman [21] offers an early theoretical analysis, and extensive work on this is reported in Wu [35], and the subsequent discussion. Wu's simulation studies strongly suggest that the bootstrap strategy encounters severe difficulty in the presence of heteroscedasticity. Also, Stine [34] has used the bootstrap to set prediction intervals in regression.

In multivariate analysis, Beran and Srivastava [4] use bootstrap techniques to set confidence intervals on the eigenvalues and eigenvectors of covariance matrices. Simulation results compare the bootstrap performance with methods available from asymptotic approximation and exact distribution theory. By inverting confidence intervals, Beran [2] shows that one can also use the bootstrap to estimate the power function of complex tests in the presence of nuisance parameters.

The bootstrap has given useful results in more exotic situations. Felsenstein [20] has used the bootstrap to assess the reliability of phylogenetic trees estimated from morphological or genetic data. Another application to tree structures is the use of the bootstrap to estimate the risk of a CART decision rule (*see* RECURSIVE PARTITIONING), described in Breiman et al. [7, Sec. 11.7]. They indicate that for some CART applications, the percentile bootstrap leads to inconsistent results. A bootstrap implementation of kriging* is described in Efron and Diaconis [17]; they also discuss a bootstrap analysis of variable selection procedures for the study of liver transplant data. Ducharme et al. [10] use the bootstrap to construct confidence cones for the mean vector of directional data*, and compare bootstrap performance with alternative methods via simulation. And Efron [13] modifies the bootstrap procedure to handle censored data*.

References

[1] Abramovitch, L. and Singh, K. (1985). *Ann. Statist.*, **13**, 116–132. (Shows that the studentized pivot is asymptotically preferable for asymptotically normal estimators.)

[2] Beran, R. (1986). *Ann. Statist.*, **14**, 151–173. (Examines the power of tests based on inverting bootstrap confidence intervals.)

[3] Beran, R. (1987). *Biometrika*, **74**, 457–468. (Compares convergence rates for different pivots, and introduces an automatic pivoting technique.)

[4] Beran, R. and Srivastava, M. S. (1985). *Ann. Statist.*, **13**, 95–115. (Uses the bootstrap to set confidence regions on functions of covariance matrices.)

[5] Bickel, P. (1987). *J. Amer. Statist. Ass.*, **82**, 191. (Current thoughts on bootstrap trends.)

[6] Bickel, P. and Freedman, D. (1981). *Ann. Statist.*, **9**, 1196–1217. (Early asymptotic work on the bootstrap.)

[7] Breiman, L., Friedman, J. H., Olshen, R. A., and Stone, C. (1984). *Classification and Regression Trees*. Wadsworth, Belmont, CA. (Application of the bootstrap to the estimation of misclassification probabilities; Chap. 11.)

[8] Davison, A., Hinkley, D., and Schechtman, E. (1986). *Biometrika*, **73**, 555–566. (Proposes balanced resampling to reduce computational effort.)

[9]   DiCiccio, T. and Tibshirani, R. (1987). *J. Amer. Statist. Ass.*, **82**, 163–170. (Develops a computationally simpler approximation to the BC_a interval.)

[10] Ducharme, G. R., Jhun, M., Romano, J. P., and Truong, K. N. (1985). *Biometrika*, **72**, 637–645. (Bootstrap circular data.)

[11] Efron, B. (1979). *Ann. Statist.*, **7**, 1–26. (Introduces the bootstrap to the world.)

[12] Efron, B. (1981). *Canad. J. Statist.*, **9**, 139–172. (Uses the bootstrap for standard errors and confidence intervals.)

[13] Efron, B. (1981). *J. Amer. Statist. Ass.*, **76**, 312–319. (Uses the bootstrap with censored data.)

[14] Efron, B. (1982). *The Jackknife, the Bootstrap, and Other Resampling Plans.* SIAM, Philadelphia. (The basic monograph on bootstrapping.)

[15] Efron, B. (1985). *Biometrika*, **72**, 45–58. (Nuisance parameters and the bootstrap.)

[16] Efron, B. (1987). *J. Amer. Statist. Ass.*, **82**, 171–185. (The BC_a interval.)

[17] Efron, B. and Diaconis, P. (1984). *Scientific American*, **248**, 116–130. (A nontechnical introduction to the bootstrap.)

[18] Efron, B. and Gong, G. (1983). *Amer. Statist.*, **37**, 36–48. (A barely technical overview of bootstrap work to that time.)

[19] Efron, B. and Tibshirani, R. (1986). *Statist. Sci.*, **1**, 54–74. (A barely technical review of Efron's recent bootstrap work.)

[20] Felsenstein, J. (1985). *Evolution*, **39**, 783–791. (Applies the bootstrap to evolutionary trees.)

[21] Freedman, D. (1981). *Ann. Statist.*, **9**, 1218–1228. (Asymptotics of the bootstrap in multiple linear regression.)

[22] Ghosh, M., Parr, W. C., Singh, K., and Babu, G. J. (1984). *Ann. Statist.*, **12**, 1130–1135. (An example of bootstrap inconsistency.)

[23] Hall, P. (1986). *Ann. Statist.*, **14**, 1431–1452. (Examines the bootstrap as a one-term Edgeworth inversion, and introduces an iterated bootstrap.)

[24] Hall, P. (1986). *Ann. Statist.*, **14**, 1453–1462. (Calculations on the number of resamples needed for confidence intervals.)

[25] Laird, N. M. and Louis, T. A. (1987). *J. Amer. Statist. Ass.*, **82**, 739–750. (Builds empirical Bayes confidence intervals from bootstrap samples.)

[26] Lo, A. (1987). *Ann. Statist.*, **15**, 360–375. (Strong convergence for the Bayesian bootstrap.)

[27] Loh, W.-Y. (1987). *J. Amer. Statist. Ass.*, **82**, 155–162. (Introduces the calibrated bootstrap.)

[28] Loh, W.-Y. and Wu, C. F. J. (1987). *J. Amer. Statist. Ass.*, **82**, 188–190. (Discussion of bootstrap accuracy.)

[29] Parr, W. C. (1985). *Statist. and Prob. Lett.*, **3**, 97–100. (A simple proof of bootstrap asymptotics.)

[30] Rubin, D. (1981). *Ann. Statist.*, **9**, 130–134. (Introduces the Bayesian bootstrap.)

[31] Schenker, N. (1985). *J. Amer. Statist. Ass.*, **80**, 360–361. (Seminal criticism of the bootstrap, leading to the BC_a method.)

[32] Silverman, B. W. and Young, G. A. (1987). *Biometrika*, **74**, 469–479. (Identifies conditions under which smoothing is not advantageous.)

[33] Singh, K. (1981). *Ann. Statist.*, **9**, 1187–1195. (Early asymptotic work on the bootstrap.)

[34] Stine, R. A. (1985). *J. Amer. Statist. Ass.*, **80**, 1026–1031. (Uses the bootstrap to estimate the probable error of forecasts in regression.)

[35] Wu, C. F. J. (1986). *Ann. Statist.*, **14**, 1261–1295. (An examination of the bootstrap in regression, with 55 pages of discussion by leading bootstrap experts.)

[36] Wynn, H. P. and Ogbonmwan, S. M. (1986). *Ann. Statist.*, **14**, 1340–1342. (Extends ideas of balanced design to bootstrap sampling.)

(JACKKNIFE METHODS
NONPARAMETRIC ESTIMATION OF
   STANDARD ERRORS
RECURSIVE PARTITIONING
RESAMPLING PROCEDURES
SAMPLE REUSE)

DAVID L. BANKS

BOSE, RAJ CHANDRA

Born: June 19, 1901, in Hoshangabad, Madhya Pradesh, India.

Died: October 30, 1987, in Fort Collins, Colorado.

Contributed to: coding theory, design and analysis of experiments, geometry, graph theory, multivariate analysis distribution theory.

Raj Chandra Bose was educated at Punjab University, Delhi University, and Calcutta University. He received M.A. degrees from the latter two institutions, in 1924 (in applied mathematics) and 1927 (in pure mathematics), respectively. At Calcutta he came under the influence of the outstanding geometer S. Mukhopadhyaya.

There followed a lectureship at Asutosh College, Calcutta. This involved very heavy teaching duities, but in his limited spare time, Bose continued to produce research, including joint publications with Mukhopadhyaya, in multidimensional and non-Euclidean geometry.

In 1932, this came to the attention of Mahalanobis*, who was in the process of forming the Indian Statistical Institute*. He needed a research worker who could apply geometrical techniques to distributional problems in multivariate analysis*, along the lines pioneered by Fisher*, and invited Bose to join the young institute.

Initially, Bose worked, as expected, on applications of geometrical methods in multivariate analysis, in particular on the distribution of Mahalanobis' D^2-statistic*. Much of this work was in calibration with Roy (e.g., ref. 6). Later, as a consequence of attending seminars given by Levi (appointed Hardinge Professor of Pure Mathematics at Calcutta University in 1936), he developed an interest in the application of geometrical methods to the construction of experiment designs*. His monumental paper [2] on the construction of balanced incomplete block designs* established his reputation in this field. Further development of these ideas constitutes a major part of Bose's work. It includes the now well-known concept of partially balanced incomplete block designs* [see ref. 4], also the demonstration (with S. S. Shrikhande and E. T. Parker) of the falsity of Euler's conjecture* (ref. 7, foreshadowed by ref. 1) on the nonexistence of $n \times n$ Graeco–Latin squares* for values of n other than a prime or a power of a prime. Reference 3 contains a broad summary of this work, including treatment of confounding* and fractional factorial designs*.

Bose took a part-time position in the Department of Mathematics at Calcutta University in 1938, and moved to the Department of Statistics upon its formation in 1941. He was head of the department from 1945 to 1949, and received a D.Litt. degree in 1947. In 1949, he became a professor in the Department of Statistics at the University of North Carolina, where he stayed until his "retirement" in 1971.

During this period, Bose's interests came to include, with growing emphasis, coding theory*—in particular, the use of geometry to construct codes. Among the outcomes of this interest there resulted, from collaboration with Ray-Chaudhuri, the well-known Bose–Chaudhuri (BCH) codes [5].

After his "retirement," Bose accepted a position, jointly with the Departments of Mathematics and Statistics (later, Mathematics alone), at Colorado State University in Fort Collins. His interests continued to change, gradually returning to more purely mathematical topics, including graph theory. He finally retired in 1980, but retained an active interest in research.

References

[1] Bose, R. C. (1938). *Sankhyā*, **3**, 323–339.

[2] Bose, R. C. (1938). *Ann. Eugen.* (*Lond.*), **9**, 358–399.

[3] Bose, R. C. (1947). *Sankhyā*, **8**, 107–166.

[4] Bose, R. C., Clatworthy, W. H., and Shrikhande, S. S. (1954). Tables of Partially Balanced Designs with Two Associate Classes. *Tech. Bull.* **107**, North Carolina Agricultural Experiment Station, Raleigh, NC. (An extensive revision, by W. H. Clatworthy, with assistance from J. M. Cameron and J. A. Speakman, appeared in 1963 in National Bureau of Standards (U.S.) *Applied Mathematics Series*, **63**.)

[5] Bose R. C. and Ray-Chaudhuri, D. K. (1960). *Inform. Control*, **3**, 68–79.

[6] Bose, R. C. and Roy, S. N. (1936). *Sankhyā*, **4**, 19–38.

[7] Bose, R. C., Shrikhande, S. S., and Parker, E. T. (1960). *Canad. J. Math.*, **12**, 189–203.

(BLOCKS, BALANCED INCOMPLETE
EULER'S CONJECTURE
GRAPH THEORY
INFORMATION THEORY AND CODING
  THEORY
PARTIALLY BALANCED DESIGNS)

BOUNDS, ESTIMATION OF (COOKE'S METHOD)

Cooke [1] proposed the following method of estimating the upper (or lower) bounds of the range of variation of a continuous random variable X, based on a random sample of size n with order statistics* $X_1 \leqslant X_2 \leqslant \cdots \leqslant X_n$.

The central idea of Cooke's suggestion is that if θ is the upper, and ϕ the lower bound, then

$$\theta = E[X_n] + \int_\phi^\theta \{F(x)\}^n \, dx, \quad (1)$$

where $F(x)$ is the CDF of X. On the right-hand side of (1), replacing $E[X_n]$ by the greatest order statistic X_n, ϕ by X_1, and $F(x)$ by the empirical distribution function* (EDF),

$$\hat{F}(x) = \begin{cases} 0, & x < X_1, \\ jn^{-1}, & X_j \leqslant x < X_{j+1}, \\ 1, & x \geqslant X_n, \end{cases}$$

we obtain *Cooke's estimator*

$$X_n + \int_{X_1}^{X_n} \{\hat{F}(x)\}^n \, dx$$

$$= 2X_n - \sum_{j=0}^{n-1} \left[(1 - jn^{-1})^n \right.$$

$$\left. - \{1 - (j+1)n^{-1}\} \right] X_{n-j}.$$

Approximately, for n large, the estimator of θ is

$$2X_n - (1 - e^{-1}) \sum_{j=0}^{n-1} e^{-j} X_{n-j}.$$

Similarly, the estimator of ϕ is

$$2X_1 - \sum_{j=1}^{n} \left[\{1 - (j-1)n^{-1}\}^n \right.$$

$$\left. - (1 - jn^{-1})^n \right] X_j$$

$$\doteq 2X_1 - (e - 1) \sum_{j=1}^{n} e^{-j} X_j.$$

Cooke compared his estimator of θ with the estimator

$$T = 2X_n - X_{n-1} \quad (2)$$

and found (2) was slightly (10–17%) less asymptotically efficient when $\nu \ll 1$ and

$$\{F(x)\}^n \sim \exp\left\{ -\left(\frac{\theta - x}{\theta - u_n} \right)^{1/\nu} \right\},$$

where $u_n = F^{-1}(1 - n^{-1})$. If ν is known, however, the modified Cooke estimator

$$X_n + \left\{ 1 - (1 - e^{-1})^{-\nu} \right\}^{-1}$$

$$\times \left\{ X_n - (1 - e^{-1}) \sum_{j=0}^{n-1} e^{-j} X_{n-j} \right\}$$

has about 30% greater efficiency than the appropriate modification of $T, T' = X_n + \nu^{-1}(X_n - X_{n-1})$.

Reference

[1]   Cooke, P. (1979). *Biometrika*, **66**, 367–374.

(EXTREME-VALUE DISTRIBUTIONS
ORDER STATISTICS
THRESHOLD PARAMETER)

BREAKDOWN

Breakdown is a measure of the proportion of contamination* that a procedure can withstand and still maintain its robustness. See, e.g., Hettmansperger [1] for a detailed discussion of this concept and its relation to influence curves.

Reference

[1]   Hettmansperger, T. P. (1984). *Statistical Inference Based on Ranks*. Wiley, New York.

(INFLUENCE CURVES)

BROWNIAN BRIDGE

The (standard) Brownian motion* process, W, is Gaussian with independent increments and covariance structure $\mathrm{Cov}(W(s), W(t)) = s$ for $s \leqslant t$. When W is Brownian motion [with $W(0) = 0$],

$$B(t) = W(t) - (t/t_0)W(t_0)$$

is called a *Brownian bridge* [with covariance structure $\mathrm{Cov}(B(s), B(t)) = s(1 - t/t_0)$ for $0 \leqslant s \leqslant t \leqslant t_0$] over the interval $[0, t_0]$. A Brownian bridge is also referred sometimes as *tied-down Brownian motion* [because $B(t_0) = B(0) = 0$]. See, e.g., Billingsley [1].

Reference

[1] Billingsley, P. (1968). *Convergence of Probability Measures*. Wiley, New York.

(BROWNIAN MOTION
WIENER MEASURE)

C

CART *See* RECURSIVE PARTITIONING

CATANA SAMPLING *See* WANDERING QUARTER SAMPLING

CENTRAL STATISTIC

Centrality, as introduced by Hartigan [2], is a useful concept for characterizing those statistics that are asymptotically normally distributed. This class of statistics includes the sample mean, as well as many other important statistics (which are not necessarily "location" estimators in the usual sense). There are, of course, many statistics that do not have normal asymptotic distributions and are not central (e.g., the sample maximum). As suggested by the name, a "central" statistic has distribution "centered" near zero.

Let $\{X_i: i \geqslant 1\}$ be independent and identically distributed random variables with common distribution F. Centrality, like asymptotic normality*, is a property involving the limit of a sequence of random variables. Therefore, we will consider a *statistic* to be determined by a sequence of functions $t_n: R^n \to R^1$, $n \geqslant 1$. Define the corresponding random variables

$$T_n \overset{\mathrm{def}}{=} t_n(X_1, X_2, \ldots, X_n), \; n \geqslant 1.$$

[For example, if $t_n(x_1, x_2, \ldots, x_n) = (x_1 + x_2 + \cdots + x_n)/n$, then $\{T_n: n \geqslant 1\}$ is a sequence of sample means.] Let S_n denote an ordered subset of $\{1, 2, \ldots, n\}$, containing $|S_n|$ elements: $(i_1, i_2, \ldots, i_{|S_n|})$. Then define the corresponding random variable

$$T(S_n) \overset{\mathrm{def}}{=} t_{|S_n|}(X_{i_1}, X_{i_2}, \ldots, X_{i_{|S_n|}}).$$

Note that t_n need not be symmetric in its n arguments, and T_n need not have finite moments.

For any random variable X and any constant $A > 0$, denote the truncated* random variable

$$_AX \overset{\mathrm{def}}{=} \begin{cases} X, & \text{if } |X| < A \\ 0, & \text{if } |X| \geqslant A \end{cases}.$$

Expectation of a random variable X is denoted by $E[X]$.

A statistic $\{t_n: n \geqslant 1\}$ is said to be *central for F with variance σ^2* if and only if

(I) $\displaystyle \lim_{A \to \infty} \limsup_{n \to \infty} A^2 \mathrm{Pr}[|T_n| \geqslant A] = 0,$

(II) $\displaystyle \lim_{A \to \infty} \limsup_{n \to \infty} A|E[_AT_n]| = 0,$

and, for each $\rho^2 \in [0, 1]$, we have

(III)

$$\lim_{A \to \infty} \limsup_{n \to \infty} \left| E\left[_A T_n \times {}_A T(S_n)\right] - \rho \sigma^2 \right| = 0,$$

for every sequence $\{S_n: n \geq 1\}$ such that $|S_n|/n \to \rho^2$ and $|S_n| \to \infty$ as $n \to \infty$.

These conditions may be interpreted as follows. Condition (I) controls the tails of the T_n distribution. Condition (II) centers the statistic. Condition (III) requires that the statistic have "meanlike" correlation behavior: The covariance between the statistic and its subsample value is approximately $E[_A T_n \times {}_A T(S_n)]$, because condition (II) centered the statistic near 0. Also, the variance of the statistic is approximately σ^2. Thus condition (III) says that the squared correlation between the statistic and its subsample value should approximately equal the limiting proportion of shared observations (ρ^2). Note that, in the case of the standardized sample mean, i.e., $t_n(x_1, x_2, \ldots, x_n) = (x_1 + x_2 + \cdots + x_n)/n^{1/2}$, the squared correlation between T_n and $T(S_n)$ is precisely $|S_n|/n$ (assuming finite variances and zero means).

Hartigan [2] shows that centrality of t_n is equivalent to asymptotic normality of T_n, in the following sense. The statistic $\{t_n: n \geq 1\}$ is central for F with variance σ^2 if and only if for each $\rho^2 \in [0, 1]$, we have

The random vector $(T_n, T(S_n))$ converges in distribution (as $n \to \infty$) to a bivariate normal* with means 0, variances σ^2, and covariance $\rho\sigma^2$, for every sequence $\{S_n: n \geq 1\}$ such that $|S_n|/n \to \rho^2$ and $|S_n| \to \infty$ as $n \to \infty$.

Note that the squared correlation parameter of the limiting normal distribution corresponds to the limiting proportion of observations shared by the statistic T_n and the subsample value $T(S_n)$.

Examples of central statistics include the sample mean [2], the sample fractiles [1], and U-statistics* [2]. If $\{t_n\}$ are symmetric functions* of their arguments, and are standardized to have $E[T_n] = 0$ for all $n \geq 1$, then a sufficient condition for centrality (and hence asymptotic normality) is simply

$$\lim_{n \to \infty} (n/m_n)^{1/2} E\left[T_n \times T_{m_n}\right] = \sigma^2,$$

for every sequence $\{m_n: n \geq 1\}$ such that

$$n \geq m_n \to \infty \quad \text{as } n \to \infty.$$

The definition of centrality and the related results on asymptotic normality have been extended to the case where $\{X_i\}$ is a strictly stationary sequence [1]. [A sequence of random variables $\{X_i\}$ is said to be *strictly stationary* if $(X_1, X_2, \ldots, X_r)$ has the same joint distribution as $(X_{1+k}, X_{2+k}, \ldots, X_{r+k})$, for all $r \geq 1$ and all k.]

References

[1] Carlstein, E. (1986). *Ann. Prob.*, **14**, 1371–1379.

[2] Hartigan, J. (1975). *Ann. Statist.*, **3**, 573–580.

(ARITHMETIC MEAN
CONVERGENCE OF SEQUENCES OF
   RANDOM VARIABLES
LIMIT THEOREMS)

E. CARLSTEIN

CHANGE-POINT PROBLEM

The change-point problem considers observations ordered by time, or some other variable, and proposes that their distribution changes, perhaps abruptly, at some unspecified point in the sequence, the change-point. Generally, the change-point is considered retrospectively with a fixed sample size.

Hinkley [10] appears to be the first to use the term *change-point* and formulates a model as follows. Let $Y_1, \ldots, Y_T$ be independent random variables and

$$Y_t = \begin{cases} \theta_0(t) + \epsilon_t, & t = 1, \ldots, \tau, \\ \theta_1(t) + \epsilon_t, & t = \tau + 1, \ldots, T, \end{cases} \tag{1}$$

where the ϵ_t are independent errors and τ is the change-point. A simple, but useful form of this model, takes $\theta_0(t) = \theta_0$ and $\theta_1(t) = \theta_1$

so that there is only a change in location or mean level after $t = \tau$. This should not be confused with intervention model* analysis where τ is assumed known. Another model assumes a regression* relationship of the form $\theta_0(t) = \mathbf{x}_t^T\beta_0$ and $\theta_1(t) = \mathbf{x}_t^T\beta_1$, which can be constrained to give the continuous model $\theta_0(t) = \alpha + \beta_0(x_t - \gamma)$, $\theta_1(t) = \alpha + \beta_1(x_t - \gamma)$ for a single covariate x, with $x_1 < \cdots < x_T$ and $x_\tau \leqslant \gamma < x_{\tau+1}$. In the first case there is an abrupt change in the regression relationship after $t = \tau$, and in the second case there is a smooth change. These regression models have been variously called "two-phase regression," Hinkley [9], "switching regression" (see REGRESSIONS, SWITCHING), and "broken-stick regression," and the last for the smooth change only. These basic models can obviously be extended to more than one change-point with different distributional assumptions for the observations. Sampling theory, likelihood and Bayesian analyses for change-point models have been given and, in addition, various ad hoc procedures and those using ranks have been proposed.

The range of applications of change-point models is large and varied, with the emphasis sometimes being on a test of "no-change" while other times it is an estimation of the change-point. Some areas of application are: epidemiology, clinical monitoring, fraud, industrial monitoring, literature analysis, and economic time series. Others abound.

FREQUENTIST LIKELIHOOD APPROACHES

For the change-point model with only a change in level specified in (1), the theory for the test of no-change offers challenging work in random walk* theory and asymptotics. The small sample analysis is, in general, complex. Consider first the case where, in terms of (1), we assume that $E[Y_t] = \theta_0$, $t = 1, \ldots, \tau$, $E[Y_t] = \theta_1$, $t = \tau + 1, \ldots, T$, and Y_t is normally distributed with unit variance. The null hypothesis of "no-change" corresponds to $\tau = T$, or equivalently, $\theta_0 =$

θ_1, whilst the alternative of "change" is given by $1 \leqslant \tau < T$ and $\theta_0 \neq \theta_1$. Let $S_t = Y_1 + \cdots + Y_t$; then the log-likelihood ratio* statistic for testing "no-change" against "change" is equivalent to

$$\max_{1 \leqslant t < T} \{S_t - tS_T/T\}^2/\{t(1 - t/T)\}, \quad (2)$$

while, if the variance is unknown but constant, this statistic is divided by the sample variance. The null hypothesis of "no-change" is rejected in favor of the alternative "change" if the statistic (2) is significantly large. The term maximized in (2) is equivalent to the square of the usual two-sample statistic to test that the $Y_1, \ldots, Y_t$ have the same mean as $Y_{t+1}, \ldots, Y_T$, given unit variances. The terms $S_t - tS_T/T$, $t = 1, \ldots, T$, constitute a series of CUSUMs* of $Y_t - \bar{Y}$, $\bar{Y} = S_T/T$, and can be usefully plotted; see Pettitt [20, Fig. 2] for example. Hawkins [8] gives an iterative method of finding the distribution of the statistic (2) which is successfully implemented for sample sizes up to $T = 50$; see also Worsley [26] for the unknown variance case. Other statistics such as those based on recursive residuals can be suggested for this problem. For a review, see James et al. [12], who also give good small sample approximations for many test statistics derived from asymptotic theory.

For nonnormal data, there have been far fewer contributions. For independent binary data, with Y_j taking values 0 or 1, Pettitt [21] suggests the statistic

$$K_T = \max_{1 \leqslant t < T} T|S_t - tS_T/T|, \quad (3)$$

for a test of no-change against change. Here the model is

$$\Pr[Y_t = 1] = 1 - \Pr[Y_t = 0] = \theta_0,$$
$$t = 1, \ldots, \tau,$$
$$\Pr[Y_t = 1] = 1 - \Pr[Y_t = 0] = \theta_1,$$
$$t = \tau + 1, \ldots, T,$$

and, as before, no-change corresponds to $\tau = T$ or $\theta_0 = \theta_1$ and change to $\tau \neq T$ and $\theta_0 \neq \theta_1$. The null hypothesis of no-change is rejected in favor of change if K_T is significantly large. For this problem, the likelihood ratio statistic is more complicated, while the statistic (3) corresponds to the maximum absolute value of the CUSUMs of $Y_t - \overline{Y}$, $\overline{Y} = S_T/T$. The null distribution of K_T conditional on $S_T = m$ is such that K_T/m $(T - m)$ has the same distribution as that of the two-sample Kolmogorov–Smirnov statistic* based on sample sizes m and $T - m$. Worsley [27, 28] investigates test statistics for Y_j's having distributions in the exponential family*. Tables of percentage points are given for exponential random variables by Worsley [28], and the binomial and Poisson cases in Worsley [27]. Power of the tests is also considered.

For the nonnull case where there is a change, estimation of and confidence regions for τ can be considered. Hinkley [10] gives some asymptotic theory based on discrete random walks to obtain confidence regions for τ, but the techniques need large ($T \geqslant 200$) samples for the approximations. Alternative small sample techniques are given by Cobb [3] and Worsley [28] and asymptotic approximations by Siegmund [25]. Estimation of τ is straightforward; the maximum likelihood estimate is the value of t which maximizes the statistic in (2); other similar, but ad hoc, estimates have been suggested, for example, Pettitt [21] compares the estimate based on K_T, (3), with the maximum likelihood estimate.

ILLUSTRATION

As an illustration we can consider a sequence of binary observations given in Table 1 of Pettitt [20]. There are $T = 40$ observations given by 0, 3(1), 5(0), 3(1), 2(0), 1, 2(0), 5(1), 0, 2(1), 0, 10(1), 0, 3(1), where $n(x)$, $x = 0$ or 1, denotes a sequence of n x's. Here $S_T = 27$ and the statistic K_T equals 179, with this value being obtained at $t = 17$. To test the null hypothesis of no-change we note, as above, K_T, conditional on $S_T = 27$, has its null (no-change) distribution the same as that of $mn\, D_{mn}$, where D_{mn} is the Kolmogorov–Smirnov statistic*, and $m = 27$, $n = 13$. From tables we find $\Pr[mn\, D_{mn} \geqslant 179] \simeq 0.02$, $m = 27$, $n = 13$, so indicating

strong evidence of a change, estimated, using K_T, to be at $t = 17$. Obviously, the probability of 1 has probably increased in the latter part of the series.

REGRESSION

The regression problem is more complex. For a discussion of some regression models see REGRESSIONS, SWITCHING and equation (4) therein, which defines a change-point model with i^* as the change-point. The context of the article is economic. The regression relationship can be specified to be continuous at the change-point or abrupt, and tests for no-change are affected by the continuity constraint. Hinkley [9] gives details of maximum likelihood* estimates and likelihood confidence regions* for the smooth change model with one regressor. Feder [6, 7] discusses problems and shows why standard techniques break down, while Beckman and Cook [1] consider the effect of the continuity constraint on test statistics and caution its use. Likelihood inference for the change-point is considered by Esterby and El-Shaarawi [5] when the relationships before and after the change-point are polynomials of, perhaps, unknown degree and continuity is not assumed. Other models, involving smooth transitions, are outside the scope of this article.

NONPARAMETRIC APPROACH

Methods involving ranks have been proposed to test the null hypothesis of no-change against a change for various alternatives. Here the model is as follows:

Y_t has distribution function

$$F_0(y), \qquad t = 1, \ldots, \tau,$$

$$\text{and } F_1(y), \qquad t = \tau + 1, \ldots, T.$$

The hypothesis of no-change corresponds to $\tau = T$ or $F_0(y) = F_1(y)$, and change to $1 \leqslant \tau < T$ and $F_0(y) \neq F_1(y)$. For a shift of mean level or location, suppose W_t is the

Mann–Whitney–Wilcoxon* two-sample statistic based on the two samples $Y_1, \ldots, Y_t$ and $Y_{t+1}, \ldots, Y_T$. Let μ_t and σ_t^2 be the null (no-change) mean and variance of W_t; then Sen and Srivastava [24] suggest

$$SS_T = \max_{1 \leqslant t < T} \left| (W_t - \mu_y)/\sigma_t \right|$$

and Pettitt [20] suggests

$$J_T = \max_{1 \leqslant t < T} |W_t - \mu_t|$$

as test statistics for no-change, rejecting for large values.

The statistic $(W_t - \mu_t)/\sigma_t$ is the standardized Mann–Whitney–Wilcoxon statistic for testing $F_0(y) = F_1(y)$ given $\tau = t$ and SS_T maximizes its absolute value over t: SS_T mimics the statistic (2) but uses ranks. Pettitt [20] derived $(W_t - \mu_t)$ as, equivalently, a CUSUM of the terms $(R_t - (T + 1)/2)$, where R_t is the rank of Y_t amongst the $Y_1, \ldots, Y_T$. The null distribution of $T^{-1}\{3/(T + 1)\}^{1/2}J_T$ converges to that of the Kolmogorov–Smirnov goodness-of-fit statistic*, $\sqrt{n}\,D$, although slowly and conservatively in the right tail; Lombard [15] gives detailed theory. The asymptotic distribution of SS_T is unknown. Sen [23] considers the use of recursive residual ranks to construct test statistics for change-point alternatives, including the abrupt location and regression changes.

Natural ad hoc estimates of the change-point are suggested by considering statistics such as J_T where the value of t which maximizes $|W_t - \mu_t|$ provides an estimate of τ. The asymptotic distribution of these estimates is generally unknown. A plot of $W_t - \mu_t$ against t is a useful monitoring device to spot changes on an ad hoc basis.

BAYESIAN ANALYSES

Bayesian analyses of change-point models generally offer explicit solutions somewhat more easily than the frequentist approach; see Booth and Smith [2] for a review and references. Although most of the Bayesian* work appears in the economics and econo-metrics literature, there are many other areas of application, especially in medicine and biometrics. Monitoring of a patient's response is an important use of these techniques. The general idea is that parameters entering the model can be removed by integration to obtain the marginal distribution of the data, given the change-point:

$$p(y_1, \ldots, y_T | \tau)$$
$$= \int \int p(y_1, \ldots, y_T | \tau, \psi_0, \psi_1)$$
$$\times p(\psi_0, \psi_1)\, d\psi_0\, d\psi_1,$$
$$1 \leqslant \tau < T, \quad (4)$$

$$p(y_1, \ldots, y_T | \tau = T)$$
$$= \int p(y_1, \ldots, y_T | \tau = T, \psi_0)\, p(\psi_0)\, d\psi_0.$$

Here ψ_0 and ψ_1 are possibly vector *nuisance* parameters, a priori independent of τ, with ψ_1 solely involved with the distribution after the change and $p(y_1, \ldots, y_T | \tau, \psi_0, \psi_1)$ is the likelihood for $1 \leqslant \tau < T$ and $p(y_1, \ldots, y_T | \tau = T, \psi_0)$ the likelihood for no-change. The posterior distribution of τ is then given by

$$p(\tau | y_1, \ldots, y_T) \propto p(y_1, \ldots, y_T | \tau)\, p(\tau),$$
$$1 \leqslant \tau \leqslant T. \quad (5)$$

Marginal posterior distributions* of ψ_0 and ψ_1 can also be found. Problems, however, arise in (5) for $\tau = T$ if the prior distribution for ψ_1 is improper, that is

$$\int p(\psi_0, \psi_1)\, d\psi_1$$

does not exist, since then the marginal distribution of ψ_0, $p(\psi_0)$, in (4) cannot be properly defined and is arbitrary up to a multiplicative constant. Booth and Smith [2] describe one way of defining the arbitrary constant so that the posterior odds of no-change ($\tau = T$) to change ($\tau < T$) can be computed. They give details of the Bayesian analysis of multivariate models, abrupt changes in multiple regression models and means in linear time series (ARMA) models.

As an example, suppose Y_t is normal with mean μ_0 and variance σ_0^2, for $t = 1, 2, \ldots, \tau$, and has mean μ_1 and variance σ_1^2, for $t =$

$\tau + 1, \ldots, T$. Here $\psi_0^T = (\mu_0, \sigma_0^2)$ and $\psi_1^T = (\mu_1, \sigma_1^2)$. If a vague improper prior for (ψ_0, ψ_1) is taken, that is

$$p(\mu_0, \sigma_0^2, \mu_1, \sigma_1^2) \propto \sigma_0^{-2}\sigma_1^{-2},$$

then $p(y_1, \ldots, y_T|\tau)$, as above, for $1 \leqslant \tau < T$, can be obtained by integration to give

$$p(y_1, \ldots, y_T|\tau)$$

$$\propto \Gamma(\tfrac{1}{2}\tau + \tfrac{1}{2})\Gamma(\tfrac{1}{2}T - \tfrac{1}{2}\tau + \tfrac{1}{2})$$

$$\times \{\tau(t - \tau)\}^{-1/2}$$

$$\times \left\{\sum_{t=1}^{\tau} (y_t - \bar{y}_\tau)^2\right\}^{-(\tau+1)/2}$$

$$\times \left\{\sum_{t=\tau+1}^{T} (y_t - \bar{y}_\tau')^2\right\}^{-(T-\tau+1)/2},$$

where

$$\bar{y}_\tau = \tau^{-1} \sum_{t=1}^{\tau} y_t,$$

$$\bar{y}_\tau' = (T - \tau)^{-1} \sum_{t=\tau+1}^{T} y_t.$$

If the prior for τ is uniform over $1, \ldots, T - 1$, then the posterior distribution for τ is proportional to $p(y_1, \ldots, y_T|\tau)$ for $\tau = 1, \ldots, T - 1$. The above-mentioned problems arise if τ has a nonzero prior probability of being T since the prior distribution of (μ_1, σ_1^2) is improper.

Additional analyses including the following: Diaz [4], scale change in gamma random variables; Raftery and Akman [22], rate change in a Poisson process*; Menzefricke [19], scale change in normal random variables; Hsu [11], robust inference for a change in a regression model, and also see this paper for earlier work by Hsu involving normal and gamma scale shifts.

FURTHER EXTENSIONS

A change-point model is proposed by Matthews and Farewell [17] for a hazard function in the analysis of failure data; see also Matthews et al. [18]. Siegmund [25], apart from obtaining good approximations for percentiles of test statistics and confidence regions for a change-point, considers a "square wave" or epidemic change-point model which is suggested by Levin and Kline [14]; see also Lombard [16] for nonparametric test procedures. Kendall and Kendall [13] consider change-points in two-dimensional space for an archaeological problem.

References

[1] Beckman, R. J. and Cook, R. D. (1979). *Technometrics*, **21**, 65–69.

[2] Booth, N. B. and Smith, A. F. M. (1982). *J. Econometrics*, **19**, 7–22.

[3] Cobb, G. W. (1978). *Biometrika*, **65**, 243–251.

[4] Diaz, J. (1982). *J. Econometrics*, **19**, 23–29.

[5] Esterby, S. R. and El-Shaarawi, A. H. (1981). *Appl. Statist.*, **30**, 277–285.

[6] Feder, P. I. (1975a). *Ann. Statist.*, **3**, 49–83.

[7] Feder, P. I. (1975b). *Ann. Statist.*, **3**, 84–97.

[8] Hawkins, D. M. (1977). *J. Amer. Statist. Ass.*, **72**, 180–186.

[9] Hinkley, D. V. (1969). *Biometrika*, **56**, 494–504.

[10] Hinkley, D. V. (1970). *Biometrika*, **57**, 1–17.

[11] Hsu, D. A. (1982). *J. Econometrics*, **19**, 89–107.

[12] James, B., James, K. L., and Siegmund, D. (1987). *Biometrika*, **74**, 71–83.

[13] Kendall, D. G. and Kendall, W. S. (1980). *Adv. Appl. Prob.*, **12**, 380–424.

[14] Levin, B. and Kline, J. (1985). *Statist. Med.*, **4**, 469–488.

[15] Lombard, F. (1983). *S. Afr. Statist. J.*, **17**, 83–105.

[16] Lombard, F. (1986). *Biometrika*, **74**, 615–624.

[17] Matthews, D. E. and Farewell, V. T. (1982). *Biometrics*, **38**, 463–468.

[18] Matthews, D. E., Farewell, V. J., and Pyke, R. (1985). *Ann. Statist.*, **13**, 583–591.

[19] Menzefricke, U. (1981). *Appl. Statist.*, **30**, 141–146.

[20] Pettitt, A. N. (1979). *Appl. Statist.*, **28**, 126–135.

[21] Pettitt, A. N. (1980). *Biometrika*, **67**, 79–84.

[22] Raftery, A. E. and Akman, V. E. (1986). *Biometrika*, **73**, 85–89.

[23] Sen, P. K. (1983). In *Recent Advances in Statistics*, M. H. Rizvi, J. Rustagi, and D. Siegmund, eds., Academic, New York, pp. 371–391.

[24] Sen, A. and Srivastava, M. S. (1975). *Ann. Statist.*, **3**, 98–108.

[25] Siegmund, D. (1986). *Ann. Statist.*, **14**, 361–404.

[26] Worsley, K. J. (1979). *J. Amer. Statist. Ass.*, **74**, 365–367.

[27] Worsley, K. J. (1983). *Biometrika*, **70**, 455–464.

[28] Worsley, K. J. (1986). *Biometrika*, **73**, 91–104.

Bibliography

Bhattacharyya, G. K. (1984). In *Handbook of Statistics, Nonparametric Statistics*, Vol. 4, P. R. Krishnaiah and P. K. Sen, eds. North-Holland, Amsterdam, pp. 89–111. (Nonparametric methods in Sec. 3, statistics such as (3).)

Sen, P. K. (1984). In *Handbook of Statistics, Nonparametric Statistics*, Vol. 4, P. R. Krishnaiah and P. K. Sen, eds. North-Holland, Amsterdam, pp. 699–739. (Nonparametric methods in Sec. 2, sequential methods mainly.)

Shaban, S. A. (1980). *Int. Statist. Rev.*, **48**, 83–93. (Annotated bibliography, 61 references.)

Wolfe, D. A. and Schectman, E. (1984). *J. Statist. Plann. Inf.*, **9**, 389–396. (Nonparametric tests.)

Zacks, S. (1983). In *Recent Advances in Statistics*, M. H. Rizvi, J. Rustagi, and D. Siegmund, eds., Academic, New York, pp. 245–269. (General review, 70 references.)

(CHANGE-POINT MODEL
REGRESSIONS, SWITCHING
ROBUST TESTS FOR CHANGE-POINT
   MODELS
TIME SERIES)

A. N. PETTITT

CHAOS

A *dynamical system* is a system which evolves forward in time by strictly deterministic rules (e.g., a system of N first-order ordinary differential equations in time). An *attractor* of a dynamical system is a bounded limit set of the forward time evolution of typical system states. The *basin of attraction* for the attractor is the set of initial conditions which are asymptotic with time to that attractor. An attractor is *chaotic* if infinitesimally separated initial conditions in the basin of the attractor typically diverge from each other exponentially in time [1, 2]. In practice, the exponential sensitivity on initial conditions means that even crude prediction of the system state too far into the future may be difficult or impossible due to unavoidable small errors in the initial conditions. In effect, the system state beyond a certain time is essentially random. Furthermore, the time dependence generated by chaotic systems can be very complex. On viewing such time dependence one often (rightly) has the feeling that a statistical description, rather than a purely deterministic description, is appropriate. For example, for chaotic attractors the orbits are ergodic on the attractor so that time averages can be replaced by averages over an invariant probability measure on the attractor. Most pseudo-random number generators (*see* GENERATION OF RANDOM VARIABLES) are examples of chaotic dynamical systems. Another aspect of chaotic dynamics is that the structure of attractors and their basins can have arbitrarily fine-scaled fractal geometry (*see* FRACTALS). Recently it has become clear that many nonlinear dynamical systems in a broad range of scientific and engineering disciplines exhibit chaotic behavior. Examples where chaotic time evolution has occurred and has had important consequences include fluid flows, chemically reacting systems, lasers, electronic circuits, etc.

References

The following two review articles provide an introduction to chaotic dynamics.

[1] Grebogi, C., Ott, E., and Yorke, J. A. (1987). *Science*, **238**, 632–638.

[2] Ott, E. (1981). *Rev. Mod. Phys.*, **53**, 655–671.

(RANDOMNESS AND PROBABILITY)

EDWARD OTT

CHERNOFF INEQUALITY

Let X be a standard normal random variable. If $g(X)$ is an absolutely continuous* function of X, and has a finite variance, then

$$\mathrm{var}(g(X)) \leqslant E\left[\{g'(x)\}^2\right] \qquad (1)$$

(Chernoff [4])

with equality if and only if $g(\cdot)$ is linear. Similar inequalities, for various distributions, have been discussed by Borovkov and Utev [1] and Cacoullos and Papathanasiou [3].

A more general inequality, derived by Borokov and Utev [2] using a simple approach based on the Cauchy-Schwarz* inequality, is as follows.

Let X be a random variable with PDF $f(x)$ and finite expectation $E[X] = \mu$. Define

$$W(x) = \{\mu - E[X|X < x]\} \times \Pr[X < x]$$

$$= -\int_{-\infty}^{x} (t - \mu) f(t) \, dt.$$

Clearly $W(x) \geqslant 0$ and $\lim_{x \to \infty} W(x) = \lim_{x \to -\infty} W(x) = 0$. If $W(x) \geqslant cf(x)$ for all x, with constant c, then for any differentiable function

$$\text{var}(g(X)) \leqslant cE\left[\{g'(X)\}^2\right].$$

The equality sign holds if and only if X is normally distributed, thus constituting a characterization of normal distributions.

Hu [5] has shown that for any positive random variable X with continuous PDF $f(x)$, CDF $F(x)$, and positive variance,

$$\text{var}(X) \leqslant E\left[\frac{F(X)\{1 - F(X)\}}{2\{f(X)\}^2}\right].$$

Equality holds if and only if X has a rectangular (uniform) distribution*. This provides a characterization of uniform distributions.

Rao and Sreehari [6] have obtained a similar type of characterization of Poisson distributions, namely: An integer nonnegative random variable, X, has a Poisson distribution if and only if

$$\sup \frac{\text{var}(h(X))}{\text{var}(X)E\left[\{h(X-1) - h(X)\}^2\right]} = 1,$$

the supremum being taken over all real-valued functions $h(\cdot)$ such that $E[\{h(X+1) - h(X)\}^2]$ is finite.

References

[1] Borovkov, A. A. and Utev, S. A. (1983). *Teor. Veroyat. Primen.*, **28**, 209–217.

[2] Borovkov, A. A. and Utev, S. A. (1983). *Teor. Veroyat. Primen.*, **28**, 606–607.

[3] Cacoullos, T. and Papathanasiou, V. (1985). *Statist. Prob Lett.*, **3**, 175–184.

[4] Chernoff, H. (1981). *Ann. Prob.*, **9**, 533–535.

[5] Hu, C.-Y. (1986). *Bull. Inst. Math. Acad. Sin.*, **14**, 21–23.

[6] Rao, B. L. S. P. and Sreehari, M. (1987). *Austral. J. Statist.*, **29**, 38–41.

(CAUCHY–SCHWARZ INEQUALITY CHARACTERIZATIONS OF DISTRIBUTIONS CRAMÉR–RAO LOWER BOUND PROBABILITY INEQUALITIES FOR SUMS OF BOUNDED RANDOM VARIABLES JENSEN'S INEQUALITY VARIANCE)

COCHRANE–ORCUTT ITERATIVE PROCEDURE *See* HILDRETH–LU SCANNING METHOD

COCHRAN'S C_0 STATISTIC

A precursor of the Mantel–Haenszel statistic* for combining data from several two-by-two tables*. Suppose there are K two-by-two tables representing two factors A and B, each at two levels, as set out below (for the kth table):

	A_1	A_2	Total
B_1	n_{11k}	n_{21k}	$n_{\cdot 1k}$
B_2	n_{12k}	n_{22k}	$n_{\cdot 2k}$
Total	$n_{1\cdot k}$	$n_{2\cdot k}$	$n_{\cdot\cdot k}$

To test the hypothesis

$$H_0: p_{1k} = p_{2k}, \qquad k = 1, 2, \ldots, K,$$

where

$$p_{jk} = \Pr[A_1|B_j, k\text{th table}], \qquad j = 1, 2,$$

Cochran [1] proposed using the statistic

$$C_0 = \frac{\sum_{k=1}^{K}\{w_k(\hat{p}_{1k} - \hat{p}_{2k})\}}{\{\sum_{k=1}^{K} w_k \bar{p}_k(1 - \bar{p}_k)\}^{1/2}},$$

where $w_k = n_{.1k} n_{.2k}/n_{..k}$, $\hat{p}_{jk} = n_{1jk}/n_{.jk}$
and

$$\bar{p}_k = (n_{.1k}\hat{p}_{1k} + n_{.2k}\hat{p}_{2k})/n_{..k}$$
$$= n_{1.k}/n_{..k}.$$

The statistic has, approximately, a unit normal distribution when H_0 is valid.

In terms of the n_{ijk}'s,

$$C_0 = \frac{\sum_{k=1}^{K}(n_{11k} - n_{1.k}n_{.1k}/n_{..k})}{\left[\sum_{k=1}^{K}\{n_{1.k}n_{2.k}n_{.1k}n_{.2k}/n_{..k}^3\}\right]^{1/2}}.$$

Comparison of C_0^2 with the Mantel–Haenszel statistic

$$\frac{\left\{\left|\sum_{k=1}^{K}(n_{11k} - n_{1.k}n_{.1k}/n_{..k})\right| - \frac{1}{2}\right\}^2}{\sum_{k=1}^{K}\left[n_{1.k}n_{2.k}n_{.1k}n_{.2k}/\{n_{..k}^2(n_{..k} - 1)\}\right]}$$

is instructive. The two statistics differ only in respect of a continuity correction* in the numerator and replacement of $n_{..k}^3$ by $n_{..k}^2(n_{..k} - 1)$ in the denominator.

Approximate power* calculations for Cochran's test are described by Woolson et al. [2]. (The hypothesis H_0 corresponds to independence of A and B in all the K tables. The test criterion is especially aimed at detecting alternatives wherein the sign of $p_{1k} - p_{2k}$ is the same for most of the k's.)

References

[1] Cochran, W. G. (1954). *Biometrics*, **10**, 417–451.
[2] Woolson, R. F., Been, J. A., and Rojas, P. B. (1986). *Biometrics*, **42**, 927–932.

(CHI-SQUARE TESTS
CONTINGENCY TABLES
MANTEL–HAENSZEL STATISTIC
TWO-BY-TWO (2 × 2) TABLES)

COMPOSITIONAL DATA

A typical sample of compositional data consists of multivariate observations x with the constraints that the components of x are nonnegative and sum to unity. Thus each observation represents a set of proportions and the appropriate sample space is a simplex. Compositional data arise in many contexts: in geology as the percentage weight compositions of rock samples in terms of the constituent oxides; in home economics as the percentages of the budget spent on different sources of expenditure; in zoology as the percentages of different species present at different sites.

Appropriate multivariate probability distributions are clearly required as models for compositional data and as the foundation for inferential procedures, in the same way that the multivariate normal* distribution is used for unrestricted multivariate data. The only familiar distributions on the simplex are the Dirichlet distributions* and these were used in much of the early work, particularly in geological applications. However, their usefulness is considerably restricted, partly because of the paucity of statistical methodology for the Dirichlet and partly because of the inability of the Dirichlet to model correlations among the coefficients of x apart from spurious correlations imposed by the basic constraints on x. (Remember that there is a characterization of a D-dimensional Dirichlet random vector x in terms of a set of D independent gamma random variables, summarized by the relationships

$$x_i = w_i \bigg/ \sum_j w_j, \qquad i = 1, \ldots, D.)$$

Recently, Aitchison [1, 2] has used a multivariate logistic transformation to escape from the simplex to unrestricted Euclidean space. For instance, with x as in the previous paragraph, define

$$y_i = \log_e(x_i/x_D), \qquad i = 1, \ldots, D - 1. \tag{1}$$

This creates a $(D - 1)$-dimensional random vector y, with sample space $\mathbb{R}^{D-1}$. As a result, the way is open to use the multivariate normal distribution as a model for y and to take advantage of the wealth of inferential tools associated with the multivariate normal. A compositional vector x constructed by inverting transformation (1) is said to have an additive *logistic-normal** distribution.

It turns out that the Dirichlet class is disjoint from that of the additive logistic-normals, but there is a single class, involving only one more parameter than the number required to describe the logistic-normals, that includes both classes as special cases. These are Aitchison's distributions* $\mathscr{A}^{D-1}$; see ref. 2, Sec. 13.4.

If the logistic-normal distribution is believed to be an adequate model, compositional data can be analyzed by "standard" techniques. Examination of the covariance structure of y reveals useful information about that of x, covariates can often be incorporated using a normal linear regression model for y, and log-contrast principal components for the elements of x can be extracted from a dimensionality reduction exercise on y. Subcompositions can be created by normalizing subsets of the components of x and a variety of useful independence concepts can be developed between disjoint subcompositions and transformations thereof.

References

[1] Aitchison, J. (1982). *J. R. Statist. Soc. B*, **44**, 139–177. (Discussion paper with a wide range of references.)

[2] Aitchison, J. (1986). *The Statistical Analysis of Compositional Data*, Chapman and Hall, London and New York. (Monograph expansion of ref. 1, also with good references and an associated statistical computer package.)

(AITCHISON DISTRIBUTIONS
DIRICHLET DISTRIBUTION
LOGISTIC-NORMAL DISTRIBUTION)

D. M. TITTERINGTON

CONCENTRATION FUNCTION AND SCATTER FUNCTION

The *concentration function* of a random variable X is

$$C_X(d) = \sup_x \Pr[x \leqslant X \leqslant x + d],$$

$$d \geqslant 0.$$

It is often called the *Lévy concentration function**. It has the properties

$$\lim_{d \to \infty} C_X(d) = 1,$$

$$C_{X_1 + X_2}(d) \leqslant \min(C_{X_1}(d), C_{X_2}(d))$$

if X_1 and X_2 are mutually independent.
The *scatter function* is

$$G_X(p) = \inf\{d : C_X(d) \geqslant p\},$$

$$0 \leqslant p \leqslant 1.$$

The concentration and scatter function, are in a sense, dual to each other as measures of variability of a random variable.

See refs. 1 and 2 and LÉVY CONCENTRATION FUNCTION for further details.

References

[1] Esséen, C. G. (1968). *Zeit. Wahrscheinlichkeitsth. Verw. Geb.*, **9**, 290–308.

[2] Kesten, H. (1969). *Math. Scand.*, **25**, 133–144.

(LÉVY CONCENTRATION FUNCTION
MEAN CONCENTRATION FUNCTION)

CONSENSUAL MATRICES

This concept is used in "expert opinion" consensus models, and in related problems, involving combination of statistics. It arises from the following background.

Suppose n experts are trying to evaluate a quantity in terms of a real scalar (or vector). Denote their initial estimates by $_1F_i$, $i = 1, \ldots, n$.

After sharing their estimates they form new estimates

$$_2F_i = \sum_{j=1}^{n} {}_1a_{ij} \cdot {}_1F_j, \qquad i = 1, \ldots, n.$$

The process is repeated, so that the $(k + 1)$th set of estimates is related to the kth set by the linear equations

$$_{k+1}F_i = \sum_{j=1}^{n} {}_ka_{ij} \cdot {}_kF_j, \qquad i = 1, \ldots, n.$$

The coefficients $_ka_{ij}$ may depend on k.

Denote the n-vector with elements $\{_k F_i\}$ by $_k \mathbf{F}$ and the $n \times n$ matrix $(_k a_{ij})$ by $_k \mathbf{A}$. Evidently,

$$_k \mathbf{F} = {_{k-1}\mathbf{A}} \cdot {_{k-2}\mathbf{A}} \cdot \cdots \cdot {_1\mathbf{A}} \cdot {_1\mathbf{F}}, \qquad \geqslant 2.$$

(Note that if $_k \mathbf{A} = \mathbf{I}$, it means that at the kth stage, each expert is not affected by the opinions of the other experts.)

The sequence of matrices

$$\{_k \mathbf{A}: k = 1, 2, \ldots\}$$

is called *consensual* if, for every pair (i, j) of experts *and* for every $_1\mathbf{F}$,

$$|_k F_i - {_k F_j}| \to 0 \quad \text{as } k \to \infty.$$

A related concept is *weak ergodicity* (see, e.g., Hajnal [8]), which requires that *for each* $j = 1, 2, \ldots$ the sequence

$$\{_k \mathbf{A}: k = j, j + 1 \ldots\}$$

is consensual. This is a stronger condition than consensuality. For example, if, for some k, $_k \mathbf{A}$ has all its rows identical, so that $_{k+1}F_i$ does not depend on i, then $\{_k \mathbf{A}: k = 1, 2, \ldots\}$ is consensual, but it need not be weakly ergodic.

Conditions for consensuality have been studied by Berger [1], Chatterjee and Seneta [2], De Groot [5], Forrest [6], Lehrer [11], and Seneta [12]. The subject is closely related to the asymptotic properties of products of random matrices; see, for example, Cohen [3], Kifer [9], and Kingman [10].

Recent (1986) discussions of applications of consensual matrices are given in Cohen et al. [4] and Genest and Zidek [7].

References

[1] Berger, R. L. (1981). *J. Amer. Statist. Ass.*, **76**, 415–418.

[2] Chatterjee, S. and Seneta, E. (1977). *J. Appl. Prob.*, **14**, 89–97.

[3] Cohen, J. E. (1980). *Proc. Natl. Acad. Sci., U.S.A.*, **77**, 3749–3752.

[4] Cohen, J. E., Hajnal, J., and Newman, C. M. (1986). *Stoch. Processes Appl.*, **22**, 315–322.

[5] De Groot, M. H. (1974). *J. Amer. Statist. Ass.*, **69**, 118–121.

[6] Forrest, P. (1985). *Synthèse*, **62**, 75–78.

[7] Genest, C. and Zidek, J. V. (1986). *Statist. Sci.*, **1**, 114–148.

[8] Hajnal, J. *Proc. Camb. Philos. Soc.*, **54**, 233–246.

[9] Kifer, Y. (1982). *Zeit. Wahrscheinlichkeitsth. Verw. Geb.* **61**, 83–95.

[10] Kingman, J. F. C. (1976). In *Lecture Notes in Mathematics*, **539**. (École d'Été de Probabilités de Saint Flour, V, P. L. Hennequin, ed.) Springer, New York, pp. 168–223.

[11] Lehrer, K. (1976). *Nous*, **10**, 327–332. (A philosophical approach.)

[12] Seneta, E. (1981). *Non-Negative Matrices*, 2nd. ed. Springer, New York.

(RANDOM MATRICES)

COPHENETIC MATRIX

In hierarchical clustering procedures a *cophenetic matrix* is constructed from the (dis)similarity matrix. The (i, j) entry of this matrix indicates the fusion level at which a pair of objects $[(i)$ and $(j)]$ appears together in the same cluster *for the first time*.

Example.   Given a dissimilarity matrix

$$
\mathbf{D} =
\begin{array}{c}
\\ 1 \\ 2 \\ 3 \\ 4 \\ 5
\end{array}
\begin{array}{ccccc}
1 & 2 & 3 & 4 & 5 \\
\left(\begin{array}{ccccc}
0.0 & & & & \\
2.0 & 0 & & & \\
6.0 & 5.0 & 0 & & \\
10.0 & 9.0 & 4.0 & 0 & \\
9.0 & 8.0 & 5.0 & 3.0 & 0
\end{array} \right)
\end{array},
$$

the single linkage procedure will result in fusion of 1 and 2 the level 2.0; the fusion of 4 and 5 at the level 3.0; the fusion of $(4, 5)$ with (3) at level 4.0 and the fusion of $(1, 2)$ with $(3, 4, 5)$ at level 5.0. The corresponding cophenetic matrix $\mathbf{C}$ is thus:

$$
\mathbf{C} =
\begin{array}{c}
\\ 1 \\ 2 \\ 3 \\ 4 \\ 5
\end{array}
\begin{array}{ccccc}
1 & 2 & 3 & 4 & 5 \\
\left(\begin{array}{ccccc}
0.0 & & & & \\
2.0 & 0.0 & & & \\
5.0 & 5.0 & 0.0 & & \\
5.0 & 5.0 & 4.0 & 0 & \\
5.0 & 5.0 & 4.0 & 3.0 & 0
\end{array} \right)
\end{array}.
$$

Large values of the correlation coefficient* between the entries of $\mathbf{D}$ and $\mathbf{C}$ indicate that the dendrogram* provides a reasonable grouping. See the references cited in DEN-

DROGRAMS and HIERARCHICAL CLUSTER ANALYSIS for further details.

(MEASURES OF SIMILARITY, DISSIMILARITY AND DISTANCE)

CRAMÉR, HARALD

Born: September 25, 1893, in Stockholm, Sweden.

Died: October 5, 1985, in Stockholm, Sweden.

Contributed to: actuarial mathematics, analytic number theory, central limit theory, characteristic functions, collective risk theory, mathematical statistics, stationary processes.

(Carl) Harald Cramér spent almost the whole of his professional life in Stockholm. He entered Stockholm University in 1912, studying chemistry and mathematics. Although he worked as a research assistant in biochemistry, his primary interest turned to mathematics, in which he obtained a Ph.D. degree in 1917, with a thesis on Dirichlet series.

He became an assistant professor in Stockholm University in 1919 and in the next seven years he published some 20 papers on analytic number theory. During this time, also, Cramér took up a position as actuary with the Svenska Life Insurance Company. This work led to a growing interest in probability and statistics, as a consequence of which Cramér produced work of great importance in statistical theory and methods over the next 60 years.

In 1929, he was appointed to a newly created professorship in "actuarial mathematics and mathematical statistics" (sponsored by Swedish Life Insurance Companies). At this time, also, he was appointed actuary to the Sverige Reinsurance Company. His work there included new developments in premium loadings for life insurance policies and ultimately, after many years, to his book, *Collective Risk Theory* [5].

The 20 years following 1929 were his most intensely productive period of research. *Random Variables and Probability Distributions* [3], published in 1937, provided a fresh, clearly expressed foundation for basic probability theory as used in the development of statistical methods. The seminal book *Mathematical Methods of Statistics* [4], written during enforced restriction of international contacts during World War II, presented a consolidation of his studies, and has been of lasting influence in the development of statistical theory and practice. During these 20 years, also, Cramér built up a flourishing institute, providing conditions wherein workers in many fields of statistics could find encouragement to develop their ideas. To this period, also, belongs the "Cramér–Wold device" [7] for establishment of asymptotic multidimensional normality.

In 1950, Cramér was appointed President of Stockholm University, and in the period until his retirement from this office in 1961 a substantial proportion of his time was occupied with administrative duties, with consequent diminution in research activities.

However, after 1961, he returned to a variety of research endeavors including participation in work for the National Aeronautics and Space Administration at the Research Triangle Institute in North Carolina, during the summers of 1962, 1963, and 1965. During this time also, in collaboration with Leadbetter he produced the book, *Stationary and Related Stochastic Processes* [6]. Collective risk theory* (see ref. 5, referred to earlier) is concerned with the progress through time of funds subject to inputs (premiums and interest) and outputs (claims), constituting a special type of stochastic process*, and Cramér's attention to this field may be regarded as a natural long-term development.

For fuller accounts of Cramér's life and work, see the obituaries by Blom [1] and Leadbetter [8]. Blom and Matérn [2] provide a bibliography of Cramér's publications.

References

[1] Blom, G. (1987). *Ann. Statist.*, **15**, 1335–1350.

[2] Blom, G. and Matérn, B. (1984). *Scand. Actu. J.*, 1–10.

[3] Cramér, H. (1937). *Random Variables and Probability Distributions*. Cambridge Tracts, **36**. Cambridge University Press, London.

[4] Cramér, H. (1945). *Mathematical Methods of Statistics*. Almqvist and Wiksell, Uppsala, Sweden; Princeton University Press; Princeton, NJ.

[5] Cramér, H. (1955). *Collective Risk Theory*. Skandia Insurance Company, Stockholm, Sweden.

[6] Cramér, H. and Leadbetter, M. R. (1967). *Stationary and Related Stochastic Processes*. Wiley, New York.

[7] Cramér, H. and Wold, H. O. A. (1936). *J. Lond. Math. Soc.*, **11**, 290–294.

[8] Leadbetter, M. R. (1988). *Int. Statist. Rev.*, **56**, 89–97.

(CRAMÉR–RAO LOWER BOUND
CRAMÉR('S) SERIES
CRAMÉR–VON MISES STATISTIC
LIMIT THEOREM, CENTRAL
LIMIT THEOREMS)

CRESSIE–READ STATISTIC

Many tests of goodness-of-fit* can be reduced to testing a hypothesis about the parameter $\pi = (\pi_1, \ldots, \pi_k)$ from the multinomial distribution

$$\Pr(\mathbf{X} = \mathbf{x}) = \frac{n!}{x_1! \cdots x_k!} \pi_1^{x_1} \cdots \pi_k^{x_k},$$

where the elements of π are probabilities that sum to 1, and the elements of $\mathbf{x}$ are nonnegative integers that sum to n. The multinomial probability vector π may contain substantial inner structure, such as that imposed by log-linear models for discrete multivariate data. It is also possible that the multinomial* random vector $\mathbf{X}$ is derivative, in that X_j represents the number of times that values of a random sample $Y_1, \ldots, Y_n$ belong to a class C_j. Here $\{C_j: j = 1, \ldots, k\}$ is a set of mutually exclusive classes exhausting the probability content of the distribu-

tion, and this distribution $F(y; \theta)$ may depend on unknown parameters $\theta = (\theta_1, \ldots, \theta_s)$.

To test the simple null hypothesis

$$H_0: \pi = \pi_0,$$

where $\pi_0 = (\pi_{01}, \ldots, \pi_{0k})$ is a prespecified probability vector, the two most commonly used statistics are Pearson's statistic X^2:

$$X^2 = \sum_{j=1}^{k} (X_j - n\pi_{0j})^2 / (n\pi_{0j}),$$

and the log-likelihood-ratio statistic G^2:

$$G^2 = 2 \sum_{j=1}^{k} X_j \ln(X_j / (n\pi_{0j})).$$

A large observed value of the statistic leads to rejection of H_0. There has been considerable controversy in the past over which statistic is more appropriate. This controversy has been documented in a chapter of Read and Cressie [14], and has largely been resolved by Cressie and Read's [3] introduction of a *family* of *power-divergence statistics*:

$$2nI^\lambda = \frac{2}{\lambda(\lambda+1)} \sum_{j=1}^{k} X_j \left\{ \left(\frac{X_j}{n\pi_{0j}} \right)^\lambda - 1 \right\},$$

$$-\infty < \lambda < \infty. \quad (1)$$

The statistics $2nI^0$ and $2nI^{-1}$ are defined to be the limits of $2nI^\lambda$ as $\lambda \to 0$ and $\lambda \to -1$, respectively. A large observed value of $2nI^\lambda$ leads to rejection of H_0.

Under the simple null hypothesis H_0 where k is fixed, *each* member of the family (1) converges in distribution to the chi-squared distribution* on $(k-1)$ degrees of freedom, as $n \to \infty$; and for a composite null hypothesis where an s-dimensional parameter θ is estimated efficiently, a further s degrees of freedom must be subtracted (Cressie and Read [3]). This distributional result is seen most easily in the case of a simple null hypothesis. Observe that (1) can be written as

$$2nI^\lambda = \frac{2n}{\lambda(\lambda + 1)}$$

$$\times \sum_{j=1}^{k} \pi_{0j}\left\{\left(1 + \frac{X_j - n\pi_{0j}}{n\pi_{0j}}\right)^{\lambda+1} - 1\right\},$$

provided $\lambda \neq 0$, $\lambda \neq -1$. Now set $V_j = (X_j - n\pi_{0j})/(n\pi_{0j})$ and expand $(1 + V_j)^{\lambda+1}$ in a Taylor series* to obtain

$$2nI^\lambda = \sum_{j=1}^{k} \pi_{0j}\left(n^{1/2}V_j\right)^2 + o_p(1),$$

where $o_p(1)$ represents a stochastic term which converges to 0 in probability, as $n \rightarrow \infty$. An identical result can be derived in the special cases $\lambda = 0$ and $\lambda = -1$. Thus

$$2nI^\lambda = X^2 + o_p(1), \qquad -\infty < \lambda < \infty.$$

It is clear now why under H_0 each family member has the same asymptotic chi-squared distribution when k is fixed. However, when k grows with n, this asymptotic equivalence no longer holds, as discussed in Cressie and Read [3]. Of course, small-sample properties of $2nI^\lambda$ vary widely as λ ranges over the real line.

Particular values of λ in (1) correspond to well-known statistics: Pearson's statistic* $X^2(\lambda = 1)$, log-likelihood-ratio statistic $G^2(\lambda = 0)$, Freeman–Tukey statistic* ($\lambda = -1/2$), modified log-likelihood-ratio statistic ($\lambda = -1$), and Neyman-modified statistic ($\lambda = -2$). Now it is clear why these statistics are indeed comparable, since they are all members of the same family. Comparisons can be found in Cochran [2], Bahadur [1], West and Kempthorne [17], Holst [5], Larntz [10], Fienberg [4], Hutchinson [6], Koehler and Larntz [9], Upton [16], Lawal [11], Kallenberg et al. [7], Koehler [8], and Rudas [15], although most debate has centered around the comparative merits of X^2 and G^2.

Cressie and Read [3] propose a statistic which is "between" $G^2(\lambda = 0)$ and $X^2(\lambda = 1)$, viz. $\lambda = 2/3$, to take advantage of the desirable properties of both. The test statistic

$$2nI^{2/3} = \frac{9}{5}\sum_{j=1}^{k} X_j\left\{\left(\frac{X_j}{n\pi_{0j}}\right)^{2/3} - 1\right\} \quad (2)$$

has been called the *Cressie–Read statistic* by Rudas [15]; a large observed value of $2nI^{2/3}$ leads to rejection of H_0. [Earlier, Moore [12] named the family (1) the Cressie–Read statistics.] As an omnibus test of goodness of fit, Cressie and Read [3] show $2nI^{2/3}$ to be

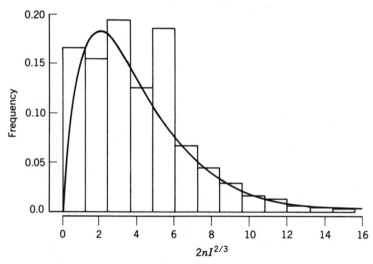

Figure 1 Exact distribution of Cressie–Read statistic and density of a chi-squared random variable on 4 degrees of freedom. (Reprinted from Read [13], with the permission of the American Statistical Association.)

the most competitive in the family (1) when the following criteria are considered jointly: finite sample approximation to the chi-squared distribution with regard to both critical value and moments, power in finite samples, Pitman and Bahadur efficiency*, approximations for large sparse multinomials, and sensitivity. More details can be found in Read [13] and Read and Cressie [14].

To give some idea of the *exact* distribution of the Cressie–Read statistic (2) under H_0, as compared to its asymptotic chi-squared distribution, the case of $n = 10$, $k = 5$, $\pi_{0j} = 1/k$; $j = 1, \ldots, k$, was computed. Figure 1 shows a histogram of the exact distribution function (for multinomial **X** and $\pi = \pi_0$) of the Cressie–Read statistic (2) along with the density of a chi-squared random variable on $k - 1 = 4$ degrees of freedom. Notice how close the two are in the upper tail, precisely where it is most important for obtaining significance levels.

References

[1] Bahadur, R. R. (1971). *Some Limit Theorems in Statistics*. SIAM, Philadelphia.

[2] Cochran, W. G. (1952). *Ann. Math. Statist.*, **23**, 315–345.

[3] Cressie, N. and Read, T. R. C. (1984). *J. R. Statist. Soc. B*, **46**, 440–464.

[4] Fienberg, S. E. (1979). *J. R. Statist. Soc. B*, **41**, 54–64.

[5] Holst, L. (1972). *Biometrika*, **59**, 137–145.

[6] Hutchinson, T. P. (1979). *Commun. Statist.-Theor. Meth.*, **8**, 327–335.

[7] Kallenberg, W. C. M., Oosterhoff, J., and Schriever, B. F. (1985). *J. Amer. Statist. Ass.*, **80**, 959–968.

[8] Koehler, K. J. (1986). *J. Amer. Statist. Ass.*, **81**, 483–493.

[9] Koehler, K. J. and Larntz, K. (1980). *J. Amer. Statist. Ass.*, **75**, 336–344.

[10] Larntz, K. (1978). *J. Amer. Statist. Ass.*, **73**, 253–263.

[11] Lawal, H. B. (1984). *Biometrika*, **71**, 415–458.

[12] Moore, D. S. (1984). *J. Statist. Plann. Inf.*, **10**, 151–166.

[13] Read, T. R. C. (1984). *J. Amer. Statist. Ass.*, **79**, 929–935.

[14] Read, T. R. C. and Cressie, N. (1988). *Goodness-of-Fit Statistics for Discrete Multivariate Data*. Springer, New York.

[15] Rudas, T. (1986). *J. Statist. Comp. Simul.*, **24**, 107–120.

[16] Upton, G. J. G. (1982). *J. R. Statist. Soc. A*, **145**, 86–105.

[17] West, E. N. and Kempthorne, O. (1971). *J. Statist. Comp. Simul.*, **1**, 1–33.

(CHI-SQUARE TESTS
CONTINGENCY TABLES
GOODNESS OF FIT)

NOEL CRESSIE
TIMOTHY R. C. READ

CROSS-VALIDATION

INTRODUCTION

Many statistical procedures can quite adequately predict the data that generated them, but do not perform nearly so well when used for predictions from new data. Cross-validation is a data-oriented method which seeks to improve reliability in this context. The lack of distributional assumptions makes the method widely applicable. Cross-validation is related to jackknifing* and bootstrap* methods. The term *sample reuse** is also used to describe cross-validation.

HISTORY

Cross-validation originated in the 1930s in attempts to improve the estimation of true multiple correlation* from the biased sample multiple correlation (Larson [6]). The cross-validatory approach consisted of calculating the regression from one sample and using a second for validation. A logical step on from this was double cross-validation where the samples were reversed and the procedure repeated.

In 1963, this approach was further developed by Mosteller and Wallace in ref. 9

where they are concerned with a set of 12 papers known to have been written by one of two authors. They looked at a set of papers of known authorship and tried to derive a discriminant function based on the words used. To do this they divided the known set into two: a screening set with which they worked out the form of the discriminant; and a calibrating set on which to test this function. They then applied the discriminant function* to the unknown set and compared the results with those from the calibrating set.

This form of assessment was further refined by various people interested in the problem of discrimination, before the first clear general statement of the refinement by Mosteller and Tukey [7] in 1968. They defined two forms, which are stated in ref. 8 as follows:

simple cross-validation:
Test the procedure on data different from those used to choose its numerical coefficients.

double cross-validation:
Test the procedure on data different both from those used to guide the choice of its form and from those used to choose its numerical coefficients.

ASSESSMENT

Having a data set $S = \{(x_i, y_i): i = 1, \ldots, n\}$, we wish to fit some model $\hat{y}(x; S)$ where $\hat{y}$ is of known form but is in some way dependent upon the data S. Cross-validation of this model provides an alternative means of determining how well the model describes the data, placing greater emphasis on prediction than goodness of fit* and thus telling us how well it predicts back to the current data.

The approach we take here is to delete a single data point (x_1, y_1), fit the model to the reduced data set $S_{\backslash 1}$, and, using the estimated model parameters, obtain a prediction $\hat{y}_1$ for the missing observation. This

is repeated for each point (x_i, y_i). We can then compute the *cross-validatory score* $C = \Sigma L(y_i, \hat{y}_i)/n$ where L is some appropriate loss function; typically $L(y, \hat{y}) = (y - \hat{y})^2$. Clearly for large data sets the computational requirements may become excessive and alternative strategies can be adopted for computing C. Rather than omitting points singly, which is equivalent to dividing the data into n groups, we can choose fewer groups and omit whole groups at a time; see Geisser [4], for example. In some cases it is possible to fit the model to the complete data set and then calculate either exactly or to an adequate approximation the effect of omitting single data points.

If two or more models are proposed for the data set S, they may be assessed by comparing the values of their cross-validatory scores: the smaller the value of C, the better the predictive power of the model. Ranking the models in terms of values of C is not necessarily the same as ranking them in terms of goodness of fit, although similar rankings are obtained in many situations.

Example 1. The top two rows of Table 1 show data from an experiment at Rothamsted Experimental Station [10]. Values of x are quantities of applied nitrogen fertilizer in kg/ha; corresponding values of y are yields of winter wheat in t/ha. Amongst models considered by Boyd et al. [2], Sparrow [12] and Wimble [18] for the yield response of cereals to nitrogen are the following three:

(i)
$$\hat{y} = a_0 + a_1 x + a_2 x^2,$$

(ii)
$$\hat{y} = b_0 + b_1 x + b_2 e^{-0.01x},$$

(iii)
$$\hat{y} = \begin{cases} c_0 + c_1(x - 75), & x \leqslant 75, \\ c_0 + c_2(x - 75), & x \geqslant 75. \end{cases}$$

These are linear models with the same number of parameters, so they may be fitted by regression* and their goodness of fit measured by the residual sum of squares, RSS: the smaller the value of RSS, the better the

Table 1

Data	x	0	30	60	90	120	150	180	210		
	y	1.67	3.77	5.42	6.29	6.24	5.90	5.94	5.77		
										RSS	C
Model	fit	1.97	3.70	5.02	5.92	6.41	6.48	6.14	5.38	0.948	
(i)	predict	2.69	3.67	4.89	5.77	6.47	6.65	6.21	4.43		0.510
Model	fit	1.70	3.89	5.25	6.01	6.32	6.28	6.01	5.54	0.329	
(ii)	predict	1.80	3.93	5.19	5.89	6.34	6.38	6.03	5.16		0.109
Model	fit	1.76	3.61	5.46	6.32	6.18	6.03	5.89	5.75	0.029	
(iii)	predict	2.03	3.55	5.48	6.33	6.16	6.06	5.87	5.73		0.016

Table 2

Model	RMS	C
mean	0.3607	0.3741
x	0.1177	0.1268
A	0.3068	0.3304
B	0.3722	0.4008
$x + A$	0.0520	0.0590
$x + B$	0.1199	0.1345
$A + B$	0.3166	0.3546
$x + A + x.A$	0.0286	0.0332
$x + B + x.B$	0.1249	0.1475
$A + B + A.B$	0.3272	0.3817
$x + A + B$	0.0515	0.0612
$x + A + x.A + B$	0.0271	0.0333
$x + B + x.B + A$	0.0537	0.0687
$A + B + A.B + x$	0.0511	0.0638
$x + A + B + x.A + x.B$	0.0283	0.0372
$x + A + B + x.A + A.B$	0.0256	0.0335
$x + A + B + x.B + A.B$	0.0534	0.0723
$x + A + B + x.A + x.B + A.B$	0.0268	0.0381
$x + A + B + x.A + x.B + A.B + x.A.B$	0.0280	0.0447

fit. The fitted values and RSS are shown in the "fit" lines of Table 1. The "predict" lines show the predicted values obtained by deleting observations, and the cross-validatory scores. Judged by either criterion, model (iii) is best and model (i) is worst.

Example 2. In a factorial* experiment wheat seedlings were inoculated with isolates of a cereal fungus (*Pseudocercosporella herpotrichoides*) at seven spore concentrations (x) and the number of leaf sheaths penetrated by the fungus was recorded (y). The isolates were classified by two factors: A, type of fungus (wheat-type or rye-type); B, susceptibility to a fungicide (sensitive or resistant). Table 2 lists all factorial models in which the effect of the fungus is at most linear in x and in which no interaction* is included without its corresponding main effects. For example, $A + x$ denotes the model in which y is linear in x, the intercept possibly changing with the different types; the addition of the term $x.A$ allows the slope to change with the different types.

The data from this experiment are not given here, but were used to calculate cross-validatory scores, shown in the right-hand column of Table 2. Notice that addition of an extra term to the model does not always decrease the value of C. The extra term may cause over-fitting to the 27 retained data points and thus poor prediction of the omitted point. In this example, the value of C increases whenever a term is added that involves the factor B.

When goodness of fit to the overall data is assessed, some consideration must be given to the dimension of the model, because the addition of a term cannot increase the residual sum of squares. Many common criteria, such as Mallows C_p-statistic* and the "percentage of variance accounted for," allow for the model dimension by using the residual mean square (RMS) in place of the residual sum of squares. The middle column of Table 2 shows the value of RMS for the 17 models.

The rankings of the 17 models by cross-validatory score and by residual mean square are similar but not identical. For example, adding the term B to the model $x + A + x.A$ increases the cross-validatory score but decreases the residual mean square.

CHOICE

The basic procedure can be used as a means of choosing one from several models $\hat{y}$, again concentrating more on their predictive properties. The definition of the model $\hat{y}$ is extended to the model $\hat{y}(x; \alpha, S)$ including an additional parameter α, which determines the form of $\hat{y}$. For instance α may be the degree of polynomial to be fitted, the set of regression terms to include, or may be a nonlinear parameter in some other model.

Having selected some reasonable values for α, the cross-validatory score $C(\alpha)$ can be computed for each; the cross-validatory choice $\alpha^\dagger$ is that value for which $C(\alpha)$ is minimum, thus indicating the best predicting model from the available choice.

When the set of possible values of α is a continuum, the cross-validatory choice $\alpha^\dagger$ is often close to the value α^0 which minimizes the residual mean square.

Example 3. In Example 1, model (ii) is an instance of the model

(iv) $\hat{y} = b_0 + b_1 x + b_2 e^{-\alpha x}$,

where α is in the semi-infinite open interval $(0, \infty)$. Cross-validatory choice gives $\alpha^\dagger = 0.013$, while $\alpha^0 = 0.015$. Similarly, model (iii) is a special case of the model

(v) $\hat{y} = \begin{cases} c_0 + c_1(x - \alpha), & x \leqslant \alpha, \\ c_0 + c_2(x - \alpha), & x \geqslant \alpha, \end{cases}$

where α is in the finite closed interval [30, 180]. Here $\alpha^\dagger = 72.3$ and $\alpha^0 = 74.0$.

Example 4. In Example 2, the values of α are sets of model terms. The cross-validatory procedure chooses the model $x + A + x.A$ as the best predictor, but the overall best-fitting model is $x + A + B + x.A + x.B$.

When α has ordered values, either or both of $C(\alpha)$ and $\text{RMS}(\alpha)$ may have local minima which are not global minima, a point to be remembered when minima are computed. In Example 4, in the increasing sequence of models

$$\text{mean, } x, x + A, x + A + B,$$
$$x + A + B + x.A,$$
$$x + A + B + x.A + x.B$$

the value of C decreases, then increases, then decreases, and finally increases again.

CHOICE AND ASSESSMENT

Stone showed [13] that choice and assessment can be integrated into a single procedure which results in the nested formulation

$$C^\dagger = \sum_i L \frac{\left\{ y_i, \hat{y}\left(x_i; \alpha^\dagger(S_{\setminus i}), S_{\setminus i}\right)\right\}}{n - 1},$$

where $\alpha^\dagger(S_{\setminus i})$ is the cross-validatory choice of α based on the reduced set $S_{\setminus i}$. Thus each observation (x_i, y_i) is omitted in turn, cross-validatory choice being applied to the remaining $n - 1$ observations to determine a value for $\hat{y}_i$.

For instance, models (iv) and (v), with cross-validatory choice of α in each case, can be compared by using $C^\dagger$. However, 56 linear submodels have to be fitted in the calculation of $C^\dagger$ for each model. Each submodel has 3 parameters and 6 data points. Smaller values of n would leave too few data points for sensible fitting of the submodels, while larger values of n increase the number of computations enormously.

APPLICATIONS

The lack of assumptions behind cross-validation means that in principle it can be used in a great variety of situations. However, it seems to work best for unstructured data, or for insufficiently specified models.

One of the most successful uses of cross-validation has been to choose the "smoothing parameter" determining the compromise between a good fit and an unacceptably tortuous fitted curve or over-complicated model. Green [5] used cross-validation in this way when he extended the usual model for yield in agricultural field trials by adding a smooth fertility trend; Wold [19] chose the number of components in a principal components analysis by cross-validation; and the method has been extensively used for fitting splines* (see refs. 3, 11, and 15–17). Titterington [14] gives a good discussion of the role of cross-validation among smoothing techniques.

Cross-validation appears to have disadvantages in some classical situations. Data from designed experiments are usually highly structured. Deletion of a single observation destroys that structure, and therefore increases the computing effort.

More dangerous, perhaps, is the use of cross-validation in regression. It is well known that, for simple models such as those in Example 1, the values of y corresponding to extreme values of x have a disproportionate influence on the fitted parameters (see ref. 1). When the data point (x_i, y_i) is omitted, the difference between the predicted value $\hat{y}_i$ and the actual value y_i is necessarily at least as great as the ith residual r_i from the model fitted to the full data. For extreme values x_i the predictions $\hat{y}_i$ are extrapolations, and so the ratios $(\hat{y}_i - y_i)/r_i$ can be expected to be larger than those for interior points x_i. Thus the disproportionate influence of the extreme data is increased by cross-validation. Stone [13] observed this problem for the cross-validatory choice of the slope of a straight line. Example 1 confirms this behavior. During cross-validation of models (i)–(v), the ratio $(\hat{y}_i - y_i)/r_i$ for internal x-values stayed remarkably constant at around 1.3, while the ratio for extreme x-values ranged from 2 to over 100.

Acknowledgment

The authors wish to thank N. Creighton and B. Fitt of Rothamsted Experimental Station for providing the data used in Example 2.

References

[1] Atkinson, A. C. (1985). *Plots, Transformations and Regression*. Oxford University Press, Oxford.

[2] Boyd, D. A., Yuen, L. T. K., and Needham, P. (1976). *J. Agric. Sci.*, **87**, 149–162.

[3] Craven, P. and Wahba, G. (1979). *Numer. Math.*, **31**, 377–403.

[4] Geisser, S. (1974). *Biometrika*, **61**, 101–107.

[5] Green, P. J. (1985). *Biometrika*, **72**, 527–538.

[6] Larson, S. C. (1931). *J. Educ. Psychol.*, **22**, 45–55.

[7] Mosteller, F. and Tukey, J. W. (1968). In *Handbook of Social Psychology*, Vol. 2, G. Lindzey and E. Aronson, eds. Addison-Wesley, Reading, MA.

[8] Mosteller, F. and Tukey, J. W. (1977). *Data Analysis and Regression*. Addison-Wesley, Reading, MA.

[9] Mosteller, F. and Wallace, D. (1963). *J. Amer. Statist. Ass.*, **58**, 275–309.

[10] Rothamsted Experimental Station (1979). Organic manuring. *Yields of the Field Experiments*, 79/W/RN/12.

[11] Silverman, B. W. (1985). *J. R. Statist. Soc. B*, **47**, 1–52.

[12] Sparrow, P. E. (1979). *J. Agric. Sci.*, **93**, 513–520.

[13] Stone, M. (1974). *J. R. Statist. Soc. B*, **36**, 111–147.

[14] Titterington, D. M. (1985). *Int. Statist. Rev.*, **53**, 141–170.

[15] Wahba, G. (1983). *J. R. Statist. Soc. B*, **45**, 133–150.

[16] Wahba, G. and Wold, S. (1975). *Commun. Statist. A*, **4**, 1–17.

[17] Wahba, G. and Wold, S. (1975). *Commun. Statist. A*, **4**, 125–141.

[18] Wimble, R. (1980). *Chemistry and Industry*, **17**, 680–683.

[19] Wold, S. (1978). *Technometrics*, **20**, 397–405.

Bibliography

Butler, R. and Rothman, E. D. (1980). *J. Amer. Statist. Ass.*, **75**, 881–889.

Geisser, S. (1975). *J. Amer. Statist. Ass.*, **70**, 320–328.

Lachenbruch, P. and Mickey, M. (1968). *Technometrics*, **10**, 1–11.

McCarthy, P. J. (1976). *J. Amer. Statist. Ass.*, **44**, 596–604.

Stone, M. (1973). *J. R. Statist. Soc. B*, **35**, 408–409.

Stone, M. (1974). *Biometrika*, **61**, 509–515.

(BOOTSTRAPPING
CURVE FITTING
GOODNESS OF FIT
JACKKNIFE METHODS
RESAMPLING PROCEDURES
SHRINKAGE ESTIMATORS
SPLINE FUNCTIONS)

R. A. BAILEY
S. A. HARDING
G. L. SMITH

D

DANIELL–KOLMOGOROV THEOREM, THE

The statistical properties of a *finite* set of random variables $X(t_1), \ldots, X(t_n)$, all defined on some probability space Ω, can be determined from their joint cumulative distribution function

$$F_{t_1, \ldots, t_n}(x_1, \ldots, x_n)$$

$$= \Pr[X(t_1) \leqslant x_1, \ldots, X(t_n) \leqslant x_n]. \quad (1)$$

As usual, we do not explicitly mention Ω in the right side of (1); it is enough to know that some suitable Ω exists, and we calculate quantities such as the covariance of $X(t_i)$ and $X(t_j)$, or the distribution of the maximum of $X(t_1), \ldots, X(t_n)$ from the functions $\{F\}$.

As well as the properties shared by all distribution functions, there are two obvious "consistency conditions" on the functions (1):

$$\lim_{x_n \to \infty} F_{t_1, \ldots, t_n}(x_1, \ldots, x_n)$$

$$= F_{t_1, \ldots, t_{n-1}}(x_1, \ldots, x_{n-1}), \quad (2)$$

$$F_{t_1, \ldots, t_n}(x_1, \ldots, x_n)$$

$$= F_{t_{\pi(1)}, \ldots, t_{\pi(n)}}(x_{\pi(1)}, \ldots, x_{\pi(n)}), \quad (3)$$

for any permutation $(\pi(1), \ldots, \pi(n))$ of $(1, \ldots, n)$.

Many applications of statistics rest on the simultaneous consideration of infinitely many random variables, $\{X(t): t \in \mathcal{T}\}$, a *stochastic process**. Typically, $\mathcal{T}$ may be a continuous time interval, and we wish to make statements about the extreme values of the process, or the time required for the process to attain a certain value. The Daniell–Kolmogorov theorem is an existence theorem, asserting that if, for all finite

n and all values $t_1, \ldots, t_n$, the distribution functions $F_{t_1, \ldots, t_n}$ satisfy (2) and (3), then there *is* a probability space Ω on which $\{X(t): t \in \mathcal{T}\}$ can be defined, whose finite-dimensional distribution functions satisfy (1).

This enormous step, from finitely many random variables to possibly uncountably many random variables, was made independently by Daniell [1] and Kolmogorov [2]. Its importance is not in the actual construction of Ω—this seldom needs to be done—but in securing the foundations of inferential statistics, when infinitely many random variables need to be simultaneously considered.

Fuller accounts can be found in, e.g.:

Kingman, J. F. C. and Taylor, S. J. (1966). *Introduction to Measure and Probability.* Cambridge University Press, London, pp. 159 and 380–382.

Loève, M. (1963). *Probability Theory*, 3rd ed. van Nostrand, New York, pp. 92–94.

References

[1] Daniell, P. J. (1918/19). Integrals in an infinite number of dimensions. *Ann. Math.*, **20**, 281–288.

[2] Kolmogorov, A. N. (1933). *Grundbegriffe der Wahrscheinlichkeitrechnung.* Ergebnisse der Mathematik. Translated as *Foundations of the Theory of Probability*, 2nd ed. (1956). Chelsea, New York.

(MEASURE THEORY IN PROBABILITY
   AND STATISTICS
STOCHASTIC PROCESSES)

JOHN HAIGH

DEFECT

A measure of the shape of a distribution introduced by Karl Pearson* [2]. It is

$$\tfrac{1}{5}\mu_6 \mu_2^{-3} - \mu_4 \mu_2^{-2},$$

where μ_r is the rth central moment. For a normal distribution* it is zero.

(*See also* Crum [1].) The term is now obsolete.

References

[1] Crum, W. L. (1923). *J. Amer. Statist. Ass.*, **18**, 607–614.

[2] Pearson, K. (1894). *Philos. Trans. R. Soc. Lond.*, **185A**, 77–110.

(EXCESS
KURTOSIS
MOMENT RATIOS
SKEWNESS)

DEFICIENCY (HODGES–LEHMANN)

If T_{1n} and T_{2n} are asymptotically unbiased estimators of a parameter, based on a random sample of size n, and, as $n \to \infty$,

$$\mathrm{var}(T_{jn}) = a_j n^{-r} + o(n^{-r}),$$
$$j = 1, 2, \ r > 0,$$

the two estimators are equally efficient* if $a_1 = a_2$. If n_1, n_2 are sample sizes such that $\mathrm{var}(T_{1n_1}) = \mathrm{var}(T_{2n_2})$, then $n_1/n_2 \to 1$ as sample sizes increase. However, it may still be the case that $n_1/n_2 > 1$ (or $n_1/n_2 < 1$) for all finite sample sizes.

Suppose T_{1n} and T_{2n} are equally efficient and

$$\mathrm{var}(T_{jn}) = an^{-r} + b_j n^{-r-q} + o(n^{-r-1}),$$
$$q > 0, \ j = 1, 2.$$

Without loss of generality we take $b_2 \geqslant b_1$.
If $\mathrm{var}(T_{1n_1}) = \mathrm{var}(T_{2n_2})$, then

$$\left(\frac{n_2}{n_1}\right)^r = \left(1 + \frac{b_2}{an_2^q}\right)\left(1 + \frac{b_1}{an_1^q}\right)^{-1}$$
$$+ o(n_1^{-1}, n_2^{-1}).$$

Writing

$$\left(\frac{n_2}{n_1}\right)^r = \left(1 + \frac{n_2 - n_1}{n_1}\right)^r$$
$$= 1 + \frac{r(n_2 - n_1)}{n_1} + o(n_1^{-1}),$$

we have

$$\frac{r(n_2 - n_1)}{n_1} \doteqdot \frac{1}{a}\left(\frac{b_2}{n_2^q} - \frac{b_1}{n_1^q}\right).$$

As sample size increases $n_1/n_2 \to 1$, and so

$$n_2 - n_1 \to \begin{cases} (ra)^{-1}(b_2 - b_1), & \text{if } q = 1, \\ \infty, & \text{if } q < 1, \\ 0, & \text{if } q > 1. \end{cases}$$

The limiting value of $(n_2 - n_1)$ is the (Hodges–Lehmann [1]) deficiency. An alternative definition, by Fisher*, is described in EFFICIENCY, SECOND ORDER.

Intuitively, the deficiency is the asymptotic number of additional observations needed, using T_{2n} in order to obtain the same variance as T_{1n}.

The most common case, by far, is $r = q = 1$, so that the deficiency is $a^{-1}(b_2 - b_1)$. Examples are given in refs. 1 and 4.

"Deficiency" should not be confused with the "deficiency distance" between two experiments defined by LeCam [5] (see also Mammen [6]), or with Chernoff and Bahadur deficiencies of tests, studied by Kallenberg [2, 3].

References

[1] Hodges, J. L. and Lehmann, E. L. (1970). *Ann. Math. Statist.*, **41**, 783–801.

[2] Kallenberg, W. C. M. (1981). *J. Multivariate Anal.*, **11**, 506–531.

[3] Kallenberg, W. C. M. (1982). *Ann. Statist.*, **10**, 583–594.

[4] Kendall, M. G. and Stuart, A. (1979). *Advanced Theory of Statistics*, 4th ed., Vol. 2. Hafner, New York; Griffin, London.

[5] LeCam, L. (1964). *Ann. Math. Statist.*, **35**, 1419–1455.

[6] Mammen, E. (1986). *Ann. Statist.*, **14**, 665–678.

(ASYMPTOTIC RELATIVE EFFICIENCY
EFFICIENCY, SECOND ORDER)

DE FINETTI, BRUNO

> **Born:** June 13, 1906, in Innsbruck, Austria.
>
> **Died:** July 20, 1985, in Rome, Italy.
>
> **Contributed to:** Bayesian inference, pedagogy, probability theory, social justice, economics.

Bruno de Finetti was born in Innsbruck of Italian parents in 1906, graduated in Mathematics from the University of Milano in 1927, and then, having attracted the attention of Corrado Gini, went to the National Institute of Statistics in Roma, where he stayed until 1931. From there he went to Trieste to work for an insurance company, staying until 1946. During this time he also held chairs in Trieste and Padova. He then moved to the University of Roma, first in Economics and later in the School of Probability in the Faculty of Science.

For statisticians, his major contributions to knowledge were his view of probability as subjective, and his demonstration, through the concept of exchangeability*, that this view could embrace standard frequency statistics as a special case. One of the aphorisms of which he was especially fond says "Probability does not exist"; by which he meant that probability does not, like the length of a table, have an existence irrespective of the observer. Nor is probability a property purely of the observer: Rather it expresses a relationship between the observer and the external world. Indeed, probability is the way in which we understand that world. It is the natural language of knowledge. This view had been held by others, but de Finetti showed that probability is the only possible language: That our statements of uncertainty must combine according to the rules of probability. He also showed that special cases, called *exchangeable*, lead an observer to appreciate chance in natural phenomena, so that frequency views are merely the special, exchangeable case of probability judgments. The unity given to disparate approaches to probability was an enormous advance in our understanding of the notions and enabled a totally coherent approach to statistics, a subject which is usually presented as a series of ad hoc statements, to be given. This is Bayesian statistics, though de Finetti was careful to distinguish between Bayesian ideas and Bayesian techniques, the latter being merely the calculations necessary to solve a problem, the former being the notions expressing the observer's view of the situation. The latter does not distinguish between sex and drawing pins—both are technically binomial (n, p)—whereas the former does.

De Finetti's view of mathematics was quite different from that dominant today, especially in the United States. His emphasis was on ideas and applications, not on technicalities and calculations. He was fond of quoting the saying of Chisini, a teacher of his, that "mathematics is the art that teaches you how *not* to make calculations." An unfortunate consequence of this is that, for those used to the modern mathematical style, he is difficult to read. Theorem, proof, and application are blended into a unity that does not allow one aspect to dominate another. He was severely critical, as are many others, of the very technical papers that one is supposed to accept because no slip has been found in an insipid and incomprehensible chain of syllogisms. Nomenclature was important to him. He insisted on "random quantity," not "random variable": for what varies?

His main contribution to applied probability was the introduction of a scoring rule*. In its simplest form, if a person gives a probability p for an event A, he is scored $(p - 1)^2$ or p^2, according as A is subsequently found to be true or false. This simple idea has led to extensions and to the appreciation of what is meant by a good probability appraiser.

Another major interest of his was teaching. He was emphatic that children should be taught to think probabilistically at an early age, to understand that the world is full of uncertainty and that probability was the way to handle and feel comfortable with it. A child should not be taught to believe that every question has a right answer; but only probabilities for different possibilities. For example, in a multiple-choice test, an examinee should give probabilities for each of the suggested solutions, not merely select one. He should not predict but forecast through a probability distribution.

He did a substantial amount of actuarial work. He was vitally interested in economics

and social justice. He believed in an economic system founded on the twin ideas of Pareto optimality and equity. This, he felt, would produce a better social system than the greed that he felt to be the basis of capitalism. His views have been described as naive but this is often a derogatory term applied by self-interested people to those who dare to disturb that self-interest. He stood as a candidate in an election (and was defeated) and was arrested when supporting what he saw to be a worthy cause.

De Finetti was kind and gentle, yet emphatic in his views. He was interested in the unity of life. Although he will best be remembered for theoretical advances, his outlook embraced the whole spectrum of knowledge.

Bibliography

Probability, Induction and Statistics: The Art of Guessing. Wiley, London, 1972.

Theory of Probability: A Critical Introductory Treatment. Wiley, London, Vol. 1, 1974; Vol. 2, 1975.

"Contro disfunzioni e storture: urgenza di riforme radicali del sistema." In *Lo Sviluppo della Società Italiana nei Prossimi Anni*. Accademia Nazionale dei Lincei, Roma, 1978, pp. 105–145.

Il Saper Vedere in Matematica. Loescher, Torino, 1967.

Matematica Logico-Intuitiva. Cremonese, Roma, 1959.

D. V. Lindley

DE WITT, JOHAN

Born: September 25, 1625 (birthplace unknown, probably outside Dordrecht, Holland).

Died: August 20, 1672, in The Hague, Holland.

Contributed to: actuarial science, mathematics, economic statistics.

The place of Johan de Witt in the history of the statistical sciences is due to his pioneering calculations of the value of annuities, based on information about mortality rates and probabilistic considerations. In Roman times no theory of probabilities existed, but a table had been produced for evaluating annuities, the basis of which is unknown. The newly developed calculus of probabilities had been given publicity by Huygens' (1629–1695) tract of 1656 on Calculations in Games of Chance, included as *De Ratiociniis in Ludo Aleae* in Van Schooten's (1615–1660) *Exercitationes Mathematicae* (1657) and three years later appearing in the original Dutch version (*Van Rekenigh in Spelen van Geluck*). No doubt De Witt was acquainted with this publication when he started his work on annuities. He collected data from the annuity registers of Holland but these were insufficient as a basis for a mortality table of potential annuitants, and so he decided to postulate the following simple model of mortality as a basis of his calculation: In each of 4 age groups considered (4–53, 54–63, 64–73, and 74–79), the rate of mortality was taken to be constant, being one and one-half times as big in the second group as in the first, twice as big in the third group as in the first, and three times as big in the third. The calculations were set out in a tract called *Waerdye van Lyf-Renten Naer Proportie van Los-Renten*, included in the *Resolutions* of the States of Holland for 1671. The title may be translated as "The Worth of Life Annuities in Proportion to Redemption Bonds." The tract demonstrated that the issue of life annuities with a rate of some 7% would be more favorable to issuer and purchasers alike than the issue of perpetual or redemption bonds (which had no termination date, but could be called in by the issuer at any time) with a rate of 4%. The conclusion was so novel that it was not widely accepted by the public at the time in spite of the simple way the argument was presented (mostly with the aid of a numerical example) in the tract. A more fundamental argument is found in the author's correspondence with Hudde (1628–1704), in which he also calculated the value of annuities based on several lives.

De Witt also supplied Pieter de la Court (1618–1685) with economic statistics for the latter's publication *Interest van Holland* which appeared in various versions from 1662 on, in some of which De Witt is mentioned incorrectly as author.

De Witt was the most prominent Dutch statesman of the third quarter of the seventeenth century, a time when the Dutch Republic had just gained its independence from Spain and, in spite of its population of only a little above one million, was the second richest country (after France) and was a major world power. At the age of 16, he and his elder brother Cornelis entered Leyden University for the study of law, finishing in 1645. Then they went to "finish their education" in the manner then customary for sons of prominent families, namely by a grand tour of France lasting one and one-half year, seeing the sights and meeting prominent people, and also gaining a doctor of law degree during a three months' stay in Angers. The brothers then spent a few months in England and returned to Holland. Johan became an apprentice with a prominent lawyer in The Hague, and in December 1647 was nominated as a representative of his home city, Dordrecht, to the governing body ("States") of the Province of Holland, whose seat was in The Hague. Because of the seniority of Dordrecht among the cities of the Province (whose representatives together with one representative of the nobility made up the "States"), its deputies were always called upon first to give their opinions in the debates. Moreover, the resident representative of Dordrecht had to take the place of the Raadspensionaris (chief executive officer) during the latter's absence. Because of these facts, together with De Witt's extraordinary abilities and devotion to his task, as well as his great integrity (a quality not common among politicians at the time), he soon became a leading figure in the "States." He was appointed Raadspensionaris in 1652, and reappointed every five years, the last time in 1668. Since the Dutch Republic was a loose federation of seven Provinces, of which the Province of Holland was by far the richest and most influential, and because of De Witt's personal qualities, his position became, in fact, one of chief executive officer of the Republic. More and more, he gave direction to its foreign, fiscal, and defense policies and their execution, even though his formal powers were very restricted and he had to operate largely by persuasion and political means. During his time in office the rulers of England several times attacked the Dutch Republic, driven largely by envy of its leading position in commerce and navigation. Moreover, the country was threatened by Louis XIV of France, who was intent on expanding his domains with the help of a powerful army. The year 1672 is known as the "Year of Disaster" in Dutch history. The English fleet attacked and Louis XIV invaded the country (which in the end was saved only by flooding the polders in the Western part of the country and by depletion of the supplies of the French army). Internally, passions ran high and many accused the De Witt brothers of treason. There was an attempt on the life of Johan on June 21, which put him out of action for some time. He submitted his resignation on August 4, which was granted with reluctance. Meanwhile, Cornelis was arrested because of a (false) accusation of plotting to assassinate the Prince of Orange. Johan came to visit his brother at the prison; and later a mob forced its way into the prison and killed both. A committee of the States of Holland, upon examining the papers of the late Raadspensionaris, found no evidence of traitorous behavior; one of the members, who had not been a political friend, when asked what had been found, replied: "Nothing but honour and virtue."

Already during his studies at Leyden, De Witt developed a strong interest in mathematics. Before his appointment as Raadspensionaris, he had drafted a treatise on conic sections, which he sent about 1657 to his former fellow student Van Schooten, then Professor of Mathematics at Leyden University. The latter was then preparing a second edition of a translation into Latin of Descartes' (1596–1650) *Géometrie* with commentaries, which was brought out by Elsevier in two volumes (1659/1661) under the title *Geometria a Renate Des Cartes*. He pro-

posed that De Witt's treatise should be included in this publication, but reformulated in closer accordance with Descartes' new approach and notation. Since he knew that De Witt was a very busy man, he offered to take upon himself the reformulation, the checking of calculations and the drafting of figures. De Witt gladly accepted, but went carefully over the reformulated version and made important further amendments. The result was the treatise named *Elementa Curvarum Linearum*. It was much praised by Huygens*; Newton* (1642–1727), when asked which books would be helpful to the study of his *Principia*, recommended it for the geometric background it supplied. It was widely used during the seventeenth century. On the other hand, there are no signs that De Witt's work on annuities influenced the development of actuarial science or practice to any large extent. James (Jacob) Bernoulli* (1654–1705) requested from Leibniz (1646–1716) a loan of his copy of De Witt's tract, but the latter could not locate it and asserted that it was not of much value. The *Waerdye* was rediscovered more or less accidentally by Bierens De Haan and reissued in 1879 by the Mathematical Society in Amsterdam. Different interpretations and evaluations of this work have appeared; the interested reader may consult the appended bibliography as well as the references quoted in these publications.

Bibliography

Bernoulli, J. (1975). *Die Werke*, Vol. 3. Birkhäuser Verlag, Basel. (Contains a facsimile copy of the *Waerdye*, and commentaries.)

Brakels, J. van (1976). Some Remarks on the Prehistory of the Concept of Statistical Probability. *Arch. History Exact Sci.*, **16**, 119–136. (The footnote on pp. 130–131 corrects a number of mistakes in Hacking and others.)

Fenaroli, G., Garibaldi, U., and Penco, M. A. (1981). Giochi, Scommesse sulle Vita, Tabelli di Mortalità: Nascita del Calcolo Probabilastico, Statistica e Teoria delle Populazioni. *Arch. History Exact Sci.*, **25**, 329–341.

van Geer, P. (1915). Johan de Witt als Wiskundige. *Nieuw Archief voor Wiskunde*, Series 2, **11**, 98–126.

Hacking, I. (1975). *The Emergence of Probability*. Cambridge University Press, Cambridge.

Rowen, H. H. (1986). *John de Witt, Statesman of the "True Freedom."* Cambridge University Press, Cambridge. (This is a short biography containing a brief bibliographical essay at the end.)

Rowen, H. H. (1975). *John de Witt, Grand Pensionary of Holland, 1625–1672.* Princeton University Press, Princeton, NJ. (This work is much more extensive and fully documented.)

(ACTUARIAL STATISTICS—LIFE
DEMOGRAPHY
HUYGENS, CHRISTIAAN
LIFE TABLES)

Hendrik S. Konijn

DORFMAN-TYPE SCREENING PROCEDURES

STANDARD DORFMAN PROCEDURES

These form a class of procedures for determining individual defective (nonconforming) items in such a way as to reduce the average amount of testing needed.

If each item is tested separately, then n items will require n tests. Sometimes, however, it is possible to test a group of n items so as to detect whether at least one of them is defective, and to proceed to individual testing only if the presence of at least one defective item is indicated. Situations where this may be possible include testing bottles of liquid for the presence of contaminants, or electric insulators for effectiveness. The method was introduced by Dorfman [1] in the context of testing blood samples for syphilis.

Reduction in the expected number of tests is to be achieved if the proportion (ω) of defective items is small, because it is then quite likely that testing will terminate with the single test of all n items in the group. If we assume random sampling from a very large (effectively infinite) population, the probability that this will happen is $(1 - \omega)^n$. There is, of course, the risk that it will not

happen; then individual testing will be needed so that $(n + 1)$ tests, in all, will be required.

The expected number of tests is

$$(1 - \omega)^n 1 + \left\{1 - (1 - \omega)^n\right\}(n + 1)$$
$$= n - \left\{n(1 - \omega)^n - 1\right\}.$$

Provided $n(1 - \omega)^n$ exceeds 1, there is a reduction in the expected number of tests. If $2(1 - \omega)^2 - 1$ is less than 0, that is, ω is greater than $1 - 1/\sqrt{2} = 0.293$, then no reduction is possible, whatever the group size. The proportionate reduction is

$$\left\{n(1 - \omega)^n - 1\right\}n^{-1}$$
$$= (1 - \omega)^n - n^{-1} \doteq 1 - n\omega - n^{-1}$$

if ω is small. This is maximized (approximately) by taking n close to $\omega^{-1/2}$.

For example, if $\omega = 0.10$ we obtain $n = 3$; if $\omega = 0.01$ we obtain $n = 10$. The corresponding proportionate reductions are 39.6% and 80.4%, respectively. The greater the value of ω, the smaller the optimal value of n. As we have seen, if ω exceeds 0.293, no advantage is gained by screening (the "optimal" n would be 1, and no repeat would be needed).

Differences between costs of a group test (c_1) and an individual item test (c_2) can be allowed for in a straightforward manner. If individual testing is needed, the total cost is $c_1 + nc_2$; if the group test is all that is needed the cost is c_1. The expected cost of the standard Dorfman procedure is

$$c_1 + nc_2\left\{1 - (1 - \omega)^n\right\}$$
$$= nc_2 - \left\{nc_2(1 - \omega)^n - c_1\right\}.$$

The proportionate reduction in cost is

$$(1 - \omega)^n - c_1(nc_2)^{-1}$$
$$= 1 - n\omega - \left(c_1 c_2^{-1}\right)n^{-1}$$

if ω is small. This is maximized by taking $n = (c_1 c_2^{-1} \omega^{-1})^{1/2}$.

There are several variants of the standard procedure, some now to be described. All are aimed at reducing the expected number of tests.

DORFMAN–STERRETT PROCEDURES

Sterrett [9] noted that if ω *is* small, then it is quite likely there will be no more than one defective item in the set of n items tested as a group. Once a defective item is found on individual testing, therefore, it is unlikely that there is another defective item among those as yet untested. Sterrett suggested that further savings might be expected if a group test is applied to these remaining items, since it might well indicate that none of them is defective, and so no further testing would be needed.

In the original proposal, reversion to group testing was to occur *whenever* a defective item was found on individual testing (except, of course, when only one item remains to be tested). For practical reasons, the number of such reversions will usually be restricted —possibly to only one or two. The advantage of the Sterrett modification is most likely to accrue when ω is small but not *very* small, so that there will be an appreciable chance of there being two defective items in a group of n, given that there is at least one defective item.

If only one reversion to group testing is permitted, the expected number of tests is

$$1 + n\omega(1 - \omega)^{n-1}n^{-1}(2 + 3 + \cdots + n + n)$$
$$+ (n+1)\left\{1 - (1 - \omega)^n - n\omega(1-\omega)^{n-1}\right\}$$
$$- \omega^2(1 - \omega)^{n-2}. \tag{1}$$

(The last term is needed because there is no reversion to group testing if there are just two defective items, and they are the last two to be tested individually.)

Expression (1) can be written:

$$n + 2 - \tfrac{1}{2}(1 - \omega)^{n-2}$$
$$\times \left\{2(n + 1) + (n^2 - 5n - 2)\omega\right.$$
$$\left. - (n^2 - 3n - 2)\omega^2\right\}.$$

Dorfman–Sterrett procedures may be modified by waiting until two (or, generally, g) items have been found to be defective on individual testing. This may be expected to be advantageous if ω, while small, is some-

what larger than for the procedure with $g = 1$. (The rationale is, that for such cases, the conditional probability of having two defective items, given that there is at least one, may be quite large.) There is a discussion of this modification in refs. 2 and 5.

CURTAILED DORFMAN PROCEDURES

When using a standard Dorfman procedure, if the presence of at least one defective item is indicated but no defective items are found among the first $(n - 1)$ items tested individually, the remaining item *must* be defective (on the assumption that there are no errors in testing). Therefore, there is no need to test the remaining item. Omission of this test further reduces the expected amount of testing. The saving in expected number of tests is $\omega(1 - \omega)^{n-1}$. This is not usually very great.

There is, however, a situation (see, e.g., Pfeifer and Enis [7]) in which this type of curtailment can be of greater importance. This is when the group test indicates not only the *presence* of defective items, but also their *number*—d, say. In this case, individual testing can cease as soon as the status of all remaining items is determined. This occurs (i) as soon as d defective items have been found—all the remaining items must be nondefective, or (ii) when $(d - x)$ defective items have been found and only x remain to be tested—these x items *must* all be defective.

For details see Kotz et al. [5], where the effects of errors in testing are also discussed.

HIERARCHAL DORFMAN PROCEDURES

Provided ω is small, a standard Dorfman procedure reduces the expected number of tests. The greater the amount of group testing in a procedure, therefore, the greater the savings to be hoped for. This reasoning underlies the Dorfman–Sterrett procedures already described. Another way of increasing the possibility for savings is by use of *hierarchal Dorfman procedures*.

If the group size n is a product of two integers $n = hn'$, then the group can be split into h separate subgroups, each containing n' items. When the group test indicates the presence of at least one defective item, standard Dorfman procedures are applied to each of the h subgroups, instead of immediate recourse to individual testing.

The expected number of tests with this procedure is

$$1 + h\left\{1 - (1 - \omega)^n\right\}$$
$$+ hn'\left\{1 - (1 - \omega)^{n'}\right\}.$$

The proportionate saving in number of tests is

$$(1 - \omega)^{n'} - n^{-1} - n'^{-1}\left\{1 - (1 - \omega)^n\right\}$$
$$= 1 - n^{-1} - (n' + h)\omega$$

if ω is small.

For given n, this is maximized by maximizing $n' + h$ subject to $n'h = n$. This is achieved by taking $n' = h = \sqrt{n}$. The corresponding proportionate expected saving; $1 - 2n^{1/2}\omega - n^{-1}$ is maximized by taking $n = \omega^{-2/3}$.

Extension to hierarchal procedures of higher order with subgroups divided into sub-subgroups, and so on, is straightforward.

Sobel and Groll [8] combined hierarchal organization with Dorfman–Sterrett procedures. Mehravari [6] further introduced the possibility that the group test might indicate the presence of *exactly* one, or *at least* one defective item, but not the actual number.

If $n = 2^k$, a process of "successive halving" can be used, each sub^{s}group being divided into two sub^{s+1}groups of equal size. This is especially effective if there is just one defective item in the original group of n items.

ERRORS IN TESTING

If testing is not perfect, the properties of procedures described in this entry will be affected. Not only do changes in the expected numbers of tests need to be considered, but also the probabilities of incorrect final decisions—both "false negatives" (defectives classed as nondefective) and "false positives" (nondefectives classed as defective).

The analysis needs to take into account, at least, probabilities of false negatives and false positives on group and individual tests. These probabilities may vary with the size of group. A specific example is studied in ref. 3. We have also noted a few relevant references in the text.

The series of papers by Johnson, Kotz, and Rodriguez (see the Bibliography) contain discussions of the effects of errors in a variety of situations, including *acceptance sampling** as well as Dorfman-type procedures.

References

[1] Dorfman, R. (1943). *Ann. Math. Statist.*, **14**, 436–440.

[2] Huang, Q.-S., Johnson, N. L., and Kotz, S. (1988). Modified Dorfman–Sterrett Screening (Group Testing) Procedures and the Effects of Faulty Inspection. *Mimeo Series No. 1750*, Institute of Statistics, University of North Carolina.

[3] Hwang, F. K. (1976). *Biometrika*, **63**, 671–673.

[4] Johnson, N. L., Kotz, S., and Rodriguez, R. N. (1987). Dorfman–Sterrett (Group Testing) Schemes and the Effects of Faulty Inspection. *Mimeo Series No. 1722*, Institute of Statistics, University of North Carolina.

[5] Kotz, S., Song, M. S., and Johnson, N. L. (1986). *Commun. Statist.-Theor. Meth.*, **15**, 831–838.

[6] Mehravari, N. (1986). *SIAM J. Alg. Discrete Math.*, **7**, 159–166.

[7] Pfeifer, C. G. and Enis, P. (1978). *J. Amer. Statist. Ass.*, **73**, 588–592.

[8] Sobel, M. and Groll, P. (1959). *Bell Syst. Tech. J.*, **38**, 1179–1253.

[9] Sterrett, A. (1957). *Ann. Math. Statist.*, **28**, 1033–1036.

Bibliography

Gastwirth, J. L. (1987). The statistical precision of medical screening procedures—Application to polygraph and AIDS antibodies test data. *Statist. Sci.*, **2**, 213–222.

Gastwirth, J. L. and Hammick, P. A. (1988). Estimation of the prevalence of a rare disease, preserving the anonymity of the subjects of a group testing. *Abstract 205-41, IMS Bull.*, **17**, 136.

Gill, A. and Gottlieb, D. (1974). The identification of a set by successive intersections. *Inf. Control.*, **24**, 20–35.

Groll, P. and Sobel, M. (1966). Binomial group-testing with an unknown proportion of defectives. *Technometrics*, **14**, 113–122.

Hwang, F. K. (1984). Robust group testing. *J. Quality Tech.*, **16**, 189–195.

Johnson, N. L. and Kotz, S. (1988). Effects of Errors in Inspection on a Binary Method for Isolating Nonconforming Items. *Austral. J. Statist.*, **30A**, 205–214.

Johnson, N. L., Kotz, S., and Rodriguez, R. N. (1985, 1987, 1988). Statistical effects of imperfect inspection sampling: I. Some basic distributions: II. Double sampling and link sampling: III. Screening (group testing). *J. Quality Tech.*, **17**, 1–31; **18**, 116–138; **20**.

Johnson, N. L., Kotz, S., and Rodriguez, R. N. (1988). Statistical effects of imperfect inspection sampling: IV. Modified Dorfman screening procedures. *J. Qual. Technol.*, (to appear).

Mundel, A. B. (1984). Group testing. *J. Quality Tech.*, **16**, 181–188.

Sobel, M. and Elashoff, R. M. (1975). Group testing with a new goal, estimation. *Biometrika*, **62**, 181–193.

(ACCEPTANCE SAMPLING
GROUP TESTING
QUALITY CONTROL, STATISTICAL)

DUDLEY METRIC

A metric for a set of random variables $\mathscr{X}(U)$ defined on a probability space (Ω, Σ, P) taking values on a complete separable metric space (U, d).

For X, Y belonging to $\mathscr{X}(U)$, the Dudley metric is defined by

$$\delta(X, Y) = \sup\{|E[g(X) - g(Y)]|: g \in \xi\},$$

where

$$\xi = \{g: |g(x)|d(y, z) + |g(y) - g(z)|$$
$$\leqslant d(y, z); x, y, z \in U\}.$$

Bibliography

Dudley, R. M. (1968). *Ann. Math. Statist.*, **39**, 1563–1572.

Zolotarev, V. M. (1983). *Theor. Prob. Appl.*, **28**, 278–302.

(PROBABILITY SPACES, METRICS AND
  DISTANCES ON)

E

EBERHARDT STATISTIC

If $X_1, X_2, \ldots, X_m$ are independent, identically distributed positive-valued random variables, then the Eberhardt statistic, A, is defined by

$$A = m \sum_{i=1}^{m} X_i^2 \bigg/ \left(\sum_{j=1}^{m} X_j\right)^2.$$

The Eberhardt statistic, which is related to the square of the coefficient of variation of the X's, is employed in some tests of the assumed randomness of location of point items in a plane, when distance measurements have been generated by T-square* or wandering-quarter* sampling. Hines and O'Hara Hines [3] and Upton and Fingleton [4] contain tables of (approximate) upper and lower critical values of A under the null hypothesis of complete randomness, and under the assumptions that boundary effects of the sampled region can be ignored and that sampling intensity is sufficiently low so that the independence of the X's can be assumed.

Hines and O'Hara Hines [3] discuss the use of the Eberhardt statistic for a test of the null hypothesis that items are located at random on the region available for searching.

References and Bibliography

[1] Diggle, P. J. (1983). *Statistical Analysis of Spatial Point Processes*. Academic, London.

[2] Eberhardt, L. L. (1967). *Biometrics*, **23**, 207–216.

[3] Hines, W. G. S. and O'Hara Hines, R. J. (1979). *Biometrika*, **66**, 73–79.

[4] Upton, G. and Fingleton, B. (1985). *Spatial Data Analysis by Example*, Vol. 1. Wiley, Chichester, England.

(SPATIAL PROCESSES
SPATIAL SAMPLING)

W. G. S. HINES

EDGEWORTH, FRANCIS YSIDRO

Born: February 8, 1845, in County Lonford, Ireland.

Died: February 13, 1926, in London, England.

Contributed to: index numbers, laws of error, estimation, economics, agricultural statistics.

Francis Ysidro Edgeworth (1845–1926) remains a relatively unheralded economist and statistician even though he is responsible for many novel and important contributions to both fields. Edgeworth's original contributions to mathematical or analytical economics include the indifference curve, the contract curve (and the related construct known as the Edgeworth box), the law of diminishing returns (which he also used in his editorial capacity to encourage brevity in journal submissions [3]), and the determination of economic equilibria. His statistical contributions include works on index numbers, the law of error, the theory of estimation, correlation, goodness of fit, and probability theory.

Edgeworth was born February 8, 1845, in County Lonford, Ireland, a descendent of a family of Anglo–Irish gentry. He received a classical education common to his time; he attended Trinity College, Dublin at age 17,

entered Oxford as a scholar of Magdalen Hall in 1867, and was admitted to Balliol College a year later. He was awarded first-class honors in Literis Humanioribus in 1869, and took his BA degree in 1873. Edgeworth was apparently primarily a "self-taught" mathematician, having received little formal training in advanced mathematics [2].

Edgeworth's earliest known publication is *New and Old Methods of Ethics*, published by Parker and Company of Oxford in 1877, in which he investigated quantitative problems arising in Utilitarianism. His only book, *Mathematical Psychics* (published in 1881), represents his first contribution to economics, and contains his treatment of contracts in a free market (and his beginning work on mathematical economics generally).

He was made Lecturer in Logic at King's College, London in 1880, and in 1890 succeeded Thorold Rogers as Tooke Professor of Economic Science and Statistics. A year later he succeeded Thorold Rogers as Drummond Professor of Political Economy at Oxford, where he remained until retiring with the title of Emeritus Professor in 1922.

He began his publications on probability and statistics while at King's College. Edgeworth was concerned with a priori probabilities, which he fit into a frequency theory of probability. (His views of "inverse probability" are distinct from Fisher's.) He wrote at length on the law of error*, emphasizing the prevalence of the normal law in nature and studying the distribution of averages. His work on index numbers* for prices, begun in 1883, drew heavily from his work on averages and error. In 1892, he published his first paper on correlation, in which he attempted to provide a framework for the concept of multiple correlation*. Edgeworth published approximately 40 statistical papers between 1893 and 1926 [5].

A large part of Edgeworth's last 35 years was occupied with the editorship of the *Economics Journal*, which was begun under his editorship in 1891. Although most of his service was devoted to economics, including serving as president of the Economic Section of the British Association for the Advancement of Science (1889 and 1922), Edgeworth also served as President of the Royal Statistical Society* from 1912 to 1914. He was made a Fellow of the British Academy in 1903.

Edgeworth was a life-long bachelor who led the life of an ascetic. He was extremely shy, with a self-effacing manner [4]. He owned few possessions, not even bothering to collect books, but preferring public libraries. (Keynes claimed that the only two material objects he knew that Edgeworth privately owned were red tape and gum.) While at Oxford, Edgeworth lived in a Fellow's room at All Soul's College. He maintained the same two barely furnished rooms on the outskirts of London for the last 50 years of his life. He died on February 13, 1926 [3].

Edgeworth's contributions to the field of mathematical statistics (74 papers and 9 reviews in all) are summarized in ref. 1.

References

[1] Bowley, A. L. (1928). *F. Y. Edgeworth's Contributions to Mathematical Statistics*. (Pamphlet published by the Royal Statistical Society. Contains an annotated bibliography.)

[2] Creedy, J. (1983). *J. R. Statist. Soc. A*, **146**, 158–162. (Speculates on mathematical connections in Edgeworth's life.)

[3] Keynes, J. M. (1951). *Essays in Biography*. Norton, New York.

[4] Schumpeter, J. A. (1954). *History of Economic Analysis*. Oxford University Press, New York.

[5] Spiegel, H. W., ed. (1952). *The Development of Economic Thought*. Wiley, New York.

Bibliography

Edgeworth, F. Y. (1925). *Papers Relating to Political Economy* (3 volumes). (Published on behalf of the Royal Economic Society by Macmillan.)

Edgeworth, F. Y. (1881). *Mathematical Psychics*. Published by Kegan Paul, Long. (Reprinted by London School of Economics, 1932.)

Henderson, J. M. and Quandt, R. E. (1971). *Microeconomic Theory: A Mathematical Approach*. McGraw-Hill, New York.

Kendall, M. G. (1968). *Biometrika*, **55**, 269–276.

Stigler, S. M. (1978). *J. R. Statist. Soc. A*, **141**, 287–322.

(Contains a detailed discussion of Edgeworth's contributions in probability and statistics, including early work on an analysis of variance for a two-way classification.)

Stigler, S. M. (1986). *The History of Statistics: The Measurement of Uncertainty before 1900*. Belknap/Harvard University Press, Cambridge, MA. (A clear and thorough exposition. Comprehensible for general readers.)

(CORRELATION
FISHER, RONALD AYLMER
FREQUENCY CURVES
INDEX NUMBERS
INVERSE PROBABILITY
LAWS OF ERROR—I, II, III
NORMAL DISTRIBUTION
PEARSON, KARL)

RAMONA L. TRADER

EFRON–MORRIS ESTIMATOR

Suppose $X_1, X_2, \ldots, X_k$ are independent normal random variables, each with the same known variance and with respective unknown means $\theta_1, \theta_2, \ldots, \theta_k$. There is no loss of generality in assuming that the common known variance is 1. In practice, the X_i's are means of samples from normal populations. The problem is to estimate $\theta_1, \ldots, \theta_k$, $k \geq 3$, and if $D_1, \ldots, D_k$ are the values of the estimates, the loss is $(D_1 - \theta_1)^2 + \cdots + (D_k - \theta_k)^2$. The maximum likelihood* estimators are $X_1, \ldots, X_k$. James and Stein [2] showed that the decision rule which estimates θ_i by $\{1 - (k - 2)/\|X\|^2\} X_i$ dominates the maximum likelihood estimator, where $\|X\|^2$ denotes $X_1^2 + \cdots + X_k^2$. Various modifications and generalizations of the James–Stein estimator* have been proposed. Efron and Morris [1] proposed an estimator, called by them a "limited translation empirical Bayes estimator," which estimates θ_i by

$$\left\{1 - \left((k - 2)/\|X\|^2\right)f\left(X_i^2/\|X\|^2\right)\right\} X_i$$

where

$$f(z) = \min\left(1, D(k - 2)^{-1/2}z^{-1/2}\right)$$
$$\text{for } z > 0,$$

with D a nonnegative constant. This type of

estimator has a slightly higher expected loss than the James–Stein estimator, but has the following advantage over the James–Stein estimator: For certain parameter values, $\max_i E[(D_i - \theta_i)^2]$ can be quite high when D_i is chosen by the James–Stein rule, compared to its value when D_i is chosen by the Efron–Morris rule.

Shinozaki [3] proposed a modification of the Efron–Morris estimator by allowing more general functions f, and gave conditions on f which imply that the estimator is minimax*.

References

[1] Efron, B. and Morris, C. (1972). *J. Amer. Statist. Ass.*, **67**, 130–139.

[2] James, W. and Stein, C. (1961). *Proc. Fourth Berkeley Symp. Math. Statist. Prob.*, Vol. 1. University of California Press, Berkeley, CA, pp. 361–379.

[3] Shinozaki, N. (1986). *J. Japan Statist. Soc.*, **16**, 191–201.

(ADMISSIBILITY
JAMES–STEIN ESTIMATORS
SHRINKAGE ESTIMATORS)

LIONEL WEISS

ESTIMATING EQUATIONS, THEORY OF

MOTIVATION AND HISTORICAL BACKGROUND

The traditional theory of point estimation as presented in most standard textbooks, e.g., Rao [25], focuses on properties of estimators which are statistics, i.e., functions of the observations alone. Restrictions such as unbiasedness* or invariance* are then imposed on such estimators and an optimal estimator sought in this restricted class, typical optimality criteria being minimum variance, minimum mean squared error, and the like. The classical theories of minimum variance unbiased (MVU) estimation* and minimum

risk invariant estimation are examples of this approach. The theory of maximum likelihood* estimation has also been developed with an emphasis on the desirable properties of the estimator itself, such as minimum asymptotic variance, efficiency, etc.

The theory of MVU estimators and in particular the emphasis on unbiasedness has been criticized in the past. Two important criticisms are: (1) that the restriction to unbiasedness may produce uniformly MVU (UMVU) estimators which are absurd (see ref. 17, p. 36); and (2) MVU estimation is not invariant under parameter transformations, i.e., if T is the MVU estimator of some parameter θ, then $g(T)$ is not necessarily the MVU estimator for $g(\theta)$, where g is a 1–1 function [$g(T)$ may not even be unbiased for $g(\theta)$]. The second of these criticisms does not apply to maximum likelihood estimation. However, maximum likelihood* estimation has been criticized, also. Neyman and Scott [24] give examples of situations where the maximum likelihood estimator (MLE) may be inefficient (asymptotically) or even inconsistent. Also the optimality properties of the MLE usually touted are asymptotic ones.

The theory of estimating equations is an approach to point estimation which tries to circumvent some of the above criticisms (of UMVU and MLE). This theory begins with the premise that any estimator may be regarded as a solution to an equation of the form $g(x, \theta) = 0$, where g is a function of the data x and the parameter θ. The second step is to impose optimality criteria on the *function* g itself rather than the estimator obtained from it. The advantages of this approach over the more traditional approach are not immediately clear. However, it turns out to be particularly advantageous in the case of nuisance parameter* problems of the Neyman–Scott variety where it overcomes the difficulties in straight maximum likelihood (Godambe [8]). Furthermore, since the pioneering paper of Godambe [7] the approach via estimating equations has found application in a wide variety of fields such as time series* (Durbin [4]), survey sampling*

(Godambe and Thompson [14]), stochastic processes* (Godambe [11], McLeish [21], and Hutton and Nelson [15]) and generalized linear models* (Morton [23] and Godambe [12]).

Historically, one of the earliest mentions of the term "equation of estimation" occurs in Fisher [5]. Later, Kimball [18] introduced the idea of a stable estimating function, i.e., one whose expectation is independent of the parameter. Kimball also defined the notion of a sufficient estimating function and used it to obtain estimating equations for the parameters of the extreme-value distribution* for which ordinary sufficiency* is inadequate. In 1960, Durbin defined the notion of a *linear* unbiased estimating equation and proved some optimality theorems with particular reference to time-series applications. In the same year Godambe noticed that the estimating equation approach provided a powerful justification for maximum likelihood estimation in the one-parameter, regular case. The argument used was in essence an extension of the Cramér–Rao bound theory to estimating functions rather than estimators. However, while the Cramér–Rao bound* is attained only for particular parametric functions and hence is limited as an optimality criterion, the maximum likelihood estimating equation is optimal according to Godambe's criterion subject only to regularity conditions. Furthermore, this justification of maximum likelihood (unlike many traditional justifications) applies in finite samples. It also has the practical and theoretical advantage of being invariant under monotonic transformations of the parameter (Fisher [6], p. 142).

OPTIMALITY OF THE MAXIMUM LIKELIHOOD EQUATION

An estimating equation $g(x, \theta) = 0$ is said to be unbiased if $E_\theta g(x, \theta) = 0$ for all θ in the parameter space. This is a natural requirement which ensures that the root of the equation is close to the true value when little random variation is present. By analogy with

the theory of MVU estimation it is natural to seek an unbiased estimating equation g for which the variance $E_\theta g^2$ is as small as possible. However, since the variance may be changed by multiplying g by an arbitrary constant (possibly depending on θ) some further standardization is necessary. Godambe [7] suggested considering the variance of the standardized estimating functions

$$g_s = g \bigg/ E_\theta \left[\frac{\partial g}{\partial \theta} \right].$$

Thus an optimal estimating function is one which minimizes

$$\mathrm{Var}_\theta(g_s) = \frac{E_\theta\left[g^2 \right]}{\left\{ E_\theta\left[\dfrac{\partial g}{\partial \theta} \right] \right\}^2}.$$

(Note that the term in the denominator on the right measures the sensitivity of g to changes in θ.) Godambe showed in the regular case that

$$\mathrm{Var}_\theta(g_s) \geqslant \frac{1}{I_E(\theta)}$$

for any unbiased estimating equation g, where $I_E(\theta)$ is the usual Fisher information*. Equality is attained if g is a multiple of the score function. Thus the optimality criterion above leads to maximum likelihood estimation in the one-parameter regular case. A linear approximation to an unbiased estimating equation $g(x, \theta) = 0$ via a Taylor expansion* about the true value of θ also indicates that the above optimality criterion approximates the variance of the estimator obtained from $g = 0$ to first order, lending an alternative justification (albeit asymptotic) to this optimality criterion.

MULTIPARAMETER PROBLEMS

Kale [16] obtains an extension of the Cramér–Rao inequality for estimating functions. He extends this to the multiparameter case using the methods of Rao [25] for estimators. Bhapkar [2] defines notions of information and efficiency for estimating functions and establishes an analog of the

Rao–Blackwell theorem* appropriate for estimating functions. He extends this to the multiparameter case. Durbin [4] had obtained a similar extension for linear estimating equations. Both Kale and Bhapkar establish optimality of the vector score function using an extension of Godambe's optimality criterion. Chandrasekar and Kale [3] consider optimality of estimating equations for a vector parameter of interest in the presence of a vector nuisance parameter. They establish the basic equivalence of three optimality criteria based on (i) nonnegative definiteness of the differences of dispersion matrices, (ii) the trace of the dispersion matrix, and (iii) the determinant of the dispersion matrix.

PROBLEMS INVOLVING NUISANCE PARAMETERS

It is perhaps in problems involving nuisance parameters that the estimating equation approach has proved most valuable. For example, it overcomes many of the objections raised by Neyman against maximum likelihood estimation. According to Barnard [1], Neyman objected to the general use of maximum likelihood not because of its failure in unusual or pathological cases, but because it gave the "wrong" answer in the simple case of estimation of the variance of a normal population. Barnard used the notion of a first-order stable pivotal to obtain an estimating equation which yields the usual unbiased estimator S^2 for σ^2. Godambe and Thompson [13] found an optimal estimating function g, minimizing

$$E_\theta \left[\{ g / E_\theta [\partial g / \partial \theta] \}^2 \right]$$

in the class of all unbiased estimating functions g which are functions of the parameter of interest alone. This also yielded S^2 as the appropriate estimate of σ^2.

More serious objections to maximum likelihood were raised by Neyman and Scott [24] in a classic paper. They showed that in problems where the number of nuisance parameters increases with the number of observations, maximum likelihood estimates of the parameter of interest could be inefficient or even inconsistent. For example, suppose (X_i, Y_i), $i = 1, \ldots, n$, are n independent pairs of observations, the ith pair being from a normal population with mean μ_i and variance σ^2. Then, treating σ^2 as the parameter of interest and the μ_i as nuisance parameters, the maximum likelihood estimate of σ^2 is inconsistent, converging to $\sigma^2/2$ as n tends to infinity. Godambe [8] showed that, under certain conditions, the optimal unbiased estimating equation leads to the conditional maximum likelihood approach and hence generates a consistent estimator of σ^2 in the above example. The optimality of the conditional score function is established, the conditioning being on a complete sufficient statistic for the nuisance parameter. Godambe also relates this approach to concepts of sufficiency and ancillarity in the presence of nuisance parameters. Other papers related to nuisance parameters problems are those of Lindsay [20] who extends Godambe's results to situations where the conditioning statistic involves the parameter of interest, Kumon and Amari [19] who adopt a differential-geometric approach, and Morton [22]. Relationships with Fisher information and ancillarity* are further explored by Godambe [9, 10].

References

[1] Barnard, G. A. (1973). *Sankhyā A*, **35**, 133–138. (Connections with Fisher's concept of a pivotal quantity are explored.)

[2] Bhapkar, V. P. (1972). *Sankhyā A*, **34**, 467–472.

[3] Chandrasekar, B. and Kale, B. K. (1984). *J. Statist. Plann. Inf.*, **9**, 45–54.

[4] Durbin, J. (1960). *J. R. Statist. Soc. B*, **22**, 139–153. (Advantages of using linear unbiased estimating equations in time-series analysis.)

[5] Fisher, R. A. (1935). *J. R. Statist. Soc.*, **98**, 39–42.

[6] Fisher, R. A. (1956). *Statistical Methods and Scientific Inference*. Oliver and Boyd, Edinburgh.

[7] Godambe, V. P. (1960). *Ann. Math. Statist.*, **31**, 1208–1211. (The seminal paper justifying maximum likelihood in finite samples.)

[8] Godambe, V. P. (1976). *Biometrika*, **63**, 277–284. (Optimality of the conditional score function in a class of nuisance parameter problems.)

[9] Godambe, V. P. (1980). *Biometrika*, **67**, 269–276. (Connections with sufficiency and ancillarity in the nuisance parameter case.)

[10] Godambe, V. P. (1984). *Biometrika*, **71**, 626–629.

[11] Godambe, V. P. (1985). *Biometrika*, **72**, 419–428. (An extension of the Gauss–Markov theorem to finite-sample estimation in stochastic processes.)

[12] Godambe, V. P. (1987). Unpublished notes, University of Waterloo. (Presents an overview of estimating functions. Explores relationships between least-squares, maximum likelihood, and minimum variance unbiased estimation within the estimating function context.)

[13] Godambe, V. P. and Thompson, M. E. (1974). *Ann. Statist.*, **2**, 277–284.

[14] Godambe, V. P. and Thompson, M. E. (1986). *Intern. Statist. Rev.*, **54**, 127–138. (Optimal estimation for both superpopulation and survey parameters in the survey sampling context.)

[15] Hutton, J. E. and Nelson, P. I. (1986). *Stoch. Processes Appl.*, **22**, 245–277. (Optimality of quasi-likelihood estimation for semimartingales.)

[16] Kale, B. K. (1962). *Skand. Actuar.*, **45**, 80–89.

[17] Kendall, M. E. and Stuart, A. (1979). *The Advanced Theory of Statistics*, 4th ed., Vol. 2. Charles Griffin, London.

[18] Kimball, B. F. (1946). *Ann. Math. Statist.*, **17**, 299–309.

[19] Kumon, M. and Amari, S. (1984). *Biometrika*, **71**, 445–459. (Differential geometry is applied to nuisance parameter problems of the Neyman–Scott variety. A new Cramér–Rao type bound is derived.)

[20] Lindsay, B. (1982). *Biometrika*, **69**, 503–512. (Extensions of the work of Godambe in [8].)

[21] McLeish, D. L. (1984). *Canad. J. Statist.*, **12**, 256–282. (Projection of the score function to produce optimal estimating equations in incomplete data problems.)

[22] Morton, R. (1981). *Biometrika*, **68**, 227–233. (A search for near optimal estimating equations based on pivot-like quantities in the nuisance parameter case.)

[23] Morton, R. (1987). *Biometrika*, **74**, 247–257. (Applications to generalized linear models and quasi-likelihood.)

[24] Neyman, J. and Scott, E. L. (1948). *Econometrica*, **16**, 1–32. (Early difficulties with maximum likelihood in nuisance parameter problems.)

[25] Rao, C. R. (1973). *Linear Statistical Inference and Its Applications*, 2nd ed. Wiley, New York.

Recent Publications of Interest

Godambe, V. P. and Heyde, C. C. (1987). Quasi-likelihood and optimal estimation. *Intern. Statist. Rev.*, **55**, 231–244. (Stochastic processes.)

Thakaneswaran, A. and Abraham, B. (1988). Estimation for non-linear time-series models using estimating equations, *J. Time Series Anal.*, **9**, 99–108. (Time series.)

(CRAMÉR-RAO LOWER BOUND
ESTIMATION, POINT
LABELS
MINIMUM VARIANCE UNBIASED
  ESTIMATION
SURVEY SAMPLING)

ANTHONY F. DESMOND

EULERIAN NUMBERS

For positive integers n and k with $1 \leqslant k \leqslant n$, the *Eulerian numbers* (see Euler [2]) $A(n, k)$ are defined by

$$A(n, k) = \sum_{j=0}^{k} (-1)^j \binom{n+1}{j} (k-j)^n.$$

Table 1 presents $A(n, k)$ for $1 \leqslant k \leqslant n \leqslant 8$.

Carlitz [1] gives a comprehensive discussion of Eulerian numbers and related polynomials. The $A(n, k)$ are related to Euler polynomials* $E_r(y)$ and Bernoulli numbers* B_r by

$$E_n(0) = -2(n+1)^{-1}(2^{n+1} - 1)B_{n+1}$$

$$= 2^{-n} \sum_{k=0}^{n} (-1)^{n+k-1} A(n, k).$$

[Note that the *Euler* numbers, $E_r(0)$—*see* EULER POLYNOMIAL—should not be confused with the *Eulerian* numbers.]

Table 1

				k				
n	1	2	3	4	5	6	7	8
1	1							
2	1	1						
3	1	4	1					
4	1	11	11	1				
5	1	26	66	26	1			
6	1	57	302	302	57	1		
7	1	120	1191	2416	1191	120	1	
8	1	247	4293	15619	15619	4293	247	1

An interpretation of $A(n, k)$ via combinatorics* is as follows. A permutation $(a_1, a_2, \ldots, a_n)$ of $(1, 2, \ldots, n)$ is said to have a *rise* at i, $i = 1, 2, \ldots, n$, if $a_{i-1} < a_i$, with the convention that $a_0 = 0$. Then the number of permutations of $(1, 2, \ldots, n)$ with exactly k rises is $A(n, k)$. It is clear from this interpretation that

$$\sum_{k=1}^{n} A(n, k) = n!.$$

An application of Eulerian numbers in statistics is for the difference-sign test for randomness* of Moore and Wallis [6] and Mann [5]. Let $X_1, X_2, \ldots, X_n$ be n observations of a time series* with X_i corresponding to the ith time point ($i = 1, 2, \ldots, n$). Let S_n be the number of rises. Then, under the null hypothesis that the X_i are independent observations from a common (assumed continuous) distribution,

$$P(S_n = k) = A(n, k)/n!.$$

Thus S_n can be used as a test statistic for a trend, with large (small) values of S_n supplying evidence of a monotone increasing (decreasing) trend*. Mann [5] has shown that

$$\{S_n - E[S_n]\}/\{\text{Var}(S_n)\}^{1/2}$$

is asymptotically standard normal as $n \to \infty$ under the null hypothesis, where

$$E[S_n] = \tfrac{1}{2}(n + 1),$$

$$\text{Var}(S_n) = (n + 1)/12.$$

See also Hensley [3] for a recent proof.

A second application, due to Kimber [4], is to the distribution function of a commonly used discordancy test statistic. Let $Y_1, Y_2, \ldots, Y_n$ be independent random variables, each, under the null hypothesis, with probability density function.

$$p(y; \mu, \sigma) = \sigma^{-1}\exp\{-(y - \mu)/\sigma\},$$
$$y > \mu.$$

Suppose μ is known and that Z_n is the maximum of the Y_i, $i = 1, 2, \ldots, n$. Consider the statistic

$$T_n = (Z_n - \mu)\bigg/\bigg(\sum_{i=1}^{n} Y_i - n\mu\bigg).$$

With the alternative hypothesis as for the null hypothesis except that a single unspecified observation comes from an exponential distribution with larger mean, T_n is a likelihood ratio* test statistic. Thus T_n is a discordancy test statistic for an upper outlier in an exponential sample. Let F_n be the distribution function of T_n under the null hypothesis. Then

$$A(n, k) = k^n F_{n+1}(1/k).$$

A third application is the following urn model of Shur [7]. We begin with one black ball and no white balls in the urn. At each stage a ball is drawn at random from the urn. The ball is then returned to the urn together with a ball of the opposite color. If n balls are drawn, then the probability that k of them are black is

$$A(n, k)/n!.$$

This is an example of a biased coin design with potential application, see Kimber [4], in treatment allocation for clinical trials.

References

[1]   Carlitz, L. (1959). *Math. Mag.*, **33**, 247–260.
[2]   Euler, L. (1755). *Institutiones Calculi Differentialis*. Petrograd.
[3]   Hensley, D. (1982). *Fibonacci Quart.*, **20**, 344–348.
[4]   Kimber, A. C. (1987). *Utilitas Mathematica*, **31**, 57–65.
[5]   Mann, H. B. (1945). *Ann. Math. Statist.*, **16**, 193–199.
[6]   Moore, G. H. and Wallis, W. A. (1943). *J. Amer. Statist. Ass.*, **38**, 153–164.
[7]   Shur, W. (1984). *Commun. Statist. A*, **13**, 877–885.

(EULER POLYNOMIAL
URN MODELS)

A. C. KIMBER

EVEN-POINT ESTIMATION

For a wide class of two-parameter discrete distributions, with parameters a and b say, one of the two maximum-likelihood* (ML) estimation equations is the first moment equation, i.e., the sample mean $\bar{x}$ is equated to the distribution mean μ expressed as a function of the parameters ($\bar{x} = \mu(a, b)$). Most non-ML estimation procedures for such distributions also utilize the first moment equation, and hence differ between themselves (and from ML) only in the form of the second estimation equation. *Even points* is one such method. Its second equation is obtained by equating the sum of the *relative* frequencies of the even values of the variable X in the sample to the sum of the even probabilities, i.e., $\sum f_{2x} = \sum p_{2x}$, $x = 0, 1, 2, \ldots$. (Note that, since $\sum p_{2x} = 1 - \sum p_{2x+1}$ and $\sum f_{2x} = 1 - \sum f_{2x+1}$, this could equally well be called the "odd-points" method.) Denoting the probability generating function* (PGF) of the distribution by $g(s) = \sum p_x s^x$, then

$$g(1) + g(-1) = \sum p_x \left[1 + (-1)^x\right]$$

$$= 2\sum p_{2x}; \qquad (1)$$

even-points may therefore be considered as a special case of estimation via the empirical PGF [1, 3].

Patel [7–9] introduced the even-points method in a detailed study of estimation properties for the Hermite distribution*. This distribution has PGF

$$g(s) = \exp\left[a(s - 1) + b(s^2 - 1)\right],$$

$$a, b > 0, \quad (2)$$

and mean

$$\mu = a + 2b.$$

From (1)

$$g(1) + g(-1) = 1 + \exp(-2a) = 2\sum p_{2x},$$
$$(3)$$

and hence the even-point estimators are

$$\tilde{a} = -\left[\log\left(2\sum f_{2x} - 1\right)\right]\Big/2,$$

$$\tilde{b} = (\bar{x} - \tilde{a})/2. \qquad (4)$$

A solution exists only if $\sum f_{2x} > 0.5$; this is consistent with the representation of the Hermite distribution as the sum of an ordinary Poisson* variable and a Poisson doublet variable [2]. Patel found that, regardless of the value of b, the even-point estimators have high joint asymptotic efficiency when $a < 0.5$, and also when $0.5 < a < 1.5$ provided b is large relative to a. In these regions the moment estimators are particularly inefficient. Hence even points provides a rapid estimation procedure complementary to moments.

A remarkable feature of (3) is the absence of b, so that (3) leads immediately to a simple explicit expression for a (and hence for b). Unfortunately, evaluating $g(-1)$ for other well-known discrete distributions (e.g., the negative binomial*) does not lead to explicit expressions for the parameter estimates, so iteration is needed. Consequently, the method has not, as yet, found favor in the context of univariate discrete distributions other than the Hermite.

In his Ph.D. thesis Papageorgiou [5] examined the extension of the even-points approach to certain bivariate discrete distributions and found that in most cases the resulting equations have to be solved iteratively. Nevertheless, for the bivariate Poisson distribution* [4] and the bivariate Hermite distribution* [6] the approach leads to simple explicit expressions for the parameter estimates. In both cases the first moment equations $\bar{x} = \mu_x$ and $\bar{y} = \mu_y$ (which are also ML equations) form two of the estimating equations. For the three-parameter bivariate Poisson distribution with PGF

$$g(u, v) = \sum \sum p_{x, y} u^x v^y$$

$$= \exp\{(a - d)(u - 1)$$

$$+ (b - d)(v - 1)$$

$$+ d(uv - 1)\}, \qquad (5)$$

evaluating the PGF at $(u, v) = (1, 1)$ and $(-1, -1)$ gives

$$g(1,1) + g(-1, -1)$$

$$= 1 + \exp\{-2(a + b - 2d)\}$$

$$= 2\left(\sum p_{2x, 2y} + \sum p_{2x+1, 2y+1}\right). \quad (6)$$

The even-point estimators are

$$\tilde{a} = \bar{x}, \qquad \tilde{b} = \bar{y},$$

$$d = (\bar{x} + \bar{y})/2 + \{\log(2A - 1)\}/4, \quad (7)$$

where A is the sum of the observed relative frequencies at the points $(2x, 2y)$ and $(2x + 1, 2y + 1)$.

For the five-parameter bivariate Hermite distribution with PGF

$$g(u, v) = \exp\left[a_1(u - 1)\right.$$

$$+ a_2(u^2 - 1) + a_3(v - 1)$$

$$\left. + a_4(v^2 - 1) + a_5(uv - 1)\right], \quad (8)$$

$g(-1, 1)$ and $g(1, -1)$ are needed, as well as $g(-1, -1)$ and $g(1, 1)$.

For the bivariate Poisson distribution, asymptotic efficiency of even-points is high either for small values of a and b or when the correlation between X and Y is high. For the bivariate Hermite distribution, efficiency is high when a_1, a_3, and a_5 are small and a_2 and a_4 are relatively large. As in the univariate case, the method provides a simple rapid estimation procedure which has high asymptotic efficiency in regions of the parameter space where estimation by moments is very inefficient.

References

[1] Kemp, A. W. and Kemp, C. D. (1986). Efficiency of empirical PGF estimation for the Hermite distribution. *1986 Proc. Statist. Comput. Sect. Amer. Statist. Ass.*, 110–112.

[2] Kemp, C. D. and Kemp, A. W. (1965). Some properties of the "Hermite" distribution. *Biometrika*, **52**, 381–394.

[3] Kemp, C. D. and Kemp, A. W. (1988). Rapid estimation for discrete distributions. *Statistician*, **37**.

[4] Loukas, S., Kemp, C. D., and Papageorgiou, H. (1986). Even point estimation for the bivariate Poisson distribution. *Biometrika*, **73**, 222–223.

[5] Papageorgiou, H. (1977). Certain Properties of Bivariate Generalized Poisson Distributions with Special Reference to the Hermite Distribution. Ph.D. thesis, University of Bradford, Bradford, England.

[6] Papageorgiou, H., Kemp, C. D., and Loukas, S. (1983). Some methods of estimation for the bivariate Hermite distribution. *Biometrika*, **70**, 479–484.

[7] Patel, Y. C. (1971). Some Problems in Estimation for the Parameters of the Hermite Distribution. Ph.D. dissertation, University of Georgia, Athens, GA.

[8] Patel, Y. C. (1976). Even point estimation and moment estimation in Hermite distribution. *Biometrics*, **32**, 865–873.

[9] Patel, Y. C. (1977). Higher moments of moment estimators and even point estimators for the parameters of the Hermite distribution. *Ann. Inst. Statist. Math. Tokyo*, **29A**, 119–130.

(EFFICIENCY, ASYMPTOTIC RELATIVE ESTIMATION, POINT
HERMITE DISTRIBUTION
METHOD OF MOMENTS)

ADRIENNE W. KEMP
C. DAVID KEMP

EXCESS

An alternative measure of kurtosis* of a distribution. It is

$$\tfrac{1}{3}\mu_4\mu_2^{-2} - 1,$$

where μ_r is the rth central moment. For a normal distribution it is zero. The term is now rarely used.

Bibliography

Crum, W. L. (1923). *J. Amer. Statist. Ass.*, **18**, 607–614.

Pearson, K. (1894). *Philos. Trans. R. Soc. Lond.*, **185A**, 77–110.

(DEFECT
KURTOSIS
MOMENT RATIOS
SKEWNESS)

F

FAULTY INSPECTION DISTRIBU-
TIONS *See* DORFMAN-TYPE SCREENING
PROCEDURES

FIBONACCI DISTRIBUTIONS

In a series of independent observations of a
binary event ($S \equiv$ "success" or F
$\equiv$ "failure"), the probability generating
function* of the number of observations (X)
needed to obtain a run of n S's is

$$G(t) = (pt)^n(1 - pt)$$

$$\times \left\{1 - t + p^n(1 - p)t^{n+1}\right\}^{-1} \quad (1)$$

(Bizley [1]).

If $p = \frac{1}{2}$ (corresponding, for example, to
tosses of an unbiased coin) then

$$G(t) = \left(\tfrac{1}{2}t\right)^n\left(1 - \tfrac{1}{2}t\right)\left\{1 - t + \left(\tfrac{1}{2}t\right)^{n+1}\right\}^{-1}$$

$$(2)$$

and

$$\Pr[X = x] = 2^{-x}F_{n, x-n},$$

$$x = n, n + 1, \ldots, \quad (3)$$

where

$$F_{n, x} = \sum_{j=1}^{n} F_{n, x-j}, \quad x = 1, 2, \ldots,$$

with $F_{n,0} = 1$ and $F_{n,j} = 0$ for $j =
-1, -2, \ldots, -n$.

If $n = 2$ then the $F_{2, x}$'s are the *Fibonacci
numbers*. Shane [2] terms the corresponding

distribution

$$\Pr[X = x] = 2^{-x}F_{2, x-2}$$

the *Fibonacci distribution*, and the general
form (3) the *poly-nacci distributions*. The distributions with general p (and n) given by
the probability generating function (1) are
called *generalized poly-nacci distributions*.

See also Taillie and Patil [3], who use a
different nomenclature. They term the distributions corresponding to 1, the nth *order
Fibonacci distribution*, and the distributions
with general p, *generalized (nth order) Fibonacci distributions*.

References

[1] Bizley, M. T. L. (1962). *J. Inst. Actu.*, **88**, 360–366.
[2] Shane, H D. (1973). *Fibonacci Quart.*, **11**, 517–522.
[3] Taillie, C. and Patil, G. P. (1986). *Commun. Statist.
A*, **15**, 951–959.

FISHER'S CHI-SQUARED APPROXIMATION

An approximation due to Fisher [1] to the
distribution of chi-squared* with ν degrees
of freedom. The formula is

$$\Pr[\chi_\nu^2 \leqslant x] \doteq \Phi(\sqrt{2x} - \sqrt{2\nu - 1}),$$

where

$$\Phi(y) = (2\pi)^{-1/2} \int_{-\infty}^{y} e^{-z^2/2} \, dz.$$

Equivalently, "χ_ν^2 is distributed approximately normally with expected value ($\sqrt{2x} -
\sqrt{2\nu - 1}$) and standard deviation 1."

It is simpler than, but not as accurate as
the Wilson–Hilferty transformation*.

Reference

[1] Fisher, R. A. (1922). *J. R. Statist. Soc. A*, **85**, 87–94.

(APPROXIMATIONS TO DISTRIBUTIONS CHI-SQUARED DISTRIBUTION)

FISHER'S TEST APPROACH TO INDEX NUMBERS   *See* TEST APPROACH TO INDEX NUMBERS, FISHER'S

FKG INEQUALITY, THE

The so-called FKG inequality concerns the problem of capturing in a mathematical definition what is meant by the notion that the coordinates of a random vector are "positively related." For the case of two random variables, X_1 and X_2, Lehmann [6] introduced *positive quadrant dependence*, which is expressed as

$$\Pr[(X_1 \geqslant x_1) \cap (X_2 \geqslant x_2)]$$
$$\geqslant \Pr[X_1 \geqslant x_1]\Pr[X_2 \geqslant x_2] \quad (1)$$

for all x_1, x_2. A stronger notion of positive dependence which has received much attention is that of *association* defined in Esary et al. [3]. A random vector $X \in R^n$ with coordinates $X = (X_1, \ldots, x_n)$ is *associated* if

$$E[f_1(X)f_2(X)] \geqslant E[f_1(X)]E[f_2(X)] \quad (2)$$

for all real-valued functions f_1 and f_2 which are nondecreasing in each coordinate variable when the other variables are held fixed. Inequality (2) (*see* DEPENDENCE, CONCEPTS OF) implies that for each k, $1 \leqslant k < n$,

$$\Pr\left[\bigcap_{I<l}^{n}(X_i \geqslant x_i)\right] \geqslant \Pr\left[\bigcap_{i=1}^{k}(X_i \geqslant x_i)\right]$$
$$\times \Pr\left[\bigcap_{i=k+1}^{n}(X_i \geqslant x_i)\right], \quad (3)$$

which in turn implies that

$$\Pr\left[\bigcap_{i>1}^{n}(X_i \geqslant x_i)\right] \geqslant \prod_{i=1}^{n}\Pr[X_i \geqslant x_i]. \quad (4)$$

This inequality has found many applications in reliability* theory where (4) often gives computable lower bounds on the probability that a system with n components in series has not failed (see, e.g., Barlow and Proschan [1] for further discussion and other applications of association). Inequality (2), which is often expressed as

$$\mathrm{cov}(f_1(X), f_2(X)) \geqslant 0 \quad (5)$$

is somewhat difficult to interpret intuitively because (2) must hold for a large class of functions. However, association has proved to be a useful tool in establishing probability inequalities. For example, see Barlow and Proschan [1], Karlin and Rinott [5], Tong [8], and Eaton [2].

In a reliability* context conditions on the distribution of a random vector X which imply that X is associated were provided in Sarkar [7]. Independently, Fortuin et al. [4] also provided conditions for association with applications to problems in physics in mind. In the physics world, inequality (5) became known as the *FKG inequality*. This nomenclature has now been adopted in much of the statistics literature as well.

Here is a statement of a sufficient condition for (5) to hold. Assume the random vector X has values in a product set. $\mathscr{X}^{(n)} = \mathscr{X}_1 \times \cdots \times \mathscr{X}_n$ where each $\mathscr{X}_i$ is a Borel subset of R^1. In most applications, $\mathscr{X}_i$ is an interval or a discrete set. It is assumed that X has a density $p(\cdot)$ with respect to a product measure $\mu^{(n)} = \mu_1 \times \cdots \times \mu_n$. For any two vectors $x \wedge y$ in $X^{(n)}$ let $x \wedge y$ ($x \vee y$, respectively) denote the vector in $\mathscr{X}^{(n)}$ whose ith coordinate is $\min\{x_i, y_i\}$ ($\max\{x_i, y_i\}$, respectively).

Theorem.   If the density $p(\cdot)$ satisfies

$$p(x)p(y) \leqslant p(x \wedge y)p(x \vee y),$$
$$\text{for all } x, y \in \mathscr{X}^{(n)}, \quad (6)$$

then (5) holds.

The article by Karlin and Rinott [5] and the monograph by Eaton [2] give self-contained proofs of the above theorem in addition to a thorough discussion concerning condition (6) and other issues related to this result and association. Some background material can also be found in Tong [8].

References

[1] Barlow, R. E. and Proschan, F. (1965). *Mathematical Theory of Reliability*. Wiley, New York.

[2] Eaton M. L. (1987). Lectures on Topics in Probability Inequalities. *CWI Tract No. 35*. Centre for Mathematics and Computer Science, Amsterdam.

[3] Esary, J. D., Proschan, F., and Walkup, D. W. (1967). *Ann. Math. Statist.*, **38**, 1466–1474.

[4] Fortuin, C. M., Kasteleyn, F. N., and Ginibre, J. (1971). *Commun. Math. Phys.*, **22**, 89–103.

[5] Karlin, S. and Rinott, Y. (1980). *J. Multivariate Anal.*, **10**, 467–498.

[6] Lehmann, E. L. (1966). *Ann. Math Statist.*, **37**, 1137–1152.

[7] Sarkar, T K. (1969). Some Lower Bounds of Reliability. *Tech. Rep. No. 124*, Dept. of Operations Research and Statistics, Stanford University, Stanford, CA.

[8] Tong, Y. L. (1980). *Probability Inequalities in Multivariate Distributions*. Academic, New York.

(ASSOCIATION, MEASURES OF DEPENDENCE, CONCEPTS OF DEPENDENCE, MEASURES AND INDICES OF MEASURES OF AGREEMENT TOTAL POSITIVITY)

M. L. EATON

FRAILTY MODELS

These are models of cohort mortality, allowing for heterogeneity in a specific manner.

For a survival function*

$$\Pr[T > t] = S_T(t), \qquad 0 < t,$$

the corresponding force of mortality (or "hazard rate") is

$$\lambda_T(t) = -\frac{d \log S_T(t)}{dt}.$$

It has long been recognized that heterogeneity among members of a cohort affects mortality experience, because the less resistant tend to die out to a relatively greater extent than the more resistant. Frailty models, as proposed by Vaupel et al. [5], allow for this by introducing a *frailty* parameter θ. For an individual, the force of mortality function is

$$\lambda(t; \theta) = \theta\lambda(t:1).$$

This is a proportional hazards* model. If the density function of θ at birth (i.e., among newborn individuals) is

$$f_\Theta(\theta; 0), \qquad 0 < \theta,$$

then the population survival function is

$$S(t) = \int_0^\infty f_\Theta(\theta; 0) \, S(t; \theta) \, d\theta,$$

where

$$S(t; \theta) = \exp\left\{ -\int_0^t \lambda(u; \theta) \, du \right\}$$

$$= \left\{ S(t; 1) \right\}^\theta.$$

The frailty distribution among survivors at age x has density function

$$f_\Theta(\theta; x) = f_\Theta(\theta; 0) S(x; \theta)/S(x),$$

$$0 < \theta.$$

As a consequence of differential mortality, the distribution of θ at age x_2 is stochastically less* than that at age x_1, if $x_2 > x_1$. Hence if a life table* were computed using force of mortality function

$$\bar{\theta}\lambda(t; 1),$$

where $\bar{\theta} = \int_0^\infty \theta f_\Theta(\theta; 0) \, d\theta$ is the expected value of θ at age 0, it would overestimate mortality at older ages. However, life-table construction is based on data which reflect mortality among *survivors* at each age. (Indeed, they could hardly relate to mortality that *might* have occurred among nonsurvivors!) Differential mortality is, therefore, allowed for automatically in fitting the data.

By assuming specific forms for $f_\Theta(\theta; 0)$, some elegant mathematical results can be derived. Vaupel et al. [5] used a gamma distribution*, while Hougaard [1] pointed out that similar results for a wide class of exponential family distributions* can be obtained, and investigated in detail the consequences of assuming an inverse Gaussian distribution*. There has been little investigation aimed at justifying these assumptions, or, indeed the assumption of a frailty multiplier, constant throughout life.

The term "frailty" was suggested in 1979 in ref. 5. In 1969, Redington [4] used a lognormal distribution* for θ, combined with a Gompertz* form for $\lambda(t:1)$, and investigated approximation of a mixture of Gompertz distributions by a single Gompertz. The general problem of the effects of heterogeneity was addressed by Perks [3] in 1932, who considered the effects of an index of "inability to withstand destruction." ("Frailty" is, of course, such an index, although Perks did not restrict himself to a multiplicative factor.) Perks noted that heterogeneity might lead to deviations from Gompertz (or Makeham–Gompertz) laws* at older ages, in suggesting the fomula

$$\lambda(t) = (A + Bc^t)/(1 + Dc^t)$$

to allow for such effects (*see* PERKS' DISTRIBUTIONS).

References

[1] Hougaard, P. (1984). *Biometrika*, **71**, 75–83.
[2] Oakes, D. (1987). *Biometrika*, **74**.
[3] Perks, W. (1932). *J. Inst. Actuar.*, **63**, 12–40.
[4] Redington, F. M. (1969). *J. Inst. Actuar.*, **95**, 243–298.
[5] Vaupel, J. W., Manton, K. G., and Stallard, E. (1979). *Demography*, **16**, 439–456.

(ACTUARIAL STATISTICS—LIFE
COHORT ANALYSIS
COX'S REGRESSION MODEL
FORCE OF MORTALITY
LIFE TABLES
MIXTURE DISTRIBUTIONS
PERKS' DISTRIBUTIONS
SURVIVAL ANALYSIS)

FREQUENCY MOMENTS *See* PROBABILITY MOMENTS AND FREQUENCY MOMENTS

G

GEARY STATISTIC *See* CLIFF–ORD TEST: SPATIAL PROCESSES

GEGENBAUER DISTRIBUTIONS

The Gegenbauer distribution was, apparently, first introduced in the literature by Plunkett and Jain [7]. It can be derived from the Hermite distribution* probability generating function* (PGF)

$$G(s) = \exp\{\gamma_1(1 - s) - \gamma_1\gamma_3(1 - s^2)\}$$

by compounding it with respect to a gamma distribution* for γ_1.

The resulting PGF of the Gegenbauer distribution is

$$G(s) = \int_0^\infty \exp\left[-\gamma_1\{1 - s + \gamma_3(1 - s^2)\}\right]$$
$$\times \{\Gamma(\lambda)\}^{-1}\gamma_1^{\lambda-1}\delta^\lambda \exp(-\gamma_1\delta)\, d\gamma_1$$
$$= (1 - \alpha - \beta)^\lambda(1 - \alpha s - \beta s^2)^{-\lambda},$$

where α and β are determined by γ_3 and the parameters λ, δ of the distribution of γ_1, according to the formulas

$$\alpha = (1 + \gamma_3 + \delta)^{-1},$$
$$\beta = \gamma_3(1 + \gamma_3 + \delta)^{-1}.$$

Note that $\alpha + \beta < 1$ since $\delta > 0$.

Expanding $G(s)$ in powers of s, we have

$$G(s) = (1 - \alpha - \beta)^\lambda \sum_{n=0}^\infty G_n^\lambda(\alpha, \beta) \frac{s^n}{n!},$$

where $G_n^\lambda(\alpha, \beta)$ is a modified Gegenbauer polynomial of order n,

$$G_n^\lambda(\alpha, \beta)$$

$$= \sum_{j=1}^{[n/2]} \frac{n!}{j!} \beta^j \frac{\Gamma(n + \lambda - j)}{\Gamma(n - 2j + 1)} \frac{\alpha^{n-2j}}{\Gamma(\lambda)}$$

(see, e.g., Abramowitz and Stegun [1], Erdelyi [4], and ULTRASPHERICAL POLYNOMIALS).
Hence

$$\Pr[X = x] = (1 - \alpha - \beta)^\lambda G_x^\lambda(\alpha, \beta)/x!.$$

Limiting cases, recursive relations among probabilities, moments and cumulants, and estimation of the parameters (α and β) were studied by Plunkett and Jain [7], and more recently by Borah [3]. Both articles feature fitting the distribution to the data on haemocytometer counts of yeast cells used by Student (*see* GOSSET, WILLIAM SEALY). The two articles used different estimation formulas: In ref. 7 the first three sample and population (factorial) moments are equated, whereas in ref. 3 the first two moments and the ratio of frequencies of 0's and 1's are used. The latter method appears to give a better fit for the data considered; both methods give better fits than those obtained with a negative binomial distribution* (Bliss [2]), a Hermite distribution* (Kemp and Kemp [5]), or a Neyman type A distribution* (Neyman [6]), according to the χ^2 criterion.

References

[1] Abramowitz, H. and Stegun, I. A. eds. (1964). Handbook of Mathematical Functions. *Appl. Math. Ser.*, **55**. National Bureau of Standards. Government Printing Office, Washington, D. C.

[2] Bliss, C. I. (1953). *Biometrics*, **9**, 176–196.

[3] Borah, M. (1984). *J. Indian Soc. Agric. Statist.*, **36**, 72–78.

[4] Erdelyi, A. (1953). *Tables of Integral Transforms*, Vol. 1. McGraw-Hill, New York.

[5] Kemp, C. D. and Kemp, A. W. (1965). *Biometrika*, **52**, 381–394.

[6] Neyman, J. (1939). *Ann. Math. Statist.*, **10**, 35–57.

[7] Plunkett, I. G. and Jain, G. C. (1975). *Biom. Z.*, **17**, 286–302.

(COMPOUND DISTRIBUTION
HERMITE DISTRIBUTIONS
NEGATIVE BINOMIAL DISTRIBUTION
NEYMAN'S TYPE A, B, AND C
   DISTRIBUTIONS
ULTRASPHERICAL POLYNOMIALS)

GENERALIZED F-DISTRIBUTION

Although there are several distributions which might be called by this name, it is applied formally (see ref. 1) to the distribution of

$$\mu + \theta \log F,$$

where μ and θ (> 0) are fixed parameters and F has an F-distribution*.

If $\mu = 0$ and $\theta = \frac{1}{2}$ we have Fisher's z-distribution*. Kalbfleisch and Prentice [1] use the distribution as a life-time distribution*.

Reference

[1] Kalbfleisch, J. D. and Prentice, R. L. (1980). *The Statistical Analysis of Failure Time Data*. Wiley, New York (pp. 28 et seq.).

(ANALYSIS OF VARIANCE
LIFE-TIME DISTRIBUTIONS
SURVIVAL ANALYSIS)

GENERIC UNIFORM LAWS OF LARGE NUMBERS

A frequently adopted approach for proving consistency of M-estimators*, and in particular of maximum likelihood estimators, involves the demonstration that the difference between the criterion function defining the estimator and the expectation of this function, both normalized by the sample size, converges to zero uniformly over the parameter space. Typically, this results in the verification of a uniform law of large numbers (ULLN), i.e., of a law of large numbers (LLN) that holds uniformly over the parameter space.

In general, ULLNs give conditions under which

$$\sup_{\theta \in \Theta} \left| n^{-1} \sum_{i=1}^{n} [f_i(x_i, \theta) - Ef_i(x_i, \theta)] \right| \to 0$$

almost surely or in probability as $n \to \infty$. Here $(x_i)_{i \in \mathbb{N}}$ is a sequence of random variables taking their values in a measurable space $(X, \mathcal{X})$, Θ is an abstract parameter (index) space, and the f_i are real-valued functions. Since Θ is an abstract space the above formulation also includes statements of the form

$$\sup_{f \in \mathcal{F}} \left| n^{-1} \sum_{i=1}^{n} [f(x_i) - Ef(x_i)] \right| \to 0,$$

where $\mathcal{F}$ is a collection of functions on X.

Recent interest in asymptotic inference in stochastic processes*, especially in nonlinear dynamic econometric* models and nonlinear time-series* models, originated the need for ULLNs that apply to dependent and nonstationary sequences (x_i). Results of this nature have been introduced recently by Andrews [1] and Pötscher and Prucha [10]. These ULLNs are called *generic* since they provide conditions under which the existence of LLNs for certain functions of x_i implies the existence of ULLNs. Since LLNs are available for wide classes of dependent and nonstationary processes these generic ULLNs can be used to generate ULLNs for such classes of processes. The technique used in deriving these results is based on the technique employed in Wald's [15] consistency proof for the maximum likelihood* estimator; compare also Hoadley [5] and Jennrich [7]. The above results also apply to triangular arrays $f_{i,n}(x_i, \theta)$.

By focusing on dependent nonstationary processes the above generic ULLNs apply to less general classes of functions than ULLNs that focus, e.g., on independent or i.i.d. processes. Examples of specific ULLNs for i.i.d. processes, available in the literature on empirical processes*, are ULLNs for Vapnik–Červonenkis (polynomial) classes; see, e.g., Gänssler [4], Pollard [9], and Yukich [16].

ULLNs are also closely related to uniformity in weak convergence, given the x_i are identically distributed with distribution P and their range space X is a separable metric space. In this case

$$\sup_{f \in \mathcal{F}} \left| n^{-1} \sum_{i=1}^{n} [f(x_i) - Ef(x_i)] \right|$$
$$= \sup_{f \in \mathcal{F}} \left| \int f \, dP_n - \int f \, dP \right|,$$

where P_n is the empirical distribution of $x_1, \ldots, x_n$. Hence if P_n converges weakly to P almost surely (which is, according to Varadarajan [14], equivalent to $|\int g \, dP_n - \int g \, dP| \to 0$ almost surely for all bounded and continuous functions g), the ULLN follows if $\mathcal{F}$ is a P-uniformity class in the sense of Billingsley and Topsøe [3]; see also Ranga Rao [11] and Topsøe [12, 13].

ULLNs can also be interpreted as LLNs in infinite-dimensional spaces. If, e.g., the functions $f_i(x_i, \cdot)$ are bounded, then an ULLN can be viewed as a LLN for random variables with values in the space $l^\infty(\Theta)$ equipped with the supremum norm. Recent references concerning LLNs in infinite-dimensional spaces include Beck et al. [2], Hoffmann-Jørgensen and Pisier [6], and Kuelbs and Zinn [8].

References

[1] Andrews, D. W. K. (1987). *Econometrica*, **55**, 1465–1471.

[2] Beck, A., Giesy, D. P., and Warren, P. (1975). *Theor. Prob. Appl.*, **20**, 126–133.

[3] Billingsley, P. and Topsøe, F. (1967). *Zeit. Wahrscheinlichkeitsth. Verw. Geb.*, **7**, 1–16.

[4] Gänssler, P. (1983). Empirical Processes. *IMS Monograph Series*, Institute of Mathematical Statistics, Hayward, CA.

[5] Hoadley, B. (1971). *Ann. Math. Statist.*, **42**, 1977–1991.

[6] Hoffmann-Jørgensen, J. and Pisier, G. (1976). *Ann. Prob.*, **4**, 587-599.

[7] Jennrich, R. I. (1969). *Ann. Math. Statist.*, **40**, 633–643.

[8]   Kuelbs, J. and Zinn, J. (1979). *Ann. Prob.*, **7**, 75–84.

[9]   Pollard, D. (1984). *Convergence of Stochastic Processes*. Springer, New York.

[10] Pötscher, B. M. and Prucha, I. R. (1988). *Econometrica* (forthcoming).

[11] Ranga Rao, R. (1962). *Ann. Math. Statist.*, **33**, 659–680.

[12] Topsøe, F. (1967). *Theor. Prob. Appl.*, **12**, 281–290.

[13] Topsøe, F. (1977). *Zeit. Wahrscheinlichkeitsth. Verw. Geb.*, **39**, 1–30.

[14] Varadarajan, V. S. (1958). *Sankhyā*, **19**, 23–26.

[15] Wald, A. (1949). *Ann. Math. Statist.*, **20**, 595–600.

[16] Yukich, J. E. (1987). *Prob. Theor. Rel. Fields*, **74**, 71–90.

(LAWS OF LARGE NUMBERS)

BENEDIKT M. PÖTSCHER
INGMAR R. PRUCHA

GIRDLE (EQUATORIAL) DATA / DISTRIBUTIONS

A random three-dimensional unit vector or axis **X** whose values tend to lie about a great circle (or equator) on the unit sphere is said to have a *girdle* or *equatorial distribution*. Data of this type arise commonly in structural geology, when measurements are made of the normal to a cylindrically folded planar surface. If the plane is assigned a facing direction, the result is a unit vector; otherwise it is a unit axis, the more usual case.

The principal parametric model for girdle axial data is the Watson, or Scheidegger–Watson, distribution* with density proportional to

$$\exp\left[\kappa(X'\lambda)^2\right], \qquad \|X\| = \|\lambda\| = 1, \qquad \kappa < 0,$$

where λ is a fixed unit axis. The distribution is rotationally symmetric about λ, and is concentrated increasingly about the great circle normal to λ as $\kappa \to -\infty$. When $\kappa > 0$, the model is sometimes appropriate for bipolar axial data (axial data forming a single cluster) which exhibit rotational symmetry about some axis λ. Both forms of the distribution are embedded in the more general

Bingham distribution (*see* DIRECTIONAL DISTRIBUTIONS) which also contains a continuum of distributions transitional between the bipolar and girdle forms of the Watson distribution*.

Statistical inference for a random sample of axial data $X_1, \ldots, X_n$ is usually based on eigenvectors and eigenvalues of the orientation matrix $\sum X_i X_i'$, both for parametric methods and large-sample theory. For details and further references see, e.g., Watson [2] or Fisher et al. [1].

References

[1]   Fisher, N. I., Lewis, T., and Embleton, B. J. J. (1987). *Statistical Analysis of Spherical Data*. Cambridge University Press, London.

[2]   Watson, G. S. (1983). *Statistics on Spheres*. Wiley, New York.

(DIRECTIONAL DATA ANALYSIS
DIRECTIONAL DISTRIBUTIONS
VON MISES' DISTRIBUTION
WATSON'S DISTRIBUTION)

N. I. FISHER

GRINGORTEN'S FORMULA

Gringorten [4] derived the formula $p_{n,i} = (i - 0.44)/(n + 0.12)$ as an approximation to the expected ith largest order statistic* in a sample of size n from a reduced type 1 extreme-value distribution*. Used as a plotting position, it has smaller bias than Weibull's formula [6] $i/(n + 1)$ or Hazen's $(i - 0.5)/n$, though the improvement over Hazen's [5] formula is slight. It has been extended to the generalized extreme-value distribution by Arnell et al. [1]. The choice of plotting positions has been reviewed by Cunnane [3], with particular reference to the type 1 extreme-value distribution. The issue appears to be more important when probability plots* are used as part of a formal estimation procedure than when they are used merely as a visual aid, and some modern authors on graphical methods (e.g.,

Chambers et al. [2]) use the Hazen formula regardless of the supposed underlying distribution.

References

[1] Arnell, N., Beran, M. A., and Hosking, J. R. M. (1986). *J. Hydrology*, **86**, 59–69.

[2] Chambers, J. M., Cleveland, W. S., Kleiner, B., and Tukey, P. A. (1983). *Graphical Methods for Data Analysis*. Duxbury, Boston, MA.

[3] Cunnane, C. (1978). *J. Hydrology*, **37**, 205–222.

[4] Gringorten, I. I. (1963). *J. Geophys. Res.*, **68**, 813–814.

[5] Hazen, A. (1914). *Trans. Amer. Soc. Civil Eng.*, **77**, 1547–1550.

[6] Weibull, W. (1939). *Ing. Vet. Akad. Handl.* (Stockholm), **151**, 1–45.

(ORDER STATISTICS
PROBABILITY PLOTTING)

R. L. SMITH

GROUP FAMILIES

A *group family* is a family $\mathscr{F}$ of probability distributions obtained by subjecting a random quantity with fixed distribution to a group, G, of transformations. The simplest example, and the one with which Fisher [5] introduced the concept, is that of a location family. (The term "group family" is due to Barndorff-Nielsen [1].)

Example 1.   Location family*. Let $(X_1, \ldots, X_n)$ have a density $f(x_1, \ldots, x_n)$. Then the joint density of

$$Y_1 = X_1 + a, \ldots, Y_n = X_n + a \quad (1)$$

has joint density

$$f(y_1 - a, \ldots, y_n - a). \quad (2)$$

Thus the group of translations (1) with $-\infty < a < \infty$ generates the location family (2).

Example 2.   Location-scale family*. If (1) is replaced by the group of transformations

$$Y_1 = bX_1 + a, \ldots, Y_n = bX_n + a,$$

$$0 < b, \quad -\infty < a < \infty, \quad (3)$$

the density of the Y's becomes the location-scale family

$$\frac{1}{b^n} f\left(\frac{x_1 - a}{b}, \ldots, \frac{x_n - a}{b}\right). \quad (4)$$

Example 3.   Simple linear regression*. If (1) is replaced by

$$Y_1 = X_1 + a + bt_1, \ldots, Y_n = X_n + a + bt_n, \quad (5)$$

where $t_1, \ldots, t_n$ are n fixed numbers, the density of Y's becomes the density of a regression model (with nonnormal errors when f is not normal)

$$f(y_1 - a - bt_1, \ldots, y_n - a - bt_n). \quad (6)$$

Other group families are more general linear models with errors that are not necessarily normal, nonparametric families, or finite-population models (see Lehmann [8]).

It follows from the fact that G is a group that quite generally the same family is generated no matter which member of $\mathscr{F}$ is used as the starting point.

Group families are closely connected to invariance concepts. (For an explanation of the concepts in this paragraph, *see* INVARIANCE CONCEPTS IN STATISTICS.) On the one hand, a group family is always invariant under the group generating it. Conversely, consider any family $\mathscr{F} = \{P_\theta, \theta \in \Omega\}$ of distributions of a random quantity X that remains invariant under a group G of transformations of X. Then $\mathscr{F}$ is a group family with respect to G if and only if the induced group $\bar{G}$ of transformations of the parameter space Ω is transitive over Ω. If $M = M(X)$ is a maximal invariant* with respect to G and $\bar{G}$ is transitive over Ω, it follows that M is an ancillary statistic. This was Fisher's reason for introducing group families. He proposed in this case to make all inferences concerning θ conditional on M (sometimes called the configuration).

The relation of group families to pivotal quantities* and hence to fiducial distributions* is discussed by Fraser [6] and in connection with his structural models* by Fraser [7]. An additional structure possessed by some group families is described by Berk [3]. A necessary and sufficient condition for the existence of a group G with respect to which a family is a group family is given in Lehmann [8, p. 26]. For the relationship between group families and exponential families, see Borges and Pfanzagl [4], Roy [9], Rukhin [10], and Barndorff-Nielsen et al. [2]. The last of these papers provides a very general treatment without the restriction that $\bar{G}$ be transitive over Ω.

References

[1] Barndorff-Nielsen, O. (1978). *Information and Exponential Families in Statistical Theory*. Wiley, New York.

[2] Barndorff-Nielsen, O., Blaesild, P., Jensen, J. L., and Jørgensen, B. (1982). *Proc. R. Soc. (Lond.) A*, **379**, 41–65.

[3] Berk, R. H. (1967). *Ann. Math. Statist.*, **38**, 1436–1445.

[4] Borges, R. and Pfanzagl, J. (1965). *Ann. Math. Statist.*, **36**, 261–271.

[5] Fisher, R. A. (1934). *J. R. Statist. Soc. A*, **144**, 285–307.

[6] Fraser, D. A. S. (1961). *Biometrika*, **48**, 261–280.

[7] Fraser, D. A. S. (1968). *The Structure of Inference*. Wiley, New York.

[8] Lehmann, E. L. (1983). *Theory of Point Estimation*. Wiley, New York.

[9] Roy, K. K. (1975). *Sankyā A*, **37**, 82–92.

[10] Rukhin, A. L. (1975). In *Statistical Distributions in Scientific Work*, Vol. 3, G. P. Patil et al., eds. Reidel, Dordrecht and Boston (pp. 149–161).

Added in Proof: Bibliography

Akahira, M. (1986). The Structure of Asymptotic Deficiency of Estimators. *Queen's Papers in Pure and Applied Mathematics*, **75**, Queen's University, Kingston, Ontario, Canada.

(ANCILLARITY
INVARIANCE CONCEPTS IN STATISTICS
INVARIANCE PRINCIPLES)

E. L. LEHMANN

H

HYBRID LOG-NORMAL DISTRIBUTIONS

This name has been applied to the distribution of X when

$$Y = \log(\rho X) + \rho X, \qquad \rho > 0,$$

is normally distributed with expected value ξ and variance σ^2. Kumazawa and Numakunai [1] proposed this distribution as a model for workers' periods of exposure to radiation, (with particular reference to data reported by the U.S. Nuclear Regulatory Commission).

As ρ increases from 0 to ∞ the shape of the distribution changes from log-normal to normal. For fixed ρ, the distribution is of log-normal type for small x, but the upper tail (large x) is of normal shape.

There is a detailed discussion of these distributions in ref. 2.

References

[1] Barndorff-Nielsen, O. (1978). *Information and Exponential Families in Statistical Theory*. Wiley, New York.

[2] Kumazawa, S. and Ohashi, Y. (1986). *Japan J. Appl. Statist.*, **15**, 1–14 (in Japanese).

(LOG-NORMAL DISTRIBUTION
NORMAL DISTRIBUTION
NUCLEAR MATERIAL SAFEGUARDS)

I

INFLUENCE DIAGRAMS

Influence diagrams were developed as a Bayesian* computer-aided modeling tool by Howard and Matheson [2]. The influence diagram provides an alternative to the decision tree (e.g., Lindley [3]) for Bayesian decision analysis*. An algorithm for solving Bayesian decision analysis problems through influence diagram manipulations was constructed by Shachter [4]. Influence diagrams also provide a graphical representation of conditional independence among random quantities. This property is useful in statistical applications, especially in experimental design (e.g., Barlow and Zhang [1]). The fault-tree* representation used in engineering risk analysis is a special case of an influence diagram.

An influence diagram is first of all a directed acyclic graph. Circle nodes in the graph denote random quantities. Arcs joining circle nodes denote *possible* statistical dependence. If there is no path (directed or undirected) in the influence diagram from one node to another, then the corresponding random quantities are *unconditionally independent*. Associated with each circle node is a conditional probability function. Conditioning is only with respect to *immediate* predecessor nodes. In Fig. 1, z is an immediate predecessor of both x and y. This fact is indicated by the direction of the arrows.

From the graph in Fig. 1, the joint probability function $p(x, y, z)$ is

$$p(z)p(x|z)p(y|z).$$

Hence from the graph the random quantity x is conditionally independent of y given z. The absence of an arc between two node random quantities means that they are *con-ditionally independent* given the states of *immediate* predecessor nodes.

Given a directed acyclic graph together with node conditional probabilities, there exists a unique joint distribution corresponding to the random quantities represented by the nodes of the graph. This is because a directed graph is acyclic if and only if there exists a list of the nodes such that any successor of a node i in the graph follows node i in the list as well.

Relative to Fig. 2, consider the problem of computing $p(x_1|x_4)$. Distinguish nodes x_1 and x_4. More generally, let J and K denote disjoint sets of nodes and x_J, x_K the associated random vectors. Then we could consider the problem of computing $p(x_J|x_K)$. An ordered list for Fig. 2 is $x_1 < x_3 < x_2 < x_4$.

To calculate $p(x_1|x_4)$ using Fig. 2, it is enough to eliminate nodes x_2 and x_3 from the graph by appropriate probabilistic and graphical manipulations which preserve the joint distribution of x_1 and x_4. The primary manipulation is arc reversal which corresponds to using Bayes' theorem*. We can start the algorithm with either node x_2 or x_3. Starting with node x_2, reverse the arc from x_2 to x_4. (Note that if we were to reverse the arc from x_1 to x_2 first we would create a directed cycle. This is *never* allowed.) After

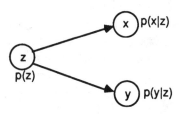

Figure 1 Simple influence diagram.

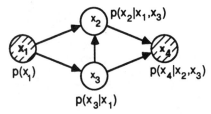

$p(x_2|x_1,x_3)$

$p(x_1)$

$p(x_4|x_2,x_3)$

$p(x_3|x_1)$

Figure 2 Influence diagram with distinguished nodes.

reversal, node x_4 has conditional probability

$$p(x_4|x_3, x_1) = \int p(x_4|x_3, x_2)$$

$$\times p(x_2|x_3, x_1)\, dx_2, \quad (1)$$

while node x_2 now has conditional probability $p(x_2|x_1, x_3, x_4)$, calculated using Bayes' theorem. After arc reversal, arrows input to x_2 are now also input to x_4 and arrows input to x_4 are also input to x_2. This is because of the possible additional dependency relations induced by calculating the expectation, as in (1), and by using Bayes' theorem. Node x_2 now has no successor nodes and has become irrelevant to our problem, see Fig. 3. Hence node x_2 together with its arcs can be deleted from the graph. Figure 4 shows the graph at this stage.

Continuing in this way, we can compute $p(x_1|x_4)$ by eliminating all nodes except x_1 and x_4 and, at the last stage, reversing the arc from x_1 to x_4. The final conditional probability attached to x_1 will be $p(x_1|x_4)$. This algorithmic approach is due to Shachter [4].

Figure 5 models the calibration* problem. A measurement y is observed. The "true"

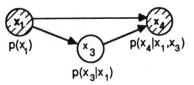

$p(x_1)$

$p(x_4|x_1,x_3)$

$p(x_3|x_1)$

Figure 4 Influence diagram with distinguished nodes after deletion of node x_2.

measurement x is related to y by

$$y = \alpha + \beta x + \epsilon,$$

where α, β, and ϵ are unknown quantities. A decision d relative to x is evaluated by a value function $v(d, x)$. A decision node is denoted by a rectangle and a value node by a diamond. Value nodes are deterministic and have no output arrows. Input arrows to a decision node denote information available at the time of decision. Input and output arrows associated with decision nodes *cannot* be reversed. To find the expected value given y, eliminate all nodes except d, y, and v, as in the previous example.

Eliminate the decision node by computing

$$\mathrm{Min}_d E[v(d, x)|y] \quad (2)$$

[where $v(d, x)$ is a loss function]. The value node now has (2) as its value. Figure 6 is the solution graph.

The influence diagram is a modeling alternative to the decision tree. Since the order of event expansion required by the decision tree is rarely the natural order in which to assess the decision maker's information, the influence diagram may be more useful. However, the influence diagram algorithm described has similarities to the algorithm used for decision trees.

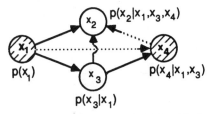

$p(x_2|x_1,x_3,x_4)$

$p(x_1)$

$p(x_4|x_1,x_3)$

$p(x_3|x_1)$

Figure 3 Influence diagram with distinguished nodes after arc reversal.

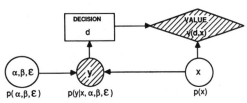

$p(\alpha,\beta,\epsilon)$

$p(y|x, \alpha,\beta,\epsilon)$

$p(x)$

Figure 5 Influence diagram for the calibration problem.

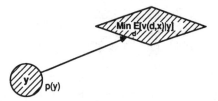

Figure 6 Influence diagram for the calibration problem after elimination of nodes.

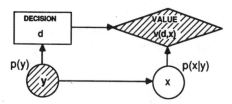

Figure 7 Influence diagram for the calibration problem after eliminating $(\alpha, \beta, \epsilon)$ and arc reversal.

As it stands, Fig. 5 cannot be represented by a decision tree. However, by integrating $(\alpha, \beta, \epsilon)$ out and then reversing the arc from x to y, we obtain the following influence diagram (Fig. 7).

Figure 7 now admits the following decision-tree representation (Fig. 8), where at each node of the decision tree we know the state of all predecessor nodes. However, this is *not* the natural ordering for our original calibration problem.

Acknowledgments

This is to acknowledge many fruitful conversations with Ross Shachter who introduced me to influence diagrams. Carlos Pereira and his MacIntosh are responsible for the final layout. Tony O'Hagan provided research facilities for this work in connection with the Bayesian Study Year at Warwick University, England.

References

[1] Barlow, R. E. and Zhang, X. (1987). Bayesian analysis of inspection sampling procedures discussed by Deming. *J. Statist. Plann. Inf.*, **16**, 285–296. (An application of influence diagram modeling to experimental designs for inspection sampling.)

[2] Howard, R. A. and Matheson, J. E. (1984). Influence diagrams. In *Readings in the Principles and Applications of Decision Analysis*, 2 volumes, R. A. Howard and J. E. Matheson, eds. Strategic Decision Group, Menlo Park, CA. (The basic ideas and symbols used in influence diagrams are developed in this reference.)

[3] Lindley, D. V. (1985). *Making Decisions*, 2nd ed. Wiley, New York. (An excellent introduction to decision analysis and decision trees.)

[4] Shachter, R. (1986). Evaluating influence diagrams. In *Reliability and Quality Control*, A. P. Basu, ed. Elsevier Science Publishers, Amsterdam, Netherlands, pp. 321–344. (A general algorithm for analyzing influence diagrams with a computer.)

(BAYESIAN INFERENCE
DECISION THEORY
FAULT-TREE ANALYSIS
GRAPH THEORY
MARKOV RANDOM FIELDS
NETWORK ANALYSIS)

RICHARD E. BARLOW

INSTITUTE OF STATISTICIANS, THE

The Institute is an international examining body of qualified professional statisticians based in the United Kingdom.

It was founded in 1948 and has two grades of qualified, corporate membership, Fellow and Member, and three grades of unquali-

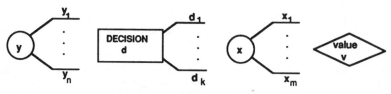

Figure 8 Decision-tree representation corresponding to the influence diagram of Fig. 7.

fied noncorporate membership, Graduate, Student, and Associate. The current membership (1985) consists of approximately 1200 Fellows and Members and 1300 Graduates, Students, and Associates located in 81 countries.

The Institute has a President and Vice-President, elected in an honorary capacity, annually, and a Council of Fellows and Members (maximum 30) elected for a term of not more than six years. It is the Chairman of Council, who, as the Chief Executive of the Institute, directs its affairs. The function of the council is to ratify formally all decisions on Institute policy as developed by the standing committees. These standing committees are Computing, Education, Executive, Finance, Meetings, Overseas, Professional Affairs, and Publications. Both Council and ordinary members serve on the standing committees and these individuals are drawn from a variety of backgrounds covering the whole spectrum of both professional and academic statistics.

The day-to-day administration of the Institute is the responsibility of the Executive Secretary and the staff of the permanent Secretariat.

The Journal of the Institute, *The Statistician**, is published five times per year with one issue being devoted to Conference proceedings. Additionally, a monthly magazine, *The Professional Statistician*, is published. This publication contains articles of general statistical interest and includes a section devoted to employment opportunities in statistics. Both publications are received by all registered members and are made available to nonmembers on subscription. The Institute also publishes a range of educational texts and monographs.

The Institute organises a number of conferences, seminars, workshops, and teaching courses every year which are open to both members and nonmembers. All members are entitled to attend these events at concessionary rates.

The Institute is regularly contacted to provide consultants or expertise on various statistical problems. Such enquiries can range from individual companies requesting the services of a statistician to national governments seeking expertise in particular specialised areas such as agricultural or medical statistics. To cater to those needs, the Institute maintains a Register of Consultants consisting of Fellows who have indicated a willingness to offer such services.

A recent innovation has been the establishment of Institute Chapters in various countries. Currently, such chapters already exist in Guyana, Hong Kong, and Nigeria, and it is anticipated that by 1988, chapters will be established in five other countries.

The Institute's examination structure is divided into four parts of stages. These are: the Ordinary Certificate in Statistics, Stages I, II, and III. Candidates who have successfully completed all of Stages II and III are awarded a Graduate Diploma in Statistics. The Graduate Diploma is equivalent to a University B.Sc. Honours Degree in Statistics and is recognised as such by various Civil Services throughout the world, by other employers, by Universities as an entrance qualification to postgraduate studies, and by the Commission of the European Communities. The examinations are held once every year worldwide.

All applicants for Fellowship and Membership, which are corporate grades, must have passed the Institute's examinations or possess equivalent, approved qualifications. Additionally, all applicants for these grades must possess relevant practical experience. On election, Fellows and Members are entitled to use the designatory letters F.I.S. and M.I.S., respectively. The Graduate grade is a noncorporate grade and is open to those individuals who, whilst possessing the necessary academic qualifications, do not have the requisite practical experience necessary for membership (currently three years minimum). Graduates are allowed to use the designatory letters Grad. I.S. Student member is a noncorporate grade open to those individuals studying for the Institute's examinations and Associate member is also a noncorporate grade open to individuals who are not sufficiently qualified for full membership but who are interested in statistics and wish to be associated with a professional body. Neither Students nor Associates are entitled to the use of designatory letters. All entrants to both the corporate and noncorporate grades of membership are required to abide by the Institute's Professional Code of Conduct.

The Institute offers to its membership the full range of services associated with a professional body such as the provision of professional indemnity insurance, regular salary surveys, etc. Additionally, the Institute responds to matters of public concern on statistics and also is represented on and consulted by outside organisations, at both national and international levels, on a wide range of statistical issues. The Institute's unique and extensive experience of worldwide examining in statistics has enabled it to monitor closely the changing needs and demands in statistical education particularly in developing nations and its response to these needs is evidenced by the increasing role it plays in the area of middle-level professional training and its involvement in the validation and moderation of statistical examinations and qualifications of an ever-increasing number of developing countries.

Further details on the Institute and its activities can be obtained from: The Executive Secretary, 50 Fitzroy Street, London, W1P 5HS, England.

(*STATISTICIAN, THE*)

PATRICK S. CLEARY

INTERNATIONAL JOURNAL OF MATHEMATICAL EDUCATION IN SCIENCE AND TECHNOLOGY

The "Aims and Scope" of this journal are described in the following terms:

"Mathematics is pervading every study and technique in our modern world, bringing

ever more sharply into focus the responsibilities laid upon those whose task it is to teach it. Most prominent among these is the difficulty of presenting an interdisciplinary approach so that one professional group may benefit from the experience of the others.

The journal exists to provide a medium by which a wide range of experience in mathematical education can be presented, assimilated, and eventually adapted to everyday needs in schools, colleges, polytechnics, universities, industry, and commerce. Contributions will be welcomed from teachers and users of mathematics at all levels on the contents of syllabuses and methods of presentation. Mathematical models arising from real situations, the use of computers, new teaching aids and techniques will also form an important feature. Discussion will be encouraged on methods of widening applications throughout science and technology.

This need for communication between teacher and user will be emphasized, and reports of relevant conferences and meetings will be included. The international experience collected in these pages will, it is hoped, provoke a discussion bringing clarity to mathematical education and better understanding of mathematical potentialities in all disciplines."

Many of the articles in the journal are of statistical or probabilistic interest. The January–February, 1985 (Vol. 16, No. 1) issue includes:

Analysis of variance and analysing variation (D. F. Stomp and M. M. Whiteside).

Optimal betting in Keno: an illustration of risk and preference analysis (J. R. Evans).

A new proof of the Neyman–Pearson lemma from the solution of a maximum receiver problem (R. C. Shiflett and L. R. Weill).

Estimation of the constant term when using ridge regression (A. J. Bertie and G. W. Cran).

There are six issues, constituting one volume, per year. Each issue contains about 150 pages. The journal is published by Taylor and Francis, London, England. The founder, and editor is A. C. Bagai, CAMET, University of Technology, Loughborough, Leicester, LE11 3TU, England.

ISOKURTIC

A term, now obsolete, used by Karl Pearson* [1] to denote "arrays of no skewness*" (and not "arrays of equal kurtosis*," as the word might taken to signify).

Reference

[1] Pearson, K. (1905). Drapers' Company Research Memoirs. *Biometric Series II*, **23**.

J

JOURNAL OF OFFICIAL STATISTICS
See STATISTISK TIDSKRIFT

K

KHINCHIN'S UNIMODALITY THEOREM

This theorem (Khinchin [4]) states that a necessary and sufficient condition for the distribution of a continuous random variable X to have a single mode at zero is that it can be expressed as ZU, where Z and U are mutually independent and U has the standard uniform distribution*. An accessible source for a proof is Feller [1]; a heuristic justification for the result is given by Johnson [2]. If the PDFs exist, then

$$f_Z(z) = -zf_X'(z).$$

A consequence of this result is that if X has a standard (unit) normal distribution, then Z is distributed as χ_3 (chi with 3 degrees of freedom) reflected about zero (*see* FOLDED DISTRIBUTIONS). In ref. 2 this fact is used as a basis for Monte Carlo generation of random variables*. Another application is a simple proof of Johnson and Rogers' [3] inequality

$$|\text{Pearson's coefficient of skewness*}|$$

$$= \left| \frac{E[X] - \text{mode}(X)}{\text{S.D.}(X)} \right| \leqslant \sqrt{3}$$

for unimodal distributions. For applications to multivariate distributions, see ref. 2.

References

[1] Feller, W. (1971). *An Introduction to Probability Theory and Its Applications*, 2nd ed., Vol. 2. Wiley, New York.

[2] Johnson, M. E. (1987). *Monte Carlo Distributions*. Wiley, New York.

[3] Johnson, N. L. and Rogers, C. A. (1951). *Ann. Math. Statist.*, **22**, 433–439.

[4] Khinchin, A. Y. (1938). *Tomskiǐ Univ. Nauch. Issled. Inst. Mat.-Mekh., Izv.*, **2**, 1–7.

(KHINCHIN'S INEQUALITY
UNIMODALITY)

KIAER, ANDERS NICOLAI

Born: September 15, 1838, in Drammen, Norway.

Died: April 16, 1919, in Christiania (now Oslo), Norway.

Contributed to: survey sampling, International Statistical Institute.

Kiaer's most lasting contribution to statistical theory was his early introduction of self-conscious sampling in population surveys. He took office as Director of the Central Bureau of Statistics of Norway at its establishment in Oslo (then Christiania) in 1876 and was head of the Bureau until 1913. He developed the practice (and to a lesser extent the theory) of sampling surveys. At the Berne International Statistical Institute meeting (1895) he introduced and clarified the meaning of "*dénombrements représentatives*" (representative investigations), i.e., based on individuals who have been selected according to a particular representative method*. He initially used this procedure in surveys for which the sample was selected at random from the returns to the Norwegian population census of 1891. For one of these surveys Kiaer linked census information with individual data collected from the income tax offices, on the basis of which he analyzed the distribution of income and wealth in Norway. He also made significant contributions

to statistical and population theory and to applications in a wide range of subject matter fields, from population and vital statistics to national income estimation and taxation. In addition, he was a pioneer in applying punch-card equipment for statistical data processing.

Kiaer's last work, which appeared posthumously in *J. Amer. Statist. Ass.*, **16**, 442–458, discussed determination of the birth rate in the United States. This article grew out of correspondence with W. F. Willcox* over the period 1918–1919—see also the obituary by Willcox on pp. 440–441 of the same issue.

A bibliography of Kiaer's work is included in *Den Representative Undersøgelsmethode*: *The Representative Method of Statistical Surveys*. Central Bureau of Statistics, Oslo, Norway (1976). (In Norwegian and English.)

Bibliography

Kruskal, W. H. and Mosteller, F. (1980). *Int. Statist. Rev.*, **48**, 169–195.

Seng, Y. P. (1951). *J. R. Statist. Soc. A*, **114**, 214–231.

(CENSUS
INTERNATIONAL STATISTICAL
    INSTITUTE
REPRESENTATIVE SAMPLING)

P. J. Bjerve
W. H. Kruskal

KOLMOGOROV, ANDREI NIKOLAYEVICH

Born: April 25, 1903, Tombov, Russia (about 400 miles southeast of Moscow).

Died: October 20, 1987, Moscow, USSR.

Contributed to: history, philosophy and foundations of mathematics, set theory, measure theory, probability theory, stochastic processes, functional analysis, information theory, statistics, approximation theory, mathematical linguistics, crystallography, theory of turbulent flow, classical dynamics.

A. N. Kolmogorov, one of the most influential mathematicians of the twentieth century, was born in Tombov, where his mother, Mariya Yakovlevna Kolmogorova, stopped on a journey to the Crimea. She died in childbirth and his aunt, Vera Yakovlevna Kolmogorova, took care of him in his childhood. His father was a learned agronomist, the son of a priest, and from his mother's side he was of gentry origin. In the fall of 1920, Kolmogorov enrolled in Moscow University. Initially, his main interest was in Russian history of the fifteenth and sixteenth centuries. In 1922, he became a student of V. V. Stepanov (an expert in trigonometric series) and the famous Russian mathematician, N. N. Luzin. In June of 1922, at the age of 19, he constructed an example of Fourier–Lebesgue series everywhere divergent, and then a Fourier–Lebesgue series divergent at each point. These were novel and unexpected results.

We now briefly survey Kolmogorov's contributions to probability theory and mathematical statistics. In 1924, he started a long and fruitful collaboration with A. Ya. Khinchin. They, jointly, obtained necessary and sufficient conditions for convergence of series whose terms are mutually independent random variables. In 1928, Kolmogorov obtained necessary and sufficient conditions for the law of large numbers* and, in 1929, he proved the law of iterated logarithm* for sums of independent random variables under very general conditions. (Earlier this law had been proved by Khinchin, first for Bernoulli and then for Poisson variables.) In the same year was obtained the remarkable result for a necessary and sufficient condition for the strong law of large numbers*—namely, the existence of a finite mathematical expectation for the summands. In 1933, his classical and famous monograph on *Foundations of the Calculus of Probability* was published in

German; it was translated into Russian in 1936 and into English in 1950 and 1956. A second Russian edition appeared in 1974. (*See* AXIOMS OF PROBABILITY.) In 1931, in his paper, "*Analytical Methods of Probability Theory*," the foundations of the modern theory of Markov processes* were laid and relations between probability theory and the theory of differential equations were revealed. In late 1939, Kolmogorov developed statistical methods in the theory of turbulent fluid-flow and, in 1941, he worked on interpolation and extrapolation of stationary time series*. This, together with (but independently from) Wiener's* work, developed later into the Kolmogorov–Wiener theory of stationary processes and later into the theory of branching processes*. His monograph, with B. V. Gnedenko, on *Limit Theorems for Sums of Independent Random Variables*, appeared in 1949. His interest in information theory began in 1956 (jointly with I. M. Gel'fand and A. M. Yaglom) and culminated in the development of algorithmic information theory* in two articles in 1965 and 1969, published in the Russian journal, *Problems of Information Transmission*, and in a survey paper in *Uspekhi Mat. Nauk*, **3**, No. 4 (1983), 27–36. (*See also IEEE Trans. Inf. Theory.* **IT-14** (1968), 662–664.)

His contributions to statistics include, among others, works on the least-squares* method, which influenced Yu. V. Linnik's work on this topic, on empirical distributions (the classical Kolmogorov statistic*

$$D_n = \sqrt{n} \, \sup_t |F_n(t) - F(t)|$$

and the Kolmogorov–Smirnov test*), on unbiased estimators, statistical forecasting, on tables of random numbers, on Wald's identity*, on measures of dependence* (generalizations of correlation coefficients), methods of medians in the theory of errors, Mendel's theory, lognormal distribution, and estimation of parameters of a process related to a model of the earth's rotation. His pedagogical work began in 1922 and flourished in his capacity as the director of the Scientific Research Institute of Mathematics at Moscow University. His famous seminars in probability theory at Moscow University inspired a great many young Soviet mathematicians and statisticians, and a great majority of the contemporary leading Soviet scientists in this area as well as a number of foreigners (including Parthasarathy, Vere-Jones, and Martin-Löëf) were among his students. He was an Academician of the USSR from 1939 and was an honorary member of over 20 foreign scientific organizations, including the Royal Statistical Society*, International Statistical Institute*, the U.S. Academy of Sciences, and the Paris Academy of Sciences. His professional activity was centered on Moscow State University (MSU) with which he was associated for over 65 years (1920–1987), being a faculty member for over 60 years (from 1925). Among his activities at MSU was the establishment of the Statistical Laboratory in 1961.

Bibliography

Bogolyubov, N. N., Gnedenko, B. V., and Sobolev, S. L. *A. N. Kolmogorov: On his 80th birthday.* In *A. N. Kolmogorov's Selected Works*, Vol. 3. Moscow, Nauka, 1987.

Kolmogorov, A. N. 1903–1987. *IMS Bull.*, **16**, 1987, pp. 324–325. (Reprinted with permission of *The Times*, (London), October 26, 1987.)

(KOLMOGOROV–KHINCHIN THEOREM
KOLMOGOROV'S INEQUALITY
KOLMOGOROV–SMIRNOV STATISTICS
KOLMOGOROV–SMIRNOV SYMMETRY
  TEST
KOLMOGOROV'S THREE-SERIES
  THEOREM
LAWS OF LARGE NUMBERS
WIENER, NORBERT)

k TH NEAREST-POINT—*k* TH NEAREST-NEIGHBOR SAMPLING

In this sampling scheme for investigating spatial point distributions over unmapped (two-dimensional) regions, sampling sites

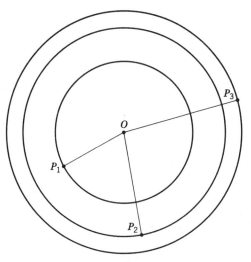

Figure 1

$O_1, O_2, \ldots, O_n$, are chosen randomly in the region of interest, for example by choosing exact coordinates of a map of the region. Each sampling site is then visited, and the k nearest items to that sampling site are then found, and the k (Euclidean) distances from these items to the corresponding sampling sites O_i are determined. If the locations of point items in the region under study are generated by a spatial Poisson process*, if boundary effects are negligible, and if $X_1, X_2, \ldots, X_k$, denote the k nearest-point distances, then the k random variables $X_1^2, X_2^2 - X_1^2, \ldots, X_k^2 - X_{k-1}^2$ are independent and identically exponentially distributed. The sampling method can be impractical for moderate or large k, since, for X_i with a fixed absolute error of measurement, the absolute error of X_i^2 grows proportionally with X_i, and the precisions of the differences of squares for large i can be unacceptably poor. A practical alternative is the wandering quarter sampling* method.

k-th nearest-neighbor sampling differs only, but importantly, in its requirement that the initial sampling positions, O_i, be the locations of randomly selected items in the population. The method is relatively impractical in many applications because of its need for a complete enumeration of all items in the population, together with their locations.

Bibliography

Diggle, P. J. (1983). *Statistical Analysis of Spatial Point Processes*. Academic, London.

Ripley, B. D. (1981). *Spatial Statistics*. Wiley, New York.

(NEAREST-POINT—NEAREST-NEIGHBOR SAMPLING
SPATIAL PROCESSES
SPATIAL SAMPLING
WANDERING-QUARTER SAMPLING)

W. G. S. Hines

L

LEARN–MERGE INVARIANCE

Learn–Merge invariance (LMI) formalizes in a Bayesian setting the requirement that it should make no difference in inference, whether passing to a coarser observation set (merging) is done before or after sampling. A prior distribution* is LM invariant if passing to the posterior distribution* (learning) and merging of the observation set are interchangeable operations, for arbitrary observations and some or all of the possible mergers [1]. LMI may be considered as a requirement of internal consistency of inference, but since merging is indeed frequently done, e.g., in the analysis of categorical data*, it is also motivated from practical viewpoints. Using a non-LMI prior could lead to the unpleasant fact that conclusions do not only depend on the prior, the data, and the merging chosen, but also on the stage at which the latter was employed.

To illustrate some of the issues further, consider the multinomial situation with k categories and unknown parameter vector $(p_1, \ldots, p_k)$, $p_i \geq 0$, $\sum_{i=1}^{k} p_i = 1$, and n independent observations tallied in a vector $\mathbf{s} = (n_1, \ldots, n_k)$ (i.e., the sufficient statistic). In a Bayesian* analysis of this problem one has to specify the prior distribution, being here any distribution over the k-simplex of all possible parameter vectors. A common choice is, for example, a *Dirichlet distribution** as parametrized by a prior mean vector $\mathbf{m} = (m_1, \ldots, m_k)$, $\sum_{i=1}^{k} m_i = 1$, and a real number $\rho \geq 0$ (for $\rho > 0$ often $\mathbf{m}/\rho$ is used as parameter vector). The posterior distribution then is again a Dirichlet distribution* with parameter

$$m_s = \frac{n + \rho_s}{\rho n + 1} \quad \text{and} \quad \rho_s = \frac{\rho}{\rho n + 1}.$$

Suppose now that some of the categories are to be merged. This situation may be handled in two ways: first, by using a "merged" prior for passing to the posterior distribution according to the "merged" data. Second, the posterior distribution as obtained with raw data may be merged according to the new categories. When having initially a Dirichlet prior with $\mathbf{m}$, ρ, it is easily established, that the above approaches both come to the same result, i.e., end up with a same posterior distribution, and this holds for arbitrary merging of categories and any possible data, in other words the Dirichlet prior is Learn–Merge invariant for all possible mergers. It is now an obvious question, which of the other possible priors would share this property. The answer proved in ref. 1 states that, given there are at least three categories, Dirichlet distributions are the only LMI priors, if LMI is required for all possible mergings. In the case of just two categories, LMI is an empty constraint which does not restrict the class of all priors. This theorem implies, for instance, that (convex) linear combinations of Dirichlet distributions are in general not LMI. For example, let $k = 4$ and there is a first Dirichlet prior with $\mathbf{m}_1 = (\frac{5}{8}, \frac{1}{8}, \frac{1}{8}, \frac{1}{8})$ and a second one with $\mathbf{m}_2 = (\frac{1}{4}, \frac{1}{4}, \frac{1}{4}, \frac{1}{4})$ both having $\rho = 0.1$ and being combined with equal weights. Suppose the observed data was $\mathbf{s} = (10, 1, 1, 8)$. If the first two and the last two of the categories are coalesced, then the posterior mean, when using a merged prior and the data $\mathbf{s}' = (11, 9)$ amounts to about $\mathbf{m}_{\mathbf{s}'} = (0.56, 0.44)$. Otherwise the merged posterior renders roughly $(0.60, 0.40)$. Note that neither the prior guess nor the data was different to be responsible for this discrepancy. Beyond the simple multinomial situation, e.g., when there is some further structure on the categories, other than Dirichlet distributions may be invariant, if LMI is relaxed to those mergers compatible with the given structure. For example, for a contingency table it appears sensible to allow only merging of each categorical variable separately. Additional LMI priors in this case, however, are just products of Dirichlet priors, implying subjective a priori independence of the categorical variables [8]. Other structures may render other sets of LMI priors [9].

The concept of LMI, originally formulated in the finite case, generalizes to arbitrary measurable spaces, thereby characterizing Dirichlet processes as proper LMI "priors." The latter were introduced by ref. 4 in the context of Bayesian analysis of nonparametric problems.

LMI is mathematically most easily handled when referring to the prior as an exchangeable distribution over the space of infinite sequences of observations, rather than as a distribution over the simplex (more generally, over all probability measures). One can do so since both sets are known to be bijectively related to each other by de Finetti's representation theorem (respectively its generalizations, [5]). In this formulation, LMI is seen to be closely related to requirements considered by the philosophers

Johnson and Carnap within their work on foundations of probability and inductive logic, respectively ([7, 2], see also ref. 10 for a historical and mathematical review). Roughly, in both their formulations it is postulated that the "probability" or "confirmation" of a category after having observed a sequence of length n, should only depend on n and the number of occurrences of this category within the sequence, but not on frequencies of other categories and not on the category itself. It may be shown [1] that this is equivalent to the special case of LMI, when invariance is required for all mergers and the prior mean vector is the equidistribution. Related philosophical and mathematical questions are treated in refs. 3 and 6.

References

[1] Böge, W. and Möcks, J. (1986). *J. Multivariate Anal.*, **18**, 83–92.

[2] Carnap, R. and Stegmüller, W. (1958). *Induktive Logik und Wahrscheinlichkeit*. Springer, Vienna.

[3] Carnap, R. and Jeffrey, R. C. eds. (1971). *Studies in Inductive Logic and Probability*, Vol. 1. University of California Press, Berkeley and Los Angeles, CA.

[4] Ferguson, T. S. (1973). *Ann. Statist.*, **1**, 209–230.

[5] Hewitt, E. and Savage, L. J. (1955). *Trans. Amer. Math. Soc.*, **80**, 470–501.

[6] Jeffrey, R. C. ed. *Studies in Inductive Logic and Probability*, Vol. 2. University of California Press, London.

[7] Johnson, W. E. and Braithwaite, R. B. (1932). *Mind*, **41**, 421–423.

[8] Möcks, J. and Stockel, M. (1972). Diploma thesis, Inst. für Angewandte Mathematik, Universität Heidelberg, German Federal Republic.

[9] Möcks, J. and Böge, W. (1981). Abstract, 14th European Meeting of Statisticians, Wroclaw, Poland, p. 226.

[10] Zabell, S. L. (1982). *Ann. Statist.*, **10**, 1091–1099.

(BAYESIAN INFERENCE
DEGREE OF CONFIRMATION
DEGREES OF BELIEF
DIRICHLET DISTRIBUTION
FOUNDATIONS OF PROBABILITY

INFERENCE, STATISTICAL
INVARIANT PRIOR DISTRIBUTIONS
LOGIC OF STATISTICAL REASONING)

JOACHIM MÖCKS
WERNER BÖGE

LIKELIHOOD DISPLACEMENT

This concept was introduced and developed by Cook and Weisberg [3] as a unified concept in regression (and more general statistical) diagnostics* for measuring the influence of individual cases (observations).

Let θ be a $p \times 1$ parameter vector partitioned as $\theta' = (\theta_1', \theta_2')$, where θ_i is $p_i \times 1$, $i = 1, 2$. Often only θ_1 may be of interest, while θ_2 represents the nuisance parameters*. Let $L(\theta; Z) = L(\theta_1, \theta_2; Z)$ be the log-likelihood for θ based on data Z. Let $L(\theta; Z_{(i)})$ denote the log-likelihood when the subscript (i) means "without case i." Then the *displacement likelihood* is defined as

$$LD_i(\theta) = 2\left[L(\hat{\theta}; Z) - L(\hat{\theta}_{(i)}, Z)\right], \quad (1)$$

where $\hat{\theta}$ is the maximum likelihood estimator (m.l.e.) of θ based on full data and $\hat{\theta}_{(i)}' = (\theta_{1(i)}', \theta_{2(i)}')$ is the m.l.e. obtained from $L(\theta; Z_{(i)})$.

The difference (1) provides a fixed metric for comparing values of likelihood displacement for different cases i. Cook and Wang [2] also proposed a conditional $LD_i(\theta_1|\theta_2)$ defined as

$LD_i(\theta_1|\theta_2)$

$$= 2\left[L(\hat{\theta}; Z) - L(\hat{\theta}_{1(i)}, g(\hat{\theta}_{1(i)}); Z)\right], \quad (2)$$

where $g(\theta_1)$ is a well-defined *implicit* function such that for a *fixed* θ_1, $L(\theta_1, g(\theta_1); Z)$ is maximized.

These authors use (2) in their development of influence measures for transformation parameters in linear models. For the case of the standard linear regression model* $Y = X\beta + \epsilon$ with Y: $n \times 1$ vector and β: $p \times 1$ vector (in the obvious notation) with the assumption that $\epsilon \sim N(0, \sigma^2 I)$, let $\hat{\beta}$ and $\hat{\sigma}^2$ be the m.l.e. of β and σ^2, respectively.

In this case

$$LD_i(\beta|\sigma^2) - n \log\left[\left(\frac{p}{n-p}\right)D_i + 1\right],$$

where D_i is the Cook D_i-statistic (distance*) defined as

$$D_i = \left(\hat{\beta} - \hat{\beta}_{(i)}\right)' X'X\left(\hat{\beta} - \hat{\beta}_{(i)}\right)/ps^2,$$

where $s^2 = e'e/(n - p)$; e being the vector of residuals between the fitted and observed values.

In this case $LD_i(\beta|\sigma^2)$ is a monotonic function of D_i and is thus equivalent to D_i for the purpose of ordering cases based on influence. See also Cook and Peña [1] for more details and additional illustrations.

References

[1] Cook, R. D. and Peña, D. (1988). *Commun. Statist. A*, **17**, 623–640.

[2] Cook, R. D. and Wang, P. C. (1983). *Technometrics*, **25**, 337–344.

[3] Cook, R. D. and Weisberg, S. (1982). *Residuals and Influence in Regression*. Chapman and Hall, London.

(INFLUENTIAL OBSERVATIONS
LIKELIHOOD
REGRESSION DIAGNOSTICS)

LINEAR MODEL SELECTION, CRITERIA AND TESTS

MODEL SELECTION CRITERIA AND MODEL SELECTION TESTS

Consider a linear model for n observed values y ($n \times 1$) with design matrix* X ($n \times p$) ($n \geq p$)

$$y = X\beta + \epsilon, \quad \epsilon \sim N(0, \sigma^2 I) \quad (1)$$

when X is of full rank [rank$(X) = p$]. Assume further that $C_n = n^{-1}X'X$ is positive definite and (with p fixed)

$$\lim_{n \to \infty} C_n = C$$

is also positive definite. Denote a series of h linear constraints on β by

$$\mathbf{r}_j - \mathbf{R}_j\beta = \mathbf{0}, \qquad j = 1, 2, \ldots, h,$$

nested so that $\mathbf{r}_j = \mathbf{G}_j\mathbf{r}_{j+1}$ and $\mathbf{R}_j = \mathbf{G}_j\mathbf{R}_{j+1}$ is $m_j \times p$ with $m_1 < m_2 < \cdots < m_h = p$, $\text{rank}(\mathbf{R}_j) = m_j$. If model (1) holds with the constraint $\mathbf{r}_j - \mathbf{R}_j\beta = \mathbf{0}$ for $j \leqslant j_0$ but not for $j > j_0$, it is denoted M_{j_0} with M_0 representing the unrestricted model $(m_0 = 0)$. Clearly, M_{j_0} has $(p - m_{j_0})$ free parameters. Because of the nesting,

$$M_0 \supset M_1 \supset M_2 \supset \cdots \supset M_h.$$

In order to decide which of the models $M_0, \ldots, M_h$ is appropriate, *model selection criteria* (MSC) are used. Three types of MSC are in common use:

$\text{MSC1}(M_j)$

$$= \ln \hat{\sigma}^2 + (p - m_j)n^{-1}f(n, 0),$$

$\text{MSC2}(M_j)$

$$= \hat{\sigma}_j^2 + (p - m_j)\sigma^2 n^{-1}f(n, 0),$$

$\text{MSC3}(M_j)$

$$= \hat{\sigma}^2 + (p - m_j)\hat{\sigma}_j^2 n^{-1}f(n, p - m_j),$$

where $f(\cdot, \cdot) > 0$, $\lim_{n \to \infty} f(n, z) = 0$ for all z, and $\hat{\sigma}_j^2 = n^{-1}$(residual sum of squares from fitting M_j by constrained least squares*).

Table 1 shows commonly used MSCs.

(i) The customary decision rule in using the MSC is as follows. Choose the model M_g if $\text{MSC}(M_g) = \min_{j=0,1\ldots,h} \text{MSC}(M_j)$.

(ii) In another strategy, formal tests of significance are applied sequentially. Start by testing M_0 against M_1, then M_1 against M_2, and so on, until the first rejection. If this occurs on testing M_g against M_{g+1}, then M_g is chosen. These tests are called *model selection tests* (MST).

If the most restricted model (M_h) is valid, the probability of a correct decision using rule (i) (minimum MSC) is

$$\alpha_h(n) = \Pr\big[\text{MSC}(M_j) \geqslant \text{MSC}(M_h),$$
$$j = 0, 1, \ldots, h - 1\big].$$

Using the Bonferroni inequality*

$$1 - \alpha_h(n) \leqslant \sum_{j=0}^{h-1} \big\{1 - \alpha_{hj}(n)\big\},$$

where

$$\alpha_{hj}(n) > \Pr\big[\text{MSC}(M_j) - \text{MSC}(M_h) \geqslant 0\big].$$

Now, for the criteria with the property $f(n, \cdot) = \ln n$ or $f(n, \cdot) = 2 \ln \ln n$, $\alpha_{hj}(n) \to 1$ for $j = 0, 1, \ldots, h - 1$ as $n \to \infty$, so that $1 - \alpha_h(n) \to 0$ as $n \to \infty$. Thus for these criteria the asymptotic probability of overestimating the size of the true model equals zero. On the other hand, $\alpha_{hj}(n) \geqslant \alpha_h(n)$, $j = 0, 1, \ldots, h - 1$, so that a simple lower bound

$$1 - \alpha_{hj}(n) \leqslant 1 - \alpha_h(n) \qquad (2)$$

holds. For the criteria with the property $\lim_{n \to \infty} \alpha_{hj}(n) = c > 0$, the left-hand side of (2) bounds the asymptotic significance level of the minimum MSC rule away from zero. This property is often called *dimension inconsistency* in time-series analysis: For general results on dimension consistency and inconsistency, see Kohn [14].

Approximations to the effective significance level of the rule (i) have been obtained by Teräsvirta and Mellin [23] using a theorem of Spitzer [22] and Shibata [20]. Table 2 contains a few values of the significance level for AIC and SBIC using these approximations together with the asymptotic upper bounds of Geweke and Meese [9, Table 2]. The bounds refer to a linear model with one regressor, where the set of alternatives consists of models involving up to 10 lags of this regressor.

The discrepancies between the asymptotic figures and finite-sample approximations for AIC are substantial. The asymptotic upper bound appears not to be very tight when the number of alternative models is low. When the number of models is large while the number of observations is relatively small the asymptotic upper bound seems to underestimate the true significance level.

Table 1 Certain Model Selection Criteria, Their Types, Critical Values, and Limits for Critical Values as $n \to \infty$

Criterion	Type	$f(n,h)$, $h = 0$, k	Critical Value	Limit ($n \to \infty$)
URV (Theil [24])	MSC3	$(1 - k/n)^{-1}$	1	1
C_p (Mallows [15])	MSC2	2	2	2
PC (Amemiya [3]) or FPE (Akaike [1])	MSC3	$2(1 - k/n)^{-1}$	$2n/(n + k)$	2
AIC (Akaike [2])	MSC1	2	$(n - p)[\exp\{2(p - k)/n\} - 1]/(p - k)$	2
SBIC (Schwarz [19] Rissanen [17])	MSC1	$\ln n$	$(n - p)[\exp\{(p - k)\ln n/n\} - 1]/(p - k)$	∞
HQ (Hannan and Quinn [11])	MSC1	$2\ln\ln n$	$(n - p)[\exp\{2(p - k)\ln\ln n/n\} - 1]/(p - k)$	∞
BEC (Geweke and Meese [9])	MSC2	$\ln n$	$\ln n$	∞
GCV (Golub et al. [10])	MSC3	$(2 - k/n)(1 - k/n)^{-2}$	$\{2 - (p + k)/n\}/(1 - p/n)$	2
S (Shibata [21])	MSC3	2	$2(1 - p/n)(1 + 2k/n)$	2
S_p (Hocking [12])	MSC3	$(2 - (k + 1)/n)(1 - k/n)^{-1}(1 - (k + 1)/n)$	$\{2 - (p + k + 1)/n\}/(1 - (p + 1)/n)$	2
T (Rice [16])	MSC3	$2(1 - 2k/n)^{-1}$	$2(1 - p/n)/(1 - 2p/n)$	2

Table 2   Approximate Significance Levels of AIC and SBIC and the Geweke and Meese [9] Asymptotic Upper Bounds (GM)

Number of Lags	$n = 50$		$n = 100$		$n = 200$	
	AIC	SBIC	AIC	SBIC	AIC	SBIC
1	0.1681	0.0539	0.1626	0.0341	0.1599	0.0222
2	0.2302	0.0651	0.2212	0.0392	0.2169	0.0246
3	0.2629	0.0685	0.2510	0.0404	0.2453	0.0250
4	0.2831	0.0697	0.2687	0.0407	0.2619	0.0251
5	0.2968	0.0703	0.2803	0.0408	0.2725	0.0251
6	0.3069	0.0705	0.2884	0.0408	0.2797	0.0251
7	0.3147	0.0706	0.2942	0.0408	0.2848	0.0251
8	0.3209	0.0707	0.2986	0.0408	0.2885	0.0251
9	0.3260	0.0707	0.3020	0.0408	0.2912	0.0251
10	0.3304	0.0707	0.3047	0.0408	0.2933	0.0251
(GM)	0.2883	0.0602	0.2883	0.0376	0.2883	0.0240

SEQUENTIAL TESTING

Consider now selecting one of the nested models $M_0 \supset M_1 \supset \cdots \supset M_h$ using customary MST. Assume that $m_j = j$, $j = 1, \ldots, h$. As mentioned above, this conditional testing procedure starts from the least restrictive model M_0 assuming that it is true. The individual statistics follow $F_{1, n-k_j}$ distributions and are mutually independent (cf. Anderson [4, 5]). Thus if M_h is true, the overall significance level of the test sequence (rule (ii)) equals

$$1 - \alpha_h(n) = 1 - \prod_{j=0}^{h-1} \tilde{\alpha}_{j+1, j}(n), \quad (3)$$

where $\tilde{\alpha}_{j+1, j}(n) = \Pr\{F_{1, n-k_j} \leqslant c_j\}$, $c_{h-1} \leqslant c_{h-2} \leqslant \cdots \leqslant c_1 \leqslant c_0$. c_j is the critical value of the test that M_j is true against the alternative that is not, given $\mathbf{r}_i = \mathbf{R}_i \beta$, $i = 1, 2, \ldots, j - 1$.

DISCUSSION

The difference between MSC and MST is that use of the former implies a simultaneous comparison of all alternatives while the latter is a sequence of bilateral comparisons. Note that as long as c_j, $j = 0, 1, \ldots, h - 1$, are finite fixed constants, the MST sequence could be called "a dimension-inconsistent

model selection procedure." The significance level (3) is then positive for all n and thus, even asymptotically, there is a positive probability of overestimating the true size of the model.

Obstensibly, the use of MSC instead of MST may seem to save the practitioner from choosing significance levels altogether. But then, many criteria with varying small-sample properties are available. Deciding between them amounts to choosing the significance level of the minimum MSC rule. There has even been discussion on expanding the possibilities of choosing the significance levels of MSC. Bhansali and Downham [8], Atkinson [6], and Bhansali [7] have considered a generalization of AIC and FPE such that $f(n, 0) = \alpha$ and $f(n, k) = \alpha(1 - k/n)^{-1}$, respectively, where selecting α is left to the user. Such a generalization makes the dimension-inconsistent criteria more flexible and brings them closer to MST as far as the possibilities of choosing the significance level of the rule are concerned.

References

[1]   Akaike, H. (1969). *Ann. Inst. Statist. Math.*, **21**, 243–247.

[2]   Akaike, H. (1974). *IEEE Trans. Autom. Control*, **AC-19**, 716–723.

[3]   Amemiya, T. (1980). *Int. Econ. Rev.*, **21**, 331–354.

[4] Anderson, T. W. (1962). *Ann. Math. Statist.*, **33**, 255–265.

[5] Anderson, T. W. (1971). *The Statistical Analysis of Time Series*. Wiley, New York.

[6] Atkinson, A. C. (1980). *Biometrika*, **67**, 413–418.

[7] Bhansali, R. J. (1983). *J. Time Ser. Anal.*, **4**, 137–162.

[8] Bhansali, R. J. and Downham, D. Y. (1977). *Biometrika*, **64**, 547–551.

[9] Geweke, J. and Meese, R. (1981). *Int. Econ. Rev.*, **22**, 55–70.

[10] Golub, G. H., Heath, M., and Wahba, G. (1979). *Technometrics*, **21**, 215–223.

[11] Hannan, E. J. and Quinn, B. G. (1979). *J. R. Statist. Soc. B*, **41**, 190–195.

[12] Hocking, R. R. (1976). *Biometrics*, **32**, 1–49.

[13] Judge, G. G., Griffiths, W. E., Hill, R. C., Lütkepohl, H., and Lee, T.-C. (1985). *The Theory and Practice of Econometrics*, 2nd ed. Wiley, New York.

[14] Kohn, R. (1983). *Econometrica*, **51**, 357–376.

[15] Mallows, C. L. (1973). *Technometrics*, **15**, 661–675.

[16] Rice, J. (1984). *Ann. Statist.*, **12**, 1215–1230.

[17] Rissanen, J. (1978). *Automatica*, **14**, 465–471.

[18] Sawa, T. (1978). *Econometrica*, **46**, 1273–1291.

[19] Schwarz, G. (1978). *Ann. Statist.*, **6**, 461–464.

[20] Shibata, R. (1976). *Biometrika*, **63**, 117–126.

[21] Shibata, R. (1981). *Biometrika*, **68**, 45–54.

[22] Spitzer, F. (1956). *Trans. Amer. Math. Soc.*, **82**, 323–339.

[23] Teräsvirta, T. and Mellin, I. (1986). *Scand. J. Statist.*, **13**, 159–171.

[24] Theil, H. (1961). *Economic Forecasts and Policy*, 2nd ed. North-Holland, Amsterdam.

(AKAIKE'S CRITERION
MULTIPLE COMPARISONS
PARTITION OF CHI-SQUARE
RYAN'S MULTIPLE COMPARISONS
    PROCEDURE
SCHEFFÉ'S SIMULTANEOUS COMPARISON
    PROCEDURE
SCHWARZ CRITERION
SIMULTANEOUS TESTING
STEPWISE REGRESSION)

T. TERÄSVIRTA
I. MELLIN

LITERATURE AND STATISTICS—II

NEW POEMS BY SHAKESPEARE?

The years 1981–1986 have witnessed the appearance, among other works, of Pollatschek and Radday's further analysis of Biblical literature [13], Brainerd's extensions of type-token models [2, 3], Ellegard's discussion on the identification of authorship [6], Kenny's expository monograph [9], the revised edition of Mosteller and Wallace's book on *The Federalist Papers* [12], Lanke's comments on Ellegard's work [11], Holmes' review of the analysis of literary style [8], and Sichel's paper on type-token characteristics [15]. All these are indicative of a new awareness of the value of statistical analysis in literature.

Most recently, popular interest has been aroused by Taylor's discovery on November 14, 1985, in a folio volume at the Bodleian Library, Oxford, of a nine-stanza poem beginning with the lines

> Shall I die? Shall I fly
> Lovers' baits and deceits,
>   sorrow breeding?

This was attributed by Taylor [17] to Shakespeare on the basis of a literary analysis, but several critics, including Robbins [14], have expressed dissenting views.

In their 1976 paper, Efron and Thisted [5] had already studied Shakespeare's vocabulary using Spevack's [16] concordance, and estimated the number of words Shakespeare might have known, but not used. It is not often that statisticians can test their results on new data, but with Taylor's discovery of the new poem, this has in fact proved possible (see Kolata [10], and the ensuing correspondence between Driver and Kolata [4] and Birkes [1]). To help resolve the question of its authorship, Thisted and Efron [18] decided to carry out a statistical analysis of the new poem based on their earlier work. Their conclusion was that "On balance, the poem is found to fit previous

Table 1   Expected and Observed Counts of Words Appearing x Times in a 429-Word Poem

Number of Occurrences x	Expected Number of Words $\hat{v}_x$	Observed Number of Words m_x
0	6.97	9
1	4.21	7
2	3.33	5
3	2.84	4
4	2.53	4
5	2.43	2
10	1.62	1
30	0.96	4
50	0.68	0
70	0.49	0
90	0.37	0
95	0.34	1
99	0.32	0

Shakespearean usage well, lending credence to belief that it was actually written by Shakespeare."

The new poem contains 429 words, 258 of them distinct, 9 of which had not appeared in Shakespeare's previous work. These unusual words were "admirations," "besots," "exiles," "inflection," "joying," "scanty," "speck," "tormentor," and "explain" (see Driver and Kolata [4] for a clarification of types, or different words). On the basis of Shakespeare's known writings, the number of distinct words, $\hat{v}_x$, expected to occur $x = 0, 1, 2, 3, 4, \ldots, 99$ times, respectively, in a poem of 429 words can be estimated (see the summary in Table 1); what Thisted and Efron [18] have done in their study is to compare these counts with the actual numbers m_x of such words appearing in the new poem.

To broaden the scope of their study, seven additional poems were considered: one by Ben Jonson, a second by Christopher Marlowe, and a third by John Donne, as well as four already included in the Shakespeare canon from *Cymbeline*, *A Midsummer Night's Dream*, *The Phoenix and the Turtle*, and Sonnets 12–15.

Three different tests were used on the collected data, one of which (the slope test) proved to be the best discriminator for detecting non-Shakespearean authorship. All the tests relied on a regression model in which the observed numbers of words $\{m_x\}$ for the particular poem under study follow the Poisson distribution* with means $\{\mu_x\}$ independently for $x = 0, 1, \ldots, 99$, where

$$\mu_x = \hat{v}_x e^{\beta_0}(x + 1)^{\beta_1}.$$

The first test was based on the total count of words occurring 99 times or less in each of the poems. For the new poem, for example, the actual number was $m_+ = \sum_{x=0}^{99} m_x = 118$, while its expectation was $v_+ = \sum_{x=0}^{99} \hat{v}_x = 94.95$, and $\mu_+ = \sum_{x=0}^{99} \mu_x$. The hypothesis tested was $H_1: \mu_+ = \hat{v}_+$; this proved the least reliable test for discriminating between Shakespearean and non-Shakespearean authorship.

The second test was concerned with the simple null hypothesis $H_2: \pi_0 = \hat{v}_0/\hat{v}_+$, where the zero count m_0 conditional on the total count m_+ in each of the poems follows a binomial distribution $B(m_+, \pi_0)$. This test proved only moderately useful in discerning Shakespearean authorship.

The third test (the slope test) of the hypothesis $H_3: \beta_1 = 0$ relied on the data $(m_1, \ldots, m_{99})$ for each poem. This is equivalent to testing H_3 conditional on (m_+, m_0) when the $(m_1, \ldots, m_{99})$ follow a multinomial distribution depending on β_1. The maximum likelihood estimates $\hat{\beta}_1$ and their standard errors $\hat{\sigma}$ were obtained (see Table 2); this test can be seen to provide the most promising method of discriminating Shakespearean authorship.

When the significant z-values for all three tests are summarized as in Table 3, the conclusion that test 3 is the most discriminating is strengthened. On the basis of these statistical tests, Thisted and Efron [18] reached the conclusion that the new poem "fits Shakespearean usage about as well as do the four Shakespeare poems." It is only fair to mention that many, including Foster [7] and Robbins [14], remain unconvinced that Shakespeare was its author.

Table 2   Estimated Slope Values $\hat{\beta}_1$, Standard Errors $\hat{\sigma}$, and z-Values for 8 Poems

Poems	$\hat{\beta}_1$	$\hat{\sigma}$	z-Value $\hat{\beta}_1/\hat{\sigma}$
Jonson	0.229	0.11	2.08**
Marlowe	−0.323	0.08	−4.04****
Donne	−0.138	0.09	−1.53*
Cymbeline	−0.047	0.10	−0.47
Midsummer	−0.050	0.12	−0.42
Phoenix	−0.127	0.09	−1.41
Sonnets	−0.034	0.09	−0.38
New Poem	−0.075	0.09	−0.83

Table 3   Summary of Significant z-Values for Tests 1, 2, and 3

Poems	Test 1	Test 2	Test 3
Jonson			**
Marlowe	***		****
Donne		***	*
Cymbeline	***		
Midsummer		*	
Phoenix	****	**	
Sonnets			
New Poem	**		

Asterisks indicate significant values as follows: *$1.5 \leqslant |z| < 2$; **$2 \leqslant |z| < 2.5$; ***$2.5 \leqslant |z| < 3$; and ****$3 \leqslant |z|$.

Most recently, Foster [7] has analyzed the word frequency, frequency of subordinating conjunctions, and 15 other statistical measures of the Peter "Funeral Elegy" signed W.S. He has concluded from these 17 different tests that W.S. is very likely to be Shakespeare. Once again statistical analysis has assisted in the identification of authorship, and justified its value in the humanities.

References

[1]   Birkes, D. (1986). Sly statistics. *Science*, **232**, 698.

[2]   Brainerd, B. (1981). Some elaborations upon Gani's model for the type-token relationship. *J. Appl. Prob.*, **18**, 452–460.

[3]   Brainerd, B. (1982). On the relation between the type-token and the species–area problems. *J. Appl. Probab.*, **19**, 785–793.

[4]   Driver, O. and Kolata, G. (1986). Shakespeare and statistics. *Science*, **231**, 1355.

[5]   Efron, B. and Thisted, R. (1976). Estimating the number of unseen species: How many words did Shakespeare know? *Biometrika*, **63**, 435–447.

[6]   Ellegard, A. (1982). Genre styles, individual styles, and authorship identification. In Text Processing, S. Allen, ed. *Proc. Nobel Symp.*, **51**, 519–537.

[7]   Foster, D. W. (1986). Elegy by W.S.—A Study in Attribution. Ph.D. thesis, University of California, Santa Barbara, CA.

[8]   Holmes, D. I. (1985). The analysis of literary style —A review. *J. R. Statist. Soc. A*, **148**, 328–341.

[9]   Kenny, A. (1982). *The Computation of Style*. Pergamon, Oxford, England.

[10]   Kolata, G. (1986). Shakespeare's new poem: An ode to statistics. *Science*, **231**, 335–336.

[11]   Lanke, J. (1985). On the art of conditioning on the right event. In *Contributions to Statistics in Honour of Gunnar Blom*, pp. 215–221.

[12]   Mosteller, F. and Wallace, D. L. (1984). *Applied Bayesian and Classical Inference—The Case of the Federalist Papers*. Springer, New York.

[13]   Pollatschek, M. and Radday, Y. T. (1981). Vocabulary richness and concentration in Hebrew Biblical literature. *Ass. Lit. Linguist. Comp. Bull.*, **8**, 217–231.

[14]   Robbins, R. (1985). ... and the counter-arguments. *TLS*, **4316**, December 20, 1985, 1449–1450.

[15]   Sichel, H. S. (1986). Word frequency distribution and type-token characteristics. *Math. Scientist*, **11**, 45–72.

[16]   Spevack, M. (1968). *A Complete and Systematic Concordance to the Works of Shakespeare*, 6 volumes. George Olms, Hildesheim, West Germany.

[17]   Taylor, G. (1985). A new Shakespeare poem? The evidence .... *TLS*, **4316**, December 20, 1985, 1447–1448.

[18]   Thisted, R. and Efron, B. (1987). Did Shakespeare write a newly-discovered poem? *Biometrika*, **74**, 445–455.

(LINGUISTICS, STATISTICS IN LITERATURE AND STATISTICS)

J. GANI

LOBATSCHEWSKI (LOBACHEVSKY) DISTRIBUTION

This is the distribution of the sum (X) of n independent random variables, each having a uniform distribution* over the interval

$[-\frac{1}{2}, \frac{1}{2}]$. The probability density function is

$$f_n(x) = \frac{1}{(n-1)!} \sum_{j=0}^{[x+n/2]} (-1)^j \binom{n}{j}$$

$$\times \left(x + \frac{1}{2}n - j\right)^{n-1};$$

$$-\frac{1}{2}n \leqslant x \leqslant \frac{1}{2}n$$

(where $[x + n/2]$ denotes the integer part of $x + n/2$). This formula was derived by the Russian mathematician I. N. Lobatschewski (1792–1856) [2], who also obtained the recursive formula

$$f_n(x) = \int_{x-1/2}^{x+1/2} f_{n-1}(t) \, dt.$$

For further details, and some historical remarks, see, e.g., Allasia [1], Maistrov [3], and Seal [4].

References

[1] Allasia, G. (1981). *Statistica* (*Bologna*), **41**, 326–332.

[2] Lobatschewski, I. N. (1842). *J. Reine Angew. Mat.*, **24**, 164–170.

[3] Maistrov, L. E. (1974). *Probability Theory: A Historical Sketch* (translated by S. Kotz). Academic, New York.

[4] Seal, H. L. (1951). *J. Inst. Actuar. Stud. Soc.*, **10**, 255–257.

(RANDOM SUM DISTRIBUTIONS
UNIFORM (RECTANGULAR)
   DISTRIBUTIONS)

LOGISTIC-NORMAL DISTRIBUTION

Suppose $(y_1, \ldots, y_d) \sim N^d(\mu, \Sigma)$ and that x is a d-dimensional random vector defined on the positive simplex,

$$S^d = \{(x_1, \ldots, x_d): x_i > 0, \, i = 1, \ldots, d,$$

$$x_1 + \cdots + x_d < 1\}.$$

If x is augmented by an additional element $x_D = x_{d+1}$, where

$$x_{d+1} = 1 - x_1 - \cdots x_d$$

and

$$x_i/x_{d+1} = \exp(y_i), \qquad i = 1, \ldots, d,$$

then x is said to have an (*additive*) *logistic-normal* distribution, written $\mathbf{x} \sim \mathscr{L}^d(\mu, \Sigma)$. (Whether x is thought of as d-dimensional or as the augmented D-dimensional version is largely optional.)

The terminology was introduced in ref. 3 but the distribution had been used much earlier; see ref. 4, p. 20, and ref. 5, for instance. In the latter and in the work of Aitchison [1–3] the transformation is used to induce, from normal distributions, distributions on the simplex. The distributions are tractable alternatives to the Dirichlet class and greatly enrich the statistical analysis of compositional data*, in which each observation is a vector of proportions or percentages.

Properties of the additive logistic-normal are discussed in refs. 3 and 2, Chap. 6. In particular, if the convention is adopted that $\mu_D = 0 = \sigma_{iD}$, $i = 1, \ldots, D$,

$$E\left[\log(x_i/x_j)\right] = \mu_i - \mu_j$$

and

$$\text{cov}\left\{\log(x_i/x_k), \log(x_j/x_l)\right\}$$

$$= \sigma_{ij} - \sigma_{il} - \sigma_{jk} + \sigma_{kl}, \quad \text{for all } i, j, k, l.$$

Further classes of distributions on S^d can be induced, from normal distributions, using transformations other than the additive logistic transformation [3].

References

[1] Aitchison, J. (1982). *J. R. Statist. Soc. B*, **44**, 139–177.

[2] Aitchison, J. (1986). *The Statistical Analysis of Compositional Data*. Chapman and Hall, London.

[3] Aitchison, J. and Shen, S. M. (1980). *Biometrika*, **67**, 261–272.

[4]   Johnson, N. L. and Kotz, S. (1972). *Distributions in Statistics. Continuous Multivariate Distributions.* Wiley, New York. (Comment from R. L. Obenchain.)

[5]   Leonard, T. (1973). *Biometrika*, **59**, 581–589.

(COMPOSITIONAL DATA
DIRICHLET DISTRIBUTION)

D. M. TITTERINGTON

LOGISTIC PROCESSES

The Malthusian theory of population growth*, which first appeared in 1798, stressed that there would be an unbounded and rapid increase in the size of a population unless some artificial restraints (such as birth control) were introduced to diminish birth rates. In three memoirs published in 1838, 1845, and 1846, Verhulst [14–16], advanced a theory of population growth which took account of forces inimical to unbounded population growth, such as limits to the supply of food and sufficiently favorable living space, and he showed that under his hypotheses the explosive growth predicted by Malthus would not occur. Rather, the population size would increase asymptotically to a finite limit. In the second of these three memoirs Verhulst introduced the word *logistic* as the name of the curve $y = K/(1 + Ae^{a-bt})$, which he proposed to describe the growth of a population subject to the vicissitudes he contemplated. The word *logistic* comes from a Greek root λογιστιχός pertaining to reckoning, reasoning, and logicality. By the eighteenth century, it was being used in a variety of mathematical contexts (in addition to its other nonscientific chores). It was used, for example, in the *Chambers' Cyclopedia* (1727–1741) as an alternative name for the logarithmic curve, and as late as 1882, Glaisher in the *Encyclopedia Britannica* used the phrase *logistic numbers* to denote what would now be called simply *ratios*. But these are only two examples; already, by the time Verhulst unfortunately selected the name *logistic* for his curve the word seems to have had too many meanings. The confusion was increased when, in 1944, in his book on the art of war, Burne [4] introduced the expression *the science of logistics* to mean "the science of moving and supplying troops." In recent times the range of meanings attached to the belabored word has increased in a Malthusian way with none of Verhulst's restraining forces to inhibit new connotations.

Further studies of population growth using the logistic curve were made by Yule [20], Pearl and Reed [13], Pearl [12], Will [19], and others. However, in 1939, Feller [6], in a ground-breaking paper, introduced a stochastic model for the development of populations in the face of "struggle" and thereby opened up a challenging area of research for probabilists and statisticians, for whom the stochastic element was an important consideration in rendering the mathematics more representative of the erratic behavior of many populations. It is thus now possible to study both the *deterministic* and the *stochastic* versions of the "logistic" theory of population growth. The published discussions of the deterministic versions are mathematically less demanding, since they typically lead to differential equations of a classical kind whose solutions are well understood. Stochastic versions, on the other hand, call for special methods of solution, of some ingenuity and complexity. For this reason this article is largely devoted to the stochastic versions.

The term "the logistic process" now usually refers to a stochastic model of population growth of the birth-and-death* type, in which the instantaneous rates for birth and death for each individual "particle" alive at a given moment are linear functions of the instantaneous population size. Let $N(t)$ be the size of the population at time t; let N_1 and N_2 be two fixed integers ($N_1 < N_2$). The probability that any individual particle will yield, in the time interval $(t, t + \delta t)$, an "offspring" is assumed to be $\lambda[N_2 - $

$N(t)$] δt, to the first order of small quantities. Similarly, the probability that an individual particle dies during $(t, t + \delta t)$, is assumed to be $\mu[N(t) - N_1] \delta t$. As is usual with birth-and-death processes*, the individuals are assumed to act independently of each other and, indeed, all the other assumptions for birth-and-death processes apply (such as their Markovian nature). Thus if we write $p_n(t)$ for $P\{N(t) = n\}$ we are led to a set of differential equations of which the following is representative (for $N_1 < n < N_2$):

$$\frac{d}{dt} p_n(t) = \lambda(n - 1)[N_2 - n + 1] p_{n-1}(t)$$
$$+ \mu(n + 1)[n + 1 - N_1] p_{n+1}(t)$$
$$- \{\lambda n[N_2 - n] + \mu n[n - N_1]\} p_n(t).$$
(1)

Obvious equations hold for the cases $n = N_1$ and $n = N_2$.

This probabilistic model is of interest in a number of distinct areas of application. In the study of *population growth* the model is one which takes account of the effect of overcrowding on the population, causing, as it often does, a struggle for available food and living space. A high value of $N(t)$, near N_2, implies a low birth rate and a high death rate. Notice that it is impossible to have $N(t) < N_1$ or $N(t) > N_2$ because the death rate vanishes when $N(t) = N_1$ and the birth rate vanishes when $N(t) = N_2$.

Another, and equally important, area of application for the logistic model is in the study of *epidemics*. We discuss here a case, frequently considered in the literature, which supposes $\mu = 0$ and $N_1 = 0$. For this special model, $N(t)$ represents the number of infected individuals in a closed population of total size N_2. Thus at any time t there are $N_2 - N(t)$ uninfected individuals and the rate of appearance of new infections is proportional to $N(t)[N_2 - N(t)]$. There is an obvious parallel here with the law of mass action in physical chemistry. Since $N(t)$ cannot decrease in this special model, it is only appropriate for certain kinds of rapidly spreading epidemics. In particular, no allowance is made for an infected individual

to become uninfected and, possibly, pass back into the pool of susceptibles.

If one introduces the usual probability generating function* (PGF)

$$\Pi(z, t) = \sum_{n=N_1}^{n=N_2} z^n p_n(t),$$
(2)

which is plainly always a polynomial for the logistic process (rather than the infinite series which arises more usually with birth-and-death processes), one can show that (1) reduces to the single partial differential equation

$$\frac{\partial \Pi}{\partial t} = \mu(1 - z) \frac{\partial}{\partial z} \left[z \frac{\partial \Pi}{\partial z} - N_1 \Pi \right]$$

$$- \lambda(1 - z) z \frac{\partial}{\partial z} \left[N_2 \Pi - z \frac{\partial \Pi}{\partial z} \right].$$
(3)

Although this is a linear equation it is of the second order and no conveniently simple solution is known.

For the special epidemic model described above, by putting $\mu = 0$ in (3), we obtain

$$\frac{\partial \Pi}{\partial t} = \lambda z(1 - z) \left[z \frac{\partial^2 \Pi}{\partial z^2} - (N_2 - 1) \frac{\partial \Pi}{\partial z} \right],$$
(4)

which is like an equation presented in a similar context by Bartlett [2]. Even this simpler equation (4) is troublesome to solve. Bailey [1] gives a detailed discussion based on Laplace transforms. He obtains formulas for the mean and variance of $N(t)$ and various other attributes of the epidemic model. Explicit formulas are presented for $N_2 = 10$, but it does seem that Bailey's solution is cumbersome if N_2 is much larger. There is a significant sequel to Bailey's paper by Whittle [18]. A different, and highly ingenious, approach to (4) using an operator technique is given in Whittle [17]. Mention should also be made of the paper by Daniels [5] which provides an approximation procedure for tackling a variety of problems to do with stochastic processes* of the logistic type.

The details of the papers just mentioned would require too much space to be presented here; the reader is referred to the original papers. It is possible that the operator technique of Whittle and the approximation methods of Daniels would yield useful information about the general logistic process described by (3). But one should also consult the important survey paper of Kendall [9] in which much useful information is obtained about the general logistic process (3) by "formal" arguments. For this, also, an exposition would require too much space and the reader is urged to consult Kendall's highly readable paper.

However, the *stationary* solution of (3) is easy to obtain; a derivation of it is provided in the book of Karlin and Taylor [8]. They show the stationary value of $P\{N(t) = N_1 + m\}$, $0 \le m \le N_2 - N_1$, to be

$$\frac{N_1}{N_1 + N_2}\binom{N_2 - N_1}{m}\left(\frac{\lambda}{\mu}\right)^m.$$

A recent paper by Hanson and Tuckwell [7] introduces an interesting modification of the classical logistic process; they suppose that random "disasters" occur, which kill a fixed proportion of the population. Another recent paper of relevance to the present subject is that of Norden [11], who uses a numerical method for the inversion of Laplace transforms, due to Bellman, to obtain information about the moments of a nonstationary logistic process at an arbitrary time t.

The reader interested in the general question of sophisticated mathematical models of ecological and epidemic processes, including logistic processes, but embracing much else, is directed to two valuable survey papers which have appeared in recent years. They are Mollison [10] and Brillinger [3].

References

[1] Bailey, N. T. J. (1950). A simple stochastic epidemic. *Biometrika*, **37**, 193–202.

[2] Bartlett, M. S. (1978). *An Introduction to Stochastic Processes*, 3rd ed. Cambridge University Press, Cambridge, England.

[3] Brillinger, D. R. (1981). Some aspects of modern population mathematics. *Canad. J. Statist.*, **9**, 173–194.

[4] Burne, A. H. (1944). *The Art of War on Land*, Methuen, London.

[5] Daniels, H. E. (1967). The distribution of the total size of an epidemic. *Proc. 5th Berkeley Symp. Math. Statist. Prob.*, University of California Press, Berkeley, CA.

[6] Feller, W. (1939). Die Grundlagen der Volterrischen Theorie des Kampfes ums Dasein in wahrscheinlichkeitstheoretischer Behandlung. *Acta Biotheoretica*, **5**, 11–40.

[7] Hanson, F. B. and Tuckwell, H. C. (1981). Logistic growth and random density independent disasters. *Theor. Pop. Biol.*, **19** 1–18.

[8] Karlin, S. and Taylor, H. M. (1975). *A First Course in Stochastic Processes*, 2nd ed. Academic, New York (see page 144).

[9] Kendall, D. G. (1949). Stochastic processes and population growth. *J. R. Statist. Soc. B*, **11**, 230–264.

[10] Mollison, D. (1977). Spatial contact models for ecological and epidemic spread. *J. R. Statist. Soc. B*, **39**, 283–326.

[11] Norden, R. H. (1984). On the numerical evaluating of the moments of states at time t in the stochastic logistic process. *J. Statist. Comput. Simul.*, **20**, 1–20.

[12] Pearl, R. (1925). *The Biology of Population Growth*. Knopf, New York (rev. ed. 1930).

[13] Pearl, R. and Reed, L. J. (1920). On the growth of the population of the United States since 1790 and its mathematical representation. *Proc. Nat. Acad. Sci. Wash.*, **6**, 275–288.

[14] Verhulst, P. F. (1838). Notice sur le loi que la population suit dans son accroissement. *Corr. Math. Phys. Publ. A. Quetelet*, **X**, 113–121.

[15] Verhulst, P. F. (1845). Recherches mathématiques sur la loi d'accroissement de la population. *Nouveaux Mém. Acad. R. Sci. Belles-Lettres de Bruxelles*, **18**, 1–38.

[16] Verhulst, P. F. (1846). Deuxième mémoire sur la loi d'accroissement de la population. *Nouveaux Mém. Acad. R. Sci. Belles-Lettres de Bruxelles*, **19**, 1–32.

[17] Whittle, P. (1952). Certain nonlinear models of population growth. *Skand. Akt.*, **35**, 211–222.

[18] Whittle, P. (1955). The outcome of a stochastic epidemic—a note on Bailey's paper. *Biometrika*, **42**, 116–122.

[19] Will, H. S. (1936). On a general solution for the parameters of any function with application to the theory of organic growth. *Ann. Math. Statist.*, **7**, 165–190.

[20] Yule, G. U. (1925). The growth of population and the factors which control it. *J. R. Statist. Soc.*, **88**, 1–58.

(BIRTH-AND-DEATH PROCESSES
GROWTH CURVES
LOGISTIC CURVE

MALTHUS, THOMAS ROBERT
POPULATION GROWTH MODELS
STOCHASTIC DEMOGRAPHY
STOCHASTIC PROCESSES)

WALTER L. SMITH

M

MARL ESTIMATOR

In 1964, Chipman [1] developed an estimator of β in the linear model $Y = X\beta + \epsilon$, which "minimized" the mean square error or risk matrix, $E[(\hat{\beta} - \beta)(\hat{\beta} - \beta)']$, among all estimators of β that were linear in Y. Chipman adopted the Bayesian context for the estimation problem so that β, as well as Y, are considered random variables. The above expectation is thus taken with respect to the joint distribution of both Y and β, and "minimum risk matrix" is taken to mean that any other estimator linear in Y has a risk matrix that exceeds the risk matrix of Chipman's estimator by a positive semidefinite matrix. Specifically, Chipman made the following assumptions concerning the moments of the random variables β, ϵ, and Y:

$$E[\beta] = \bar{\beta}, \quad E[\epsilon] = 0,$$
$$E[Y|\beta] = X\beta, \quad E[Y|\epsilon] = X\bar{\beta} + \epsilon,$$
$$\text{cov}(\beta) = U, \quad \text{cov}(\epsilon) = V,$$
$$\text{cov}(Y|\beta) = V, \quad \text{cov}(Y|\epsilon) = XUX'.$$

He then examined all estimators of β linear in Y, as

$$\hat{\beta} = AY + b.$$

Using results on optimum inverses developed by Foster [3], Chipman found that $\hat{\beta}$ minimizes the risk matrix when $b = (I - AX)\bar{\beta}$ and $A = UX'(XUX' + V)^+$, where $+$

denotes the Moore–Penrose generalized inverse*. It follows that the optimum $\hat{\beta}$ in the case where U is nonsingular is given by

$$\hat{\beta} = (X'V^{-1}X + U^{-1})^{-1}(X'V^{-1}Y + U^{-1}\bar{\beta}).$$

Swamy and Mehta [6, 7] later named Chipman's estimator the *minimum average risk linear estimator*, or MARL estimator, due to the fact that $E[(\hat{\beta} - \beta)(\hat{\beta} - \beta)'] = E_\beta[E[(\hat{\beta} - \beta)(\hat{\beta} - \beta)'|\beta]]$, i.e., risk is essentially averaged over the distribution of β, and hence Chipman's estimator minimizes average risk in the class of linear estimators.

Unaware of Chipman's work, Duncan and Horn [2] developed the MARL estimating formula under the assumption of nonsingular U in their discussion of what they term "a wide—sense random—regression theory." Also, LaMotte [4], in his discussion of Bayes affine estimators, establishes the MARL estimator formula by minimizing what he calls "total mean square error," which is essentially the trace of the risk matrix defined above. In LaMotte's terminology, "Bayes affine estimator" refers to an estimator of the form $\hat{\beta} = AY + b$, whereas "Bayes linear estimator" refers to $\hat{\beta} = AY$, the b constant being deleted.

As originally defined, Chipman's estimator required full knowledge of the mean vector and covariance matrix of β. Mittelhammer and Conway [5] examined the case where only the mean and covariance matrix of certain linear combinations of β, i.e., $\mathbf{R}\beta$, are known. Here, $\mathbf{R}$ is a known $j \times k$ matrix of rank j used to define the particular linear combinations of the β vector about which prior information exists. They show that, in the class of linear estimators $\hat{\beta} = \mathbf{AY} + \mathbf{b}$ for which average risk is defined, the estimator

$$\hat{\beta} = \left(\mathbf{X'V^{-1}X} + \mathbf{R'\Psi^{-1}R} \right)^{-1}$$
$$\times \left(\mathbf{X'V^{-1}Y} + \mathbf{R'\Psi^{-1}\mu} \right)$$

minimizes average risk, where $\mu = E[\mathbf{R}\beta]$ and $\text{cov}(\beta) = \Psi$. They name the estimator the *minimum average risk–rankable linear estimator, or* MARRL estimator. In the special case where $\mathbf{R} = \mathbf{I}$, the MARRL estimator becomes Chipman's MARL estimator.

Mittelhammer and Conway go on to derive some admissibility* results regarding the MARRL estimator. They find that, in the class of linear estimators, the MARRL estimator is quadratic risk admissible if the number of independent rows of $\mathbf{R}$ are greater than or equal to the number of entries in the β vector, less two. They also find that if ϵ is multinormally distributed, then the MARRL estimator is quadratic risk admissible in the class of all estimators of β provided the same condition on the number of independent rows of $\mathbf{R}$ holds as above.

The main attraction of the MARL (or MARRL) estimation technique is that it allows the incorporation of prior information into the estimation of β in a rather straightforward, closed-form manner that avoids the often complex apparatus of a full Bayesian approach to inference. Only prior information on the first two moments of elements of the β vector are actually required, thus a full specification of the prior distribution on β is not necessary. However, optimality of the estimator is restricted to the linear class of estimators, and the estimator does not generally minimize overall Bayes risk.

References

[1] Chipman, J. S. (1964). On least squares with insufficient observations. *J. Amer. Statist. Ass.*, **59**, 1078–1011.

[2] Duncan, D. B. and Horn, S. D. (1972). Linear dynamic recursive estimation from the viewpoint of regression analysis. *J. Amer. Statist. Ass.*, **67**, 815–821.

[3] Foster, M. (1961). An application of the Wiener–Kolmogorov smoothing theory to matrix inversion. *SIAM J. Appl. Math.*, **9**, 387–392.

[4] LaMotte, L. R. (1978). Bayes linear estimators. *Technometrics*, **20**, 281–290.

[5] Mittelhammer, R. C. and Conway, R. K. (1985). Extending Chipman's MARL estimator to cases of ignorance in one or more parameter dimensions. *Commun. Statist. Theor. Meth.*, **14**, 1879–1888.

[6] Swamy, P. A. V. B. and Mehta, J. S. (1977). A note on minimum average risk estimators for coefficients in linear models. *Commun. Statist. A*, **6**, 1181–1186.

[7] Swamy, P. A. V. B. and Mehta, J. S. (1976). Minimum average risk estimators for coefficients in linear models. *Commun. Statist. A*, **5**, 803–818.

(BAYESIAN INFERENCE
ESTIMATION, POINT
GENERAL LINEAR MODEL
LEAST SQUARES
MULTIVARIATE ANALYSIS)

ROGER K. CONWAY
RON C. MITTELHAMMER

MEDIAL CORRELATION COEFFICIENT

Let X and Y be random variables with medians $\tilde{x}$ and $\tilde{y}$, respectively. The *medial correlation coefficient* is

$$m = \Pr[(X - \tilde{x})(Y - \tilde{y}) > 0]$$
$$- \Pr[(X - \tilde{x})(Y - \tilde{y}) < 0].$$

For any symmetrical bivariate distribution, this is the same as the Kendall's τ, given by

$$\tau(X, Y) =$$
$$4 \int_{-\infty}^{\infty} \int_{-\infty}^{\infty} [H(x, y) \, dH(x, y)] - 1,$$

where $H(\cdot, \cdot)$ is the joint cumulative distribution function of X and Y.

(CORRELATION COEFFICIENT
DEPENDENCE, MEASURES AND
   INDICES OF
KENDALL'S TAU)

META-ANALYSIS

Meta-analysis is the use of statistical methods in reviews of related research studies. For example, meta-analyses often involve the use of statistical methods to combine estimates of treatment effects from different research studies that investigate the effects of the same or related treatments. Meta-analysis is distinguished from primary analysis (the original analysis of a data set) and secondary analysis (reanalysis of another's data) by the fact that meta-analyses do not usually require access to raw data, but only to summary statistics. Thus the data points for meta-analyses are summary statistics, and a sample of studies in meta-analysis is analogous to a sample of subjects or respondents in primary analysis.

Meta-analysis in the social and medical sciences began as an attempt to utilize better the evidence from increasingly large numbers of independent experiments. Important scientific and policy questions often stimulate dozens of research studies designed to answer essentially the same question. Meta-analysis provides a procedure for extracting much of the information in these studies to provide a broader base for conclusions than is possible in any single study and to increase statistical power.

Meta-analyses are much like original research projects and involve the same general stages of research procedures such as problem formulation, data collection*, data evaluation, data analysis and interpretation, and presentation of results [4]. Subjective decisions about procedure are necessary at each stage, and the use of an overall plan or protocol is often useful to avoid biases by constraining procedural variations. Such a protocol describes the details of procedures in each state of the meta-analysis.

Problem formulation is the process of stating precise questions or hypotheses in operational terms that make it possible to decide whether or not a given study examines the hypothesis of interest. Even studies that are superficially similar usually differ in the details of treatment, controls, procedures, and outcome measures, and consequently problem formulation involves the specification of the range of acceptable variation in these details. For example, consider studies of the effects of a drug. The treatment consists of administration of the drug but what dosage level, schedule and modality of administration, etc., should count as instances of the "same" treatment? Should control conditions involve placebos, alternative treatments, or the conventional treatment (and if so, are there variations in the conventional treatment)? Outcomes like death rates seem unambiguous, but calculation of death rates may not be identical across studies. For example, should deaths from all causes be included or should only deaths from the disease under treatment be considered (and if so what about deaths due to side effects of treatment)? The breadth of the question or hypothesis of the meta-analysis has implications for the range of acceptable variation in constructs of treatment, control, outcome, and so on.

Data collection in meta-analysis consists of assembling the studies that may provide relevant data and extracting an estimate of effect magnitude from each study. Sometimes (as in studies of the efficacy of a new drug) all of the studies that were conducted are immediately available, but more frequently studies must be obtained from the published or unpublished literature in a substantive research area. The method used to obtain the sample of studies requires careful attention because some procedures used to search for studies may introduce biases. For example, published journal articles may be a selected sample of the set of all research studies actually conducted, overrepresenting studies that yield statistically significant results [7, Chap. 14; 11]. Statistical corrections for selection are sometimes possible [7] but exhaustive enumeration (and then perhaps sampling) of studies is usually advisable.

Data collection also involves the selection of an estimate of effect magnitude from each study. Typical indices of effect magnitude compare a treatment group with a control group via indices such as raw mean differences, raw or transformed differences in incidence proportions, risk ratios, odds ratios, or correlation coefficients. The standardized mean difference $\delta = (\mu_1 - \mu_2)/\sigma$, where σ is a within-group standard deviation, is frequently used in the social and behavioral sciences as a scale-free index of treatment effect or as an index of overlap between distributions (i.e., a one-dimensional Mahalanobis distance*).

Data evaluation is the process of deciding which possibly relevant studies should be included in the data analysis. Sometimes the criteria for study inclusion will be straightforward (such as all randomized clinical trials*) but frequently inclusion criteria that are more subjective (such as well-controlled and possibly nonrandomized trials). Because serious biases can arise due to decisions about study inclusion, protocols for decisions about inclusion of studies may be useful in just the same way as they are necessary in decisions for inclusion of patients in clinical trials. Such protocols may also enhance the reliability of decisions about study quality, which may otherwise be quite low [4, 9]. Some investigators also suggest the use of "blind" ratings of study quality in which raters are not aware of the *findings* of the studies they rate. Empirical methods for detecting outliers* or influential observations* may also have a role in data evaluation. Such methods sometimes reveal problem studies that were initially overlooked [7].

Data analysis is the process of combining estimates of effect magnitude across studies. The details of these procedures are discussed in COMBINATION OF DATA. The analyses vary depending on the conceptualization of between-study variability used in the data analysis model. Different models lead to different interpretations of results and can lead to very different estimates of the precision of results. The models differ primarily in whether they treat between-study differences as fixed or random and consequently whether between-study variability should be incorporated into the uncertainty of the combined result.

Let $\delta_1, \ldots, \delta_k$ be effect magnitude parameters from k studies and let $d_1, \ldots, d_k$ and $S_1, \ldots, S_k$, respectively, be the corresponding sample estimates and their approximate

standard errors. The simplest and most common procedures treat $\delta_1, \ldots, \delta_k$ as if $\delta_1 = \cdots = \delta_k = \delta$ and estimate the common effect δ. The procedures most frequently used to estimate δ involve a weighted mean $\bar{d}$ in which the ith weight is proportional to $1/S_i^2$. The weighted sum of squares $\Sigma(d_i - \bar{d})^2/S_i^2$ about the weighted mean $\bar{d}$ is often used as a statistic to test the consistency of the δ_i across studies. (*see* COMBINATION OF DATA, Hedges and Olkin [7] or Cochran [2]). A test of homogeneity of effects may reveal that the assumption of a common δ is unrealistic. In this situation one of three approaches to the situation is usually used.

The fixed-effects approach treats between-study variation in effect magnitudes as if it were a function of known explanatory variables that arise as characteristics of studies (e.g., aspects of particular treatments, controls, procedures, or outcome measures). Fixed-effects analyses usually estimate the vector $\boldsymbol{\beta}$ of unknown coefficients in a linear model of the form

$$\delta_i = \mathbf{x}_i'\boldsymbol{\beta}, \qquad i = 1, \ldots, k,$$

where $\mathbf{x}_i$ is a vector of study characteristics [7]. The estimates of $\boldsymbol{\beta}$ provide insight into the relationship between study characteristics and effect magnitude. Tests of the goodness of fit* of the model (analogous to the test of homogeneity given above) can help to determine the degree to which the data are consistent with the model. For example, Becker and Hedges [1] used a linear model to estimate the historical trend in gender differences in cognitive abilities by using the year that a study was conducted to explain between-study variations in the magnitude of sex differences in cognitive performance.

An alternative to fixed-effects models are simple random-effects models in which the effect magnitudes $\delta_1, \ldots, \delta_k$ are modeled via $\delta_i = \delta + \eta_i$, where $\eta_1, \ldots, \eta_k$ are treated as independent random variables with a mean of zero and a variance of σ_η^2. This conceptualization is particularly appealing when the treatments used in studies exhibit substantial uncontrolled variability that may plausibly be treated as "random." Random-effects analyses of this sort usually concentrate on estimation of δ and the variance component σ_η^2 (*see* COMBINATION OF DATA, Hedges and Olkin [7] or Cochran and Cox [3]). For example, Schmidt and Hunter [10] used a random-effects model to study variation in (population) correlation coefficients that were used as indicators of the validity of psychological tests used in personnel selection.

More complex random-effects or mixed models involve the use of both explanatory variables and random effects. The effect magnitudes might be modeled via

$$\delta_i = \mathbf{x}_i'\boldsymbol{\beta} + \eta_i, \qquad i = 1, \ldots, k,$$

where $\mathbf{x}_i$ is a vector of known study characteristics, $\boldsymbol{\beta}$ is a vector of coefficients, and $\eta_1, \ldots, \eta_k$ are independent, identically distributed random variables with zero mean and variance σ_η^2. Such models are sometimes more realistic when effect magnitudes are more variable than would be expected from fixed-effects models with no random contribution.

The presentation of the results of a meta-analysis should include a description of the formal protocol as well as any other steps of problem formulation, data collection, data evaluation, and data analysis that are likely to affect results. It is usually helpful to provide a brief tabular summary which presents relevant characteristics of each of the individual studies along with the computed index of effect magnitude for each study.

References

[1] Becker, B. J. and Hedges, L. V. (1984). *J. Educ. Psychol.*, **76**, 583–587.

[2] Cochran, W. G. (1954). *Biometrics*, **10**, 101–129.

[3] Cochran, W. G. and Cox, G. M. (1957). *Experimental Design*, 2nd ed. Wiley, New York. (Chapter 14 is a discussion of the analysis of series of experiments.)

[4] Cooper, H. M. (1984). *The Integrative Literature Review*. Sage, Beverly Hills, CA. (Includes a discussion of problem formulation, data collection, and data evaluation.)

[5] Fleiss, J. L. (1973). *Statistical Methods for Rates and Proportions*. Wiley, New York. (Chapter 10 is a discussion of combining analyses of contingency tables.)

[6] Glass, G. V., McGaw, B., and Smith, M. L. (1981). *Meta-Analysis in Social Research*. Sage, Beverly Hills, CA. (A good introduction to the perspective of meta-analysis although its treatment of statistics is somewhat dated.)

[7] Hedges, L. V. and Olkin, I. (1985). *Statistical Methods for Meta-Analysis*. Academic, New York. (A comprehensive review of statistical methods for meta-analysis using standardized mean differences and correlation coefficients.)

[8] Light, R. J. and Pillemer, D. (1984). *Summing Up: The Science of Reviewing Research*. Harvard University Press, Cambridge, MA. (A good treatment of conceptual issues in combining research results.)

[9] Rosenthal, R. (1984). *Meta-Analytic Procedures for Social Research*. Sage, Beverly Hills, CA. (A very clear introduction to selected statistical procedures for meta-analysis.)

[10] Schmidt, F. L. and Hunter, J. (1977). *J. Appl. Psychol.*, **62**, 529–540.

[11] Sterling, T. C. (1959). *J. Amer. Statist. Ass.*, **54**, 30–34.

(CLINICAL TRIALS
COMBINATION OF DATA
FIXED-, RANDOM-, AND MIXED-EFFECTS
   MODELS
LINEAR MODEL SELECTION, CRITERIA
   AND TESTS)

LARRY V. HEDGES

MODIFIED MAXIMUM LIKELIHOOD ESTIMATION

This entry should be read in conjunction with MAXIMUM LIKELIHOOD ESTIMATION. There are situations where a maximum likelihood* (ML) equation $d \log L/d\theta = 0$ has no explicit solution. This is particularly true in the case of censored* samples (Schneider [10] and Tiku et al. [19]). The only way then to compute the maximum likelihood estimator (MLE) of θ is to solve this equation by iterative methods. Due to the implicit nature of iterations, however, it is difficult to make any analytical study of the resulting MLE, especially for small samples. It has therefore been suggested that such a maximum likelihood equation be modified so that the modified equation has an explicit solution [called the *modified maximum likelihood estimator* (MMLE)]. Plackett [9] was the first to suggest a modification to the ML equation based on a type II censored sample

$$y_{(a)}, y_{(a+1)}, \ldots, y_{(b)},$$

$$a = r_1 + 1, \ b = n - r_2, \ r_1 \geq 0, \ r_2 \geq 0, \quad (1)$$

where $y_{(1)} \leq y_{(2)} \leq \cdots \leq y_{(n)}$ are the order statistics* corresponding to random observations y_i, $i = 1, 2, \ldots, n$, obtained by censoring the r_1 smallest and r_2 largest observations. If the underlying distribution is normal $N(\mu, \sigma^2)$, the ML equations are

$$\frac{\partial \log L}{\partial \mu} = \frac{1}{\sigma} \left\{ \sum_{i=a}^{b} z_i - r_1 g_1(z_a) + r_2 g_2(z_b) \right\}$$

$$= 0 \quad (2)$$

and

$$\frac{\partial \log L}{\partial \sigma} = \frac{1}{\sigma} \left\{ -(n - r_1 - r_2) \right.$$

$$+ \sum_{i=a}^{b} z_i^2 - r_1 z_a g_1(z_a)$$

$$\left. + r_2 z_b g_2(z_b) \right\} = 0, \quad (3)$$

where $z_i = (y_{(i)} - \mu)/\sigma$. The functions $g_1(z)$ and $g_2(z)$ are given by

$$g_1(z) = f(z)/F(z),$$
$$g_2(z) = f(z)/\{1 - F(z)\}, \quad (4)$$

where

$$f(z) = (2\pi)^{-1/2} \exp(-z^2/2) \text{ and}$$

$$F(z) = \int_{-\infty}^{z} f(z) \, dz.$$

Of course, (2) and (3) have no explicit solutions. Plackett suggested the following modification:

Write

$$t_i = E[y_{(i)} - \mu]/\sigma \quad (5)$$

and expand $\hat{z} = (y - \hat{\mu})/\hat{\sigma}$ in a Taylor series*, in terms of t_i and the derivatives of $f(z)$ evaluated at t_i. Ignoring all the third- and higher-order derivatives in this expansion, Plackett obtained the MML equations.

These equations are rather cumbersome and will not be reproduced here. They do, however, admit explicit solutions which, when approximated further, are of the type

$$\hat{\mu} = \sum_{i=a}^{b} k_i y_{(i)}$$

$$\hat{\sigma} = \sum_{i=a}^{b} l_i y_{(i)}. \quad (6)$$

(Plackett called them *linearized maximum likelihood estimators*.)

The expressions for k_i and l_i are rather tedious. Plackett gave tables to facilitate the computations for a few $(0, r_2)$ and $(r_1, 0)$ combinations.

Tiku [13] modified (2) and (3) by using linear approximations

$$g_1(z) \simeq \alpha_1 - \beta_1 z$$
$$g_2(z) \simeq \alpha_2 + \beta_2 z; \quad (7)$$

(α_1, β_1) and (α_2, β_2) are obtained such that (7) give close approximations. In fact, (α_1, β_1) and (α_2, β_2) are obtained simply by replacing r by r_1 and r_2, respectively, in the equations

$$\beta = -f(t)\{t - [f(t)/q]\}/q$$
$$\alpha = [f(t)/q] - \beta t, \quad q = r/n. \quad (8)$$

The value of t is determined by $1 - F(t) = q$. Thus the following MML equations are obtained:

$$\frac{\partial \log L^*}{\partial \mu} = \frac{1}{\sigma} \left\{ \sum_{i=a}^{b} z_i - r_1(\alpha_1 - \beta_1 z_a) \right.$$

$$\left. + r_2(\alpha_2 + \beta_2 z_b) \right\} = 0 \quad (9)$$

and

$$\frac{\partial \log L^*}{\partial \sigma} = \frac{1}{\sigma} \left\{ -(n - r_1 - r_2) \right.$$

$$- r_1 z_a(\alpha_1 - \beta_1 z_a)$$

$$\left. + r_2 z_b(\alpha_2 + \beta_2 z_b) \right\} = 0; \quad (10)$$

(9) and (10) admit explicit solutions. For example, if $r_1 = r_2 = r$, then $\alpha_1 = \alpha_2 = \alpha$ and $\beta_1 = \beta_2 = \beta$, and the MMLE are given

by

$$\hat{\mu} = \left\{ \sum_{i=r+1}^{n-r} y_{(i)} + r\beta \left[y_{(r+1)} + y_{(n-r)} \right] \right\} / m,$$

$$m = n - 2r + 2r\beta, \quad (11)$$

and

$$\hat{\sigma} = \frac{1}{2} \left\{ B + \sqrt{(B^2 + 4AC)} \right\} \bigg/ 2\sqrt{\{A(A-1)\}},$$

$$A = n - 2r, \quad (12)$$

where

$$B = r\alpha \left[y_{(n-r)} - y_{(r+1)} \right],$$

$$C = \sum_{i=r+1}^{n-r} y_{(i)}^2 + r\beta \left[y_{(r+1)}^2 + y_{(n-r)}^2 \right] - m\hat{\mu}^2.$$

$$(13)$$

Tiku [15] showed that for large n, $g_1(z) \doteq \alpha_1 - \beta_1 z$ and $g_2(z) \doteq \alpha_2 + \beta_2 z$ in which case the ML and MML equations above are nearly identical. Tiku [16] generalized this method to multisample situations in the framework of experimental designs. Tiku [16] showed that for large n, $\sqrt{m}\,\hat{\mu}/\sigma$ is approximately a normal $N(0, 1)$ variate and $(A - 1)\hat{\sigma}^2/\sigma^2$, an independent chi-squared variate with $\nu = A - 1$ degrees of freedom. Tan [11] and Tan and Balakrishnan [12] studied the distributions of $\hat{\mu}$ and $\hat{\sigma}^2$ (and $t = \sqrt{m}\,\hat{\mu}/\hat{\sigma}$) from a Bayesian* point of view. For example, they showed that the highest posterior density interval for μ is given by

$$\left[\hat{\mu} + D\frac{\hat{\sigma}}{m} - \frac{\hat{\sigma}}{\sqrt{m}} t_{\delta/2}(\nu), \right.$$

$$\left. \hat{\mu} + \hat{D}\frac{\hat{\sigma}}{m} + \frac{\hat{\sigma}}{\sqrt{m}} t_{\delta/2}(\nu) \right], \quad (14)$$

where $D = r_2\alpha_2 - r_1\alpha_1$. If $r_1 = r_2$, then $D = 0$.

Tiku and Stewart [18] discussed the estimation of μ and σ from a random sample of size n from a truncated normal distribution*

$$\frac{1}{\sqrt{2\pi}\,\sigma} e^{-\frac{1}{2}(y-\mu)^2/\sigma^2} \bigg/ [F(z'') - F(z')],$$

$$y' < y < y''; \quad (15)$$

y' and y'' are known, and $z' = (y' - \mu)/\sigma$ and $z'' = (y'' - \mu)/\sigma$. Here, the maximum

likelihood equations are

$$\frac{\partial \log L}{\partial \mu} = \frac{1}{\sigma} \{ z_i - g_1(z') + g_2(z'') \} = 0$$

$$(16)$$

and

$$\frac{\partial \log L}{\partial \sigma} = \frac{1}{\sigma} \left\{ -n + \sum_{i=1}^{n} z_i^2 - z'g_1(z') \right.$$

$$\left. + z''g_2(z'') \right\} = 0, \quad (17)$$

where

$$z_i = (y_i - \mu)/\sigma,$$

$$g_1(z') = f(z')/\{F(z'') - F(z')\},$$

and

$$g_2(z'') = f(z'')/\{F(z'') - F(z')\};$$

$f(z)$ and $F(z)$ are the same as in (4). These equations have no explicit solutions and are difficult to solve by iterative methods. The MML equations are obtained from linear approximations exactly similar to (7). Thus the MMLEs of μ and σ are obtained (Tiku and Stewart [18]) from

$$\hat{\mu} = K + D\sigma \quad (\sigma \text{ to be replaced by } \hat{\sigma}) \quad (18)$$

and $\hat{\sigma}$ is the positive root of the equation

$$\sigma^2 - \left[(\alpha_2 y'' - \alpha_1 y') \right.$$

$$- (1 + \beta_1 + \beta_2)KD \big] \sigma$$

$$- \left[\frac{1}{n} \sum_{i=1}^{n} y_i^2 + \beta_1 y'^2 + \beta_2 y''^2 \right.$$

$$\left. - (1 + \beta_1 + \beta_2)K^2 \right] = 0, \quad (19)$$

where

$$K = (\bar{y} + \beta_1 y' + \beta_2 y'')/(1 + \beta_1 + \beta_2),$$

$$D = (\alpha_2 - \alpha_1)/(1 + \beta_1 + \beta_2). \quad (20)$$

Tiku and Stewart [18] generalized this method to experimental designs where the observations come from truncated normal distributions.

Mehrotra and Nanda [6] discussed a modification of (2) and (3) which is a simple variant of a second method proposed by Tiku [13], namely, (α_1, β_1) and (α_2, β_2) in

(7) are obtained such that $E(\partial \log L/\partial \mu) = 0$ and $E(\partial \log L/\partial \sigma) = 0$. Assuming that σ is known, Mehrotra and Nanda replaced $g_2(z_b)$ in (2) by its expected value and obtained the MMLE of μ (σ known) from the resulting equation,

$$\hat{\mu} = \left\{ \sum_{i=1}^{n-r} y_{(i)} + \sigma \sum_{i=r+1}^{n} E[V_{(i)}] \right\} \Big/ (n-r),$$

$$r = r_2, \quad (21)$$

$V_{(i)}$, $i = 1, 2, \ldots, n$, being the order statistics of a random sample of size n from a normal $N(0, 1)$ distribution. Assuming now that μ is known, they modified (3) in a similar fashion and obtained the MMLE of σ^2 (μ known),

$$\hat{\sigma}^2 = \sum_{i=1}^{n-r} \{y_{(i)} - \mu\}^2 \Big/ \left\{ \sum_{i=1}^{n-r} E[V_{(i)}^2] \right\}. \quad (22)$$

Mehrotra and Nanda did not, however, discuss the simultaneous estimation of μ and σ.

Bhattacharya [1] showed that the Tiku and Mehrotra–Nanda methods produce asymptotically unbiased and efficient estimators of μ and σ in any location–scale distribution $f[(y-\mu)/\sigma]$. Tiku et al. [19] worked out MMLEs of μ and σ for a few nonnormal distributions.

Persson and Rootzén [8] proposed a different modification of (2) and (3). Like Mehrotra and Nanda [6], they assumed that $r_1 = 0$ and replaced $F(z_b)$ by its asymptotic value, namely,

$$\lim_{n \to \infty} \{F(z_b)\} = 1 - q, \quad q = r_2/n. \quad (23)$$

Realize that if t_q is determined by the equation $F(t_q) = 1 - q$, then

$$\lim_{n \to \infty} \{g_2(z_b)\} = f(t_q)/q. \quad (24)$$

Replacing $g_2(z_b)$ in (2) and (3) by its asymptotic value, Persson and Rootzén obtained the MMLE of μ and σ as solutions of the resulting equations. They are essentially similar to the solutions of (9) and (10) with $r_1 = 0$. (*See also* PERSSON-ROOTZÉN ESTIMATOR.)

Lee et al. [5] used Tiku's method to obtain the MMLE of σ in the Rayleigh distribution*

$$\frac{2}{\sigma} y e^{-y^2/\sigma}, \quad 0 < y < \infty. \quad (25)$$

The MMLE, based on (1), is given by

$$\hat{\sigma} = \left\{ (B^2 + 4AC) - B(B^2 + 8AC)^{1/2} \right\} / (8A)^2, \quad (26)$$

where

$$A = n - r_1 - r_2, \quad B = r_1 \alpha y_{(a)},$$

$$C = 2 \sum_{i=a}^{b} y_{(i)}^2 + r_1 \beta y_{(a)}^2 + 2r_2 y_{(b)}^2, \quad (27)$$

$$\beta = -\frac{2e^{-h^2}}{1 - e^{-h^2}} \left(1 - \frac{2h^2}{1 - e^{-h^2}} \right) \quad (28)$$

$$\alpha = \frac{f(h)}{q_1} + \beta h, \quad \left(q_1 = \frac{r_1}{n} \right),$$

$f(z) = z e^{-z^2}$ and h is determined by the equation $F(h) = 1 - e^{-h^2} = q_1$. Lee et al. [5] gave a method for improving the performance of $\hat{\sigma}$ in small samples. This method essentially sharpens the linear approximations (7) through a second iteration.

Oakes [7] considered the estimation of parameters of a gamma distribution

$$\frac{1}{\sigma \Gamma(p)} e^{-y/\sigma} \left(\frac{y}{\sigma} \right)^{p-1}, \quad 0 < y < \infty, \quad (29)$$

from the type II censored sample (1) with $r_1 = 0$ and $r_2 = r$. The ML equations work in terms of

$$F(z) = \int_0^z f(z) \, dz,$$

$$f(z) = \{1/\Gamma(p)\} e^{-z} z^{p-1}.$$

Oakes replaced $F(z)$ by a Kaplan–Meier* (product–limit) estimator and suggested an algorithm for solving the resulting equations. Oakes noted, however, that his estimators are consistent but rather inefficient.

MML estimation has also been attempted for a few three-parameter distributions. For example, for the lognormal* distribution

$$\frac{1}{\sqrt{2\pi}\,\sigma(y-\eta)}$$

$$\times\exp\left\{-\frac{1}{2\sigma^2}\left[\log(y-\eta)-\mu\right]^2\right\},$$

$$\eta < y < \infty, \quad (30)$$

Tiku [14] used the equations (9) and (10) with y_i replaced by $\log(y_i-\eta_0)$ and obtained the MMLE of μ and σ for a given $\eta=\eta_0$. The MMLEs of μ, σ, and η are then obtained from (9) and (10) and the MML equation $\partial \log L^*/\partial \eta = 0$ by iteration; the expression for $\partial \log L^*/\partial \eta = 0$ is too lengthy to be reproduced here. Cohen and Whitten [2, 3] obtained different MMLEs for the three-parameter lognormal and Weibull* distributions. These equations are, however, also too lengthy to be reproduced here.

Example. To compare the ML and MML estimates, consider the following data from an accelerated life-test experiment involving specimens of electrical insulation (Lawless [4, p. 226]): 10 specimens were put on test and the test was terminated at the time of the eighth failure. The eight observed log-failure times are

6.00   6.43   6.77   7.07   7.40

7.66   8.10   8.40   —   —.

Assuming normality $N(\mu, \sigma^2)$, the ML and MML estimates are (Tiku et al. [19, p. 50]).

	μ	σ
Cohen ML	7.59	1.069
(corrected for bias)		
Tiku MML	7.5895	1.0696
Persson and Rootzén	7.57	0.99
Lawless	7.60	1.07
(based on a normal probability plot*)		

Tiku et al. [19] give several other examples and show that the ML and MML estimates are numerically very close.

Of course, the MML methods mentioned above utilize the entire distribution and seek close approximations to the likelihood function. There are, however, methods which uti-

lize only the first two moments of a distribution, namely, the methods of least squares* and quasi-likelihood (*see* QUASI-LIKELIHOOD FUNCTIONS).

References

[1] Bhattacharya, G. K. (1985). *J. Amer. Statist. Ass.*, **80**, 398–404.

[2] Cohen, A. C. and Whitten, B. (1980). *J. Amer. Statist. Ass.*, **75**, 399–404. (Gives various methods of modifying the ML equations for a three-parameter log-normal distribution.)

[3] Cohen, A. C. and Whitten, B. (1982). *Commun. Statist. A*, **11**, 2631–2656. (Gives various methods of modifying the ML equations for a three-parameter Weibull distribution.)

[4] Lawless, J. F. (1982). *Statistical Models and Methods for Lifetime Data*. Wiley, New York. (Discusses in particular approximate ML estimates for censored samples.)

[5] Lee, K. R., Kapadia, C. H., and Dwight, B. B. (1980). *Statist. Hefte*, **21**, 14–29. (Gives a technique to improve the performance of Tiku's method for small samples.)

[6] Mehrotra, K. G. and Nanda, P. (1974). *Biometrika*, **61**, 601–606. (Derives MML estimators by a method which is a simple variant of Tiku's method.)

[7] Oakes, D. (1986). *Biometrics*, **42**, 177–188. (Attempts MML estimation for a gamma distribution.)

[8] Persson, T. and Rootzén, H. (1977). *Biometrika*, **64**, 123–128. (Obtains MML estimators by equating the CDF at a sample point to its asymptotic value.)

[9] Plackett, R. L. (1958). *Ann. Math. Statist.*, **29**, 131–142. (Uses asymptotic properties of maximum likelihood equations to obtain MML estimators.)

[10] Schneider, H. (1986). *Truncated and Censored Samples from Normal Populations*. Marcel Dekker, New York. (Discusses various methods of estimation for type I and type II censored samples.)

[11] Tan, W. Y. (1985). *J. Statist. Plann. Inf.*, **11**, 329–340. (Studies the Bayesian distributions of Tiku's MMLE based on symmetric censored samples.)

[12] Tan, W. Y. and Balakrishnan, N. (1986). *J. Statist. Comut. Simul.*, **24**, 17–31. (Studies the Bayesian distributions of Tiku's MMLE based on asymmetric censored samples.)

[13] Tiku, M. L. (1967). *Biometrika*, **54**, 155–165. (Derives MML estimators for a normal distribution from a censored sample.)

[14] Tiku, M. L. (1968). *J. Amer. Statist. Ass.*, **63**, 134–140. (Derives MML estimators for a three-parameter log-normal distribution from censored samples.)

[15] Tiku, M. L. (1970). *Biometrika*, **57**, 207–211. (Establishes the asymptotic equivalence of the ML and MML estimators for a normal population.)

[16] Tiku, M. L. (1973). *Biometrics*, **29**, 25–33. (Studies the efficiency properties of Tiku's estimators in single and multi-sample situations.)

[17] Tiku, M. L. (1982). *Biom. J.*, **24**, 613–627. (Derives in particular the large-sample distributions of Tiku's MML estimators based on symmetrically censored normal samples.)

[18] Tiku, M. L. and Stewart, D. W. (1977). *Commun. Statist. A*, **6**, 1485–1501. (Extends the method of MML estimation to experimental designs under type I censoring.)

[19] Tiku, M. L., Tan, W. Y., and Balakrishnan, N. (1986). *Robust Inference*. Marcel Dekker, New York. (Discusses in detail the applications of Tiku's MML estimators in constructing robust inference procedures.)

(CENSORED DATA
ITERATED MAXIMUM LIKELIHOOD
   ESTIMATES
MAXIMUM LIKELIHOOD ESTIMATION
ORDER STATISTICS)

M. L. TIKU

MOMENT APPROXIMATION PROCEDURES

The approximation of a binomial distribution by a normal distribution is a primary example of a moment approximation procedure. The Wilson–Hilferty* and Patnaik approximations for central and noncentral chi-square distributions are also examples of moment approximation procedures [14, 15]. While the exact permutation distributions* of many statistics are well defined, an attempt to use these exact distributions to obtain P-values* is usually prohibitive because of computational requirements. Since some techniques such as multiresponse permutation procedures* [11, 12] do not possess an invariance principle [5], approximations based on known asymptotic distributions are nonexistent in such instances. Furthermore,

some statistics such as the Pearson chi-square test statistic and the likelihood ratio test* statistic for testing goodness of fit* and independence in contingency tables* possess different asymptotic distributions (e.g., normal and chi-square distributions) which depend on the limit conditions [9, 10]. Moment approximation procedures provide a useful method for obtaining P-values for these types of situations. If the first three or four exact moments of a statistic can be obtained for a specific permutation distribution, then the inferences of a moment approximation procedure are based on an approximate distribution whose lower moments match the obtained three or four exact moments. Since continuous distributions such as the Pearson type I, III, and VI distributions [8] are fully specified by either the first three or four moments, such distributions are good candidates for an approximate distribution.

For example, let Y_i be the ith of c frequencies and let $p_i > 0$ be the ith of c known proportions where $\sum_{i=1}^{c} Y_i = n$, $\sum_{i=1}^{c} p_i = 1$, and $c \geqslant 2$. Then the Pearson chi-square goodness-of-fit test statistic* given by

$$T = \sum_{i=1}^{c} \left\{ (Y_i - E_i)^2 / E_i \right\}$$

is used to test the null hypothesis (H_0) specifying that $E_i = np_i$ for $i = 1, \ldots, c$. Under H_0, the exact permutation distribution of $Y_1, \ldots, Y_c$ is given by

$$\Pr[Y_1 = y_1, \ldots, Y_c = y_c | n, p_1, \ldots, p_c]$$

$$= n! \prod_{i=1}^{c} (p_i^{y_i} / y_i!),$$

and the exact mean, variance, and skewness of T are given by

$$\mu_T = c - 1,$$

$$\sigma_T^2 = 2(c - 1) + \left[3 - (c + 1)^2\right] / n$$

$$+ \sum_{i=1}^{c} E_i^{-1},$$

and

$$\gamma_T = \left\{ 8(c-1) - 2(3c-2)(3c+8)/n \right.$$

$$+ 2(c+3)(c^2 + 6c - 4)/n^2$$

$$+ \left[22 - (3c+22)/n \right] \sum_{i=1}^{c} E_i^{-1}$$

$$\left. + \sum_{i=1}^{c} E_i^{-2} \right\} / \sigma_T^3.$$

Then the distribution of the standardized statistic

$$Z = (T - \mu_T)/\sigma_T$$

may be approximated by the Pearson type III distribution with density function given by

$$f(x) = \frac{(2/\gamma)^{4/\gamma^2}}{\Gamma(4/\gamma^2)} \left[(2 + \gamma x)/\gamma \right]^{(4-\gamma^2)/\gamma^2}$$

$$\times e^{-2(2+\gamma x)/\gamma^2},$$

where $-2/\gamma < x < \infty$ and $\gamma = \gamma_T$ [6, 8]. In particular, let T_0 denote an observed value of T. Then the P-value* [i.e., $P(T \geqslant T_0|H_0)$] is approximated by

$$\int_{Z_0}^{\infty} f(x) \, dx,$$

where $Z_0 = (T_0 - \mu_T)/\sigma_T$. In addition to being fully specified by the skewness parameter (γ), the choice of the Pearson type III distribution is also because the known limiting distributions of Z [10] are special cases [i.e., the standardized χ_m^2 distribution when $\gamma = (8/m)^{1/2}$ and the $N(0,1)$ distribution when $\gamma = 0$]. As a result, this moment approximation procedure is able to provide reasonable approximations for cases near the limiting chi-square and normal distributions of T and also for the various exact distributions of T occurring in routine applications which resemble neither a chi-square nor a normal distribution. While μ_T, σ_T^2, and γ_T are obtained in a straightforward manner, the corresponding exact moments under H_0 of the likelihood ratio* test statistic given by

$$G = 2 \sum_{i=1}^{c} Y_i \ln(Y_i/E_i)$$

are not computationally feasible except for very small data sets. Thus statistic G demonstrates an example where a moment approximation procedure is not feasible.

In cases where exact moments are not computationally feasible, moment approximation procedures are not attainable. A randomized test* (also termed a *simulation* or *Monte Carlo** test) is always feasible in such cases. However, an additional type I statistical error is introduced in a randomization test as a consequence of the inherent randomization process (i.e., the random sampling variability associated with a randomization test) required to approximate a P-value. In contrast with the P-value approximation associated with a randomized test, a moment approximation procedure's P-value does not involve an additional type I statistical error. The moment approximation procedure's P-value does however involve a technical error due to approximating a discrete distribution with a continuous distribution (i.e., analogous to a binomial distribution's approximation by a normal distribution). A major feature is that a moment approximation procedure's technical error is almost never an overwhelming error, whereas a randomization test's type I statistical error is by definition devastating (i.e., reverses the conclusion which would conceptually be obtained with an exact permutation test) some portion of the time [13]. Unpublished comparisons have been made between moment approximation procedures and randomized tests for cases where both methods are applicable and exact P-values are also attainable. These comparisons indicate that exact and moment approximation P-values are always quite close to each other while randomization test P-values are occasionally either much too large or much too small (as expected with a randomized sampling process). If a choice exists between using either a moment approximation procedure or a randomization test, then the moment approximation procedure is preferred.

Moment approximation procedure computer algorithms exist for statistical methods such as completely randomized and randomized block permutation techniques based on distance functions [1, 4, 7] and both the Pearson chi-square and Goodman–Kruskal tau* analyses for two-way contingency tables [2, 3].

References

[1] Berry, K. J. and Mielke, P. W. (1985). *Commun. Statist. Simul. Comp.*, **14**, 229–248. (Three-moment approximation procedure algorithm given for matched pairs permutation tests.)

[2] Berry, K. J. and Mielke, P. W. (1986). *Educ. Pyschol. Meas.*, **46**, 169–173. (Three-moment approximation procedure algorithm given for the Pearson chi-square statistic analysis of r by c contingency tables.)

[3] Berry, K. J. and Mielke, P. W. (1986). *Educ. Pyschol. Meas.*, **46**, 645–649. (Four-moment approximation procedure algorithm given for the Goodman–Kruskal tau statistic analysis of r by c contingency tables.)

[4] Berry, K. J., Mielke, P. W., and Wong, R. K. W. (1986). *Commun. Statist. Simul. Comp.*, **15**, 581–589. (Four-moment approximation procedure algorithm given for multiresponse permutation procedures.)

[5] Brockwell, P. J., Mielke, P. W., and Robinson, J. (1982). *Aust. J. Statist.*, **24**, 33–41. (Asymptotic noninvariance is established for multiresponse permutation procedures which satisfy the congruence principle.)

[6] Harter, H. L. (1969). *Technometrics*, **11**, 177–187. (Description and tables are given for Pearson type III distribution.)

[7] Iyer, H. K., Berry, K. J. and Mielke, P. W. (1983). *Commun. Statist. Simul. Comp.*, **12**, 479–499. (Three-moment approximation procedure algorithm given for multiresponse randomized block permutation procedures.)

[8] Johnson, N. L. and Kotz, S. (1970). *Distributions in Statistics: Continuous Univariate Distributions* —1. Wiley, New York. (Descriptions of Pearson type I, III, and VI distributions are given.)

[9] Koehler, K. J. (1986). *J. Amer. Statist. Ass.*, **81**, 483–493. (Simulations establish that neither the asymptotic chi-square nor normal distributions provide all-purpose fits for the Pearson chi-square and likelihood ratio test analyses of sparse contingency tables.)

[10] Koehler, K. J. and Larntz, K. (1980). *J. Amer. Statist. Ass.*, **75**, 336–344. (The Pearson chi-square and likelihood ratio test statistics for goodness of fit are implicitly shown to require a continuum of distributions between the chi-square and normal distributions when sparse data are involved.)

[11] Mielke, P. W. (1984). In *Handbook of Statistics*, Vol. 4, P. R. Krishnaiah and P. K. Sen, eds. North-Holland, Amsterdam, pp. 813–830. (Moment approximation procedures are described for a variety of permutation tests involving completely randomized and randomized block designs.)

[12] Mielke, P. W. (1986). *J. Statist. Plann. Inf.*, **13**, 377–387. (Median-based permutation tests which satisfy the congruence principle do not possess an asymptotic invariance principle and thus require efficient moment approximation procedures for inference purposes.)

[13] Mielke, P. W. and Medina, J. G. (1983). *J. Climate Appl. Meteor.*, **22**, 1290–1295. (Emphasizes additional type I statistical error problems with inferences based on simulation and also presents estimation methods which nevertheless require simulation-based inferences since a moment approximation procedure is not feasible.)

[14] Patnaik, P. B. (1949). *Biometrika*, **36**, 202–232. (A noncentral chi-square distribution is approximated with a central chi-square distribution by equating moments.)

[15] Wilson, E. B. and Hilferty, M. M. (1931). *Proc. Natl. Acad. Sci. U.S.A.*, **17**, 684–688. (A central chi-square distribution is approximated with a normal distribution by equating moments.)

(CONTINGENCY TABLES
GOODMAN–KRUSKAL TAU AND GAMMA
GOODNESS OF FIT
MULTIRESPONSE PERMUTATION
  PROCEDURES
RANDOMIZED TESTS)

PAUL W. MIELKE, JR.

MORAN STATISTIC *See* CLIFF–ORD TEST: SPATIAL PROCESSES

N

NEAREST-POINT–NEAREST-NEIGHBOR SAMPLING

A method of sampling spatial distributions of hidden or unmapped (point) items. The measurements produced provide information both about the probability mechanism generating the locations of the items and about the density of the items. Unlike quadrant sampling*, which requires prior decisions about the approximate density of the items, the distances used in nearest-point–nearest-neighbor sampling are determined by the population sampled.

In this sampling scheme, sampling sites $O_1, O_2, \ldots, O_n$, are chosen randomly in the region of interest, for example by choosing exact coordinates on a map of the region. Each sampling site is then visited, and the location of the nearest item to that sampling site is then found, say at P_i for sampling site O_i. Two variants of nearest-point–nearest-neighbor—T-square sampling and wandering-quarter sampling*—can be used to obtain multivariate distance measurements with distributional properties which are computa-

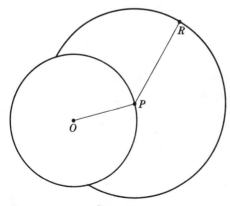

Figure 1

tionally simpler than those arising with nearest-point–nearest-neighbor sampling.

Nearest-point–nearest-neighbor sampling, T-square sampling, and wandering quarter sampling are alternatives to kth nearest-point or kth nearest-neighbor* sampling.

Bibliography

Diggle, P. J. (1983). *Statistical Analysis of Spatial Point Processes*. Academic, London.

Ripley, B. D. (1981). *Spatial Statistics*. Wiley, New York.

(kth NEAREST-POINT—kth NEAREST NEIGHBOR SAMPLING
SPATIAL PROCESSES
SPATIAL SAMPLING
T-SQUARE SAMPLING
WANDERING-QUARTER SAMPLING)

W. G. S. HINES

NORMALIZATION CONSTANT

If the distribution of a variable X is given in the form

$$\Pr[X = x_j] = k_1 g_j, \qquad g_j > 0,$$

or if it has a probability density function of the form

$$k_2 f(x), \qquad f \geq 0,$$

the constants k_1, k_2 must be such that

$$\sum_j k_1 g_j = 1 \quad \text{or} \quad \int_{-\infty}^{\infty} k_2 f(x)\, dx = 1,$$

that is,

$$k_1 = \left(\sum_j g_j \right)^{-1} \quad \text{or} \quad k_2 = \left(\int_{-\infty}^{\infty} f(x)\, dx \right)^{-1}$$

as the case may be.

k_1 and k_2 are called *normalization constants*.

O

OPTIMUM DESIGN OF EXPERIMENTS

INTRODUCTION

In a linear regression* model with response Y, the expected value of Y is given by $E(Y) = \beta_0 + \beta_1 x$. If the errors of observation are additive with constant variance σ^2, the variance of the least-squares* estimate of the slope parameter β_1 from a sample of N observations (x_i, y_i) is $\text{var}(\hat{\beta}_1) = \sigma^2/\sum_{i=1}^{N}(x_i - \bar{x})^2$, where $\bar{x} = \sum_{i=1}^{N} x_i/N$. A simple problem in the design of experiments is the choice of the N values of x_i to minimize this variance. The values at which it is possible to experiment form a *design region* $\mathcal{X}$. For single-variable regression this will be a set of values on the real line, which can be scaled to lie between -1 and $+1$. The variance of $\hat{\beta}_1$ is minimized by the design which puts $N/2$ of the trials at the highest value of x and $N/2$ at the lowest, provided N is even. For this design the variance of the predicted response $\text{var}(\hat{y}(x)) = \text{var}(\hat{\beta}_0 + \hat{\beta}_1 x)$ has its maximum value of $2\sigma^2/N$ at the design points. This is the smallest maximum that can be achieved over $\mathcal{X}$ for any design. For all other designs the maximum variance will be larger.

This simple example illustrates many of the features of optimum design theory which is concerned with the extension of these ideas to the case when β is a vector of k parameters. These features include the dependence of the design on an assumed model and design region $\mathcal{X}$ and the equivalence of one criterion, here minimizing the variance of the parameter estimate, to another seemingly unrelated one, here minimizing the maxi-

mum variance of prediction. This result is an example of the general equivalence theorem stated in the next section. Another property is that the optimum design in the example is concentrated at two sets of experimental conditions—often only k distinct sets are needed. In the general case, the maximum variance of prediction with the optimum design is equal to $k\sigma^2/n$, which maximum occurs at the points of the experimental design.

When several parameters are of interest the simple variance of $\hat{\beta}_1$ in our example is replaced by the covariance matrix of the vector $\hat{\beta}$. Many functionals of the matrix have been used to define optimality criteria, which are known by alphabetic names. We describe A-, D-, E-, and G-optimality in detail. In addition some further criteria are defined. The theory of these designs is associated with Kiefer. A survey of Kiefer's contribution to optimum experimental design is given by Wynn [57]. Kiefer's collected papers on design (Brown et al. [14]) cover both designs for regression models and classical experimental designs where interest is in the use of blocking to improve the precision of the estimation of contrasts in unstructured treatments.

References to the early history of the subject are given in Kiefer [30], particularly Sect. 4. The earliest work cited is that of Smith [51] who, in a remarkable 85-page paper, gave optimum designs for single-factor polynomials up to the sixth degree. Historically important papers which appeared after Kiefer [30] include Karlin and Studden [29] and Whittle [55]. Nalimov et al. [41] give a survey of applications of the theory which provides references to Russian work on optimum experimental design.

The theory of the optimum design of experiments provides not only a systematization of the theory of experimental designs but also a method for the construction of designs in nonstandard situations through algorithms based on the General Equivalence Theorem. The review papers of Atkinson [2] and Steinberg and Hunter [52] provide a context for the contribution of the theory to the development of experimental design.

CONVEX DESIGN THEORY

In its original form (Kiefer [30]) optimum design theory is concerned with the linear model $E(Y) = \beta^T f(x)$. The errors in the observations are independent with constant variance σ^2. The k elements of the parameter vector β are estimated by least squares from the results of N trials and have variance

$$\text{Var}(\hat{\beta}) = \sigma^2 \left\{ \sum_{i=1}^{N} f(x_i) f(x_i)^T \right\}^{-1},$$

$$i = 1, \ldots, N.$$

It is here assumed that the $k \times k$ information matrix $\sum_{i=1}^{N} f(x_i) f(x_i)^T$ is of full rank. The fitted value at x is

$$\hat{y}(x) = \hat{\beta}^T f(x)$$

with

$$\text{var}\{\hat{y}(x)\}$$

$$= \sigma^2 f(x)^T \left\{ \sum_{i=1}^{N} f(x_i) f(x_i)^T \right\}^{-1} f(x).$$

The standard version of the optimum design of experiments is concerned with the choice of x_i, $i = 1, \ldots, N$, to minimize various functions of $\text{Var}(\hat{\beta})$ or $\text{var}\{\hat{y}(x)\}$.

Both in the theory and in the construction of optimum experimental designs it is convenient to replace the N trial design by a measure ξ over the design region $\mathscr{X}$. For an exact N trial design, that is, one for which the weights at the experimental points are integer multiples of $1/N$, the measure is

denoted ξ_N. If the integer restriction is removed the design is denoted ξ and is referred to as a continuous or approximate design, since it may not be exactly realizable in practice. The information matrix of the approximate design is defined as

$$M(\xi) = \int f(x) f(x)^T \xi(dx).$$

For an exact design evidently

$$M(\xi_N) = N^{-1} \sum_{i=1}^{N} f(x_i) f(x_i)^T.$$

It is also convenient, instead of $\text{Var}(\hat{\beta})$ and $\text{var}\{\hat{y}(x)\}$, to consider their standardized analogs $M^{-1}(\xi)$ and $d(x, \xi) = f(x)^T M^{-1}(\xi) f(x)$. An advantage is the replacement of functions of integers by functions of continuous variables.

One design criterion which has been much studied is that of *D-optimality* in which the determinant $\det M(\xi)$ is maximized. This minimizes the generalized variance* of the parameter estimates. That is, if the errors in the observations are normally distributed, the D-optimum design minimizes the volume of a fixed level confidence region* for β.

Another criterion, *G-optimality*, is concerned with the variance of the predicted response. If we let

$$\bar{d}(\xi) = \max_{x \in \mathscr{X}} d(x, \xi),$$

a G-optimum design is one which minimizes $\bar{d}(\xi)$, that is, it minimizes the maximum variance of the predicted response over the design region. As was implied in the Introduction, these two optimum designs are identical. More formally, the celebrated General Equivalence Theorem (Kiefer and Wolfowitz [32]) states the equivalence of the following three requirements on the optimum measure ξ^* (*see also* WHITTLE EQUIVALENCE THEOREM):

(i) ξ^* maximizes $\det M(\xi)$,

(ii) ξ^* minimizes $\bar{d}(\xi)$, and

(iii) $\bar{d}(\xi^*) = k$, the number of linearly independent parameters in the model.

This result establishes the equivalence of the

D and G criteria in the approximate theory for constant σ^2. However, for exact designs there may be values of N for which the two optimum designs are not identical.

Two further criteria which have a statistical interpretation in terms of the information matrix $\mathbf{M}(\xi)$ are A- and E-optimality. Both have been much employed in the construction of block designs. In A-optimality $\text{tr}\{\mathbf{M}^{-1}(\xi)\}$, the average variance of the parameter estimates, is minimized. In E-optimality the variance of the least well-estimated contrast $\mathbf{a}^T\beta$ is minimized, subject to the constraint $\mathbf{a}^T\mathbf{a} = 1$. This is equivalent to minimizing the maximum eigenvalue of $\mathbf{M}^{-1}(\xi)$. In D-optimality the product of the eigenvalues is minimized. In terms of the eigenvalues $\lambda_1, \ldots, \lambda_k$ of $\mathbf{M}(\xi)$, the three criteria are:

A Minimize the sum of the variances of the parameter estimates: $\min \sum_{i=1}^{k}(1/\lambda_i)$.

D Minimize the generalized variance of the parameter estimates: $\min \prod_{i=1}^{k}(1/\lambda_i)$.

E Minimize the variance of the least well-estimated contrast $\mathbf{a}^T\beta$ with $\mathbf{a}^T\mathbf{a} = 1$: $\min \max_i (1/\lambda_i)$.

All three may be regarded as special cases of the more general criterion of choosing designs to minimize

$$\Phi_p(\xi) = \left\{ k^{-1} \sum_{i=1}^{k} \lambda_i^p \right\}^{1/p},$$

$$0 < p < \infty.$$

For A-, D-, and E-optimality the values of p are, respectively, 1, 0, and ∞ when the limiting operations are properly defined. Kiefer [31] uses this family to study the variation in structure of the optimum design as the optimality criterion changes in a smooth way.

There are several useful extensions to D-optimality. If interest is not in all k parameters but only in s linear combinations of β which are the elements of $\mathbf{A}^T\beta$ with $s < k$, the analog of D-optimality is generalized

D-optimality or D_A-*optimality* (Sibson [49]) in which $\log \det \mathbf{A}^T\mathbf{M}^{-1}(\xi)A$ is minimized. The analog of the variance function $d(\mathbf{x}, \xi)$ is

$$d_A(\mathbf{x}, \xi) = \mathbf{x}^T\mathbf{M}^{-1}(\xi)\mathbf{A}\{\mathbf{A}^T\mathbf{M}^{-1}(\xi)\mathbf{A}\}^{-1}$$

$$\times \mathbf{A}^T\mathbf{M}^{-1}(\xi)\mathbf{x}.$$

A special case of D_A-optimality is D_s-optimality in which $\mathbf{A}^T = \mathbf{I}_s(\neq \mathbf{O})$ so that designs are sought for which a subset of the parameters is estimated as precisely as possible, the remaining parameters being treated as nuisance parameters*.

A different extension of D-optimality was used by Atkinson and Cox [4] to generate designs which give reasonably precise information on the parameters of several potential models. The extension of the criterion to generalized D-optimality is that, for nonnegative weights w_i, designs are sought to minimize

$$\sum w_i \log \det \mathbf{A}_i^T\mathbf{M}_i^{-1}(\xi)\mathbf{A}_i.$$

This criterion, called S-*optimality* by Läuter [36], makes possible the design of experiments for the simultaneous estimation of parameters in a variety of models whilst estimating subsets for discrimination.

In addition to extensions to D-optimality, extensions can also be found to variance-based criteria. Instead of G-optimality in which the maximum value of $d(\mathbf{x}, \xi)$ is of interest, designs can be found to minimize a weighted average of the variance, either over $\mathscr{X}$ or over some other region of interest (Welch [53]). This criterion has been called Q-*optimality*. Finally, we would not wish to leave this survey of alphabetical criteria without mention of T-*optimum designs* (Atkinson and Fedorov [5, 6]) which are used for discriminating between two or more rival models. The criterion applies whether the models are separate or partially nested, so that some are special cases of others.

All the foregoing criteria can be formulated as special cases of the following optimization problem which provides a general equivalence theorem for each criterion. For

Table 1   Functions Appearing in the General Equivalence Theorem for a Variety of Optimality Criteria

Criterion	Ψ	ϕ	Δ
D	$\log \det \mathbf{M}^{-1}$	$\mathbf{f}(\mathbf{x})^T \mathbf{M}^{-1} \mathbf{f}(\mathbf{x})$	k
Linear	$\operatorname{tr} \mathbf{A} \mathbf{M}^{-1}$	$\mathbf{f}(\mathbf{x})^T \mathbf{M}^{-1} \mathbf{A} \mathbf{M}^{-1} \mathbf{f}(\mathbf{x})$	$\operatorname{tr} \mathbf{A} \mathbf{M}^{-1}$
E	$\min \lambda_\alpha(\mathbf{M}) = \lambda_{\min}$	$\mathbf{f}(\mathbf{x})^T \mathbf{r} \mathbf{r}^T \mathbf{f}(\mathbf{x})$ [a]	$\lambda_{\min}$
Generalized D	$\log \det \mathbf{A}^T \mathbf{M}^{-1} \mathbf{A}$	$\mathbf{f}(\mathbf{x})^T \mathbf{M}^{-1} \mathbf{A} \{\mathbf{A}^T \mathbf{M}^{-1} \mathbf{A}\}^{-1} \mathbf{A}^T \mathbf{M}^{-1} \mathbf{f}(\mathbf{x})$	$s = \operatorname{rank} \mathbf{A}$
Generalized G	$\max_{\mathbf{x} \in Z} w(\mathbf{x}) d(\mathbf{x}, \xi) = C(\xi)$	$\mathbf{f}(\mathbf{x})^T \mathbf{M}^{-1} \int_Z w(\mathbf{x}) \mathbf{f}(\mathbf{x}) \mathbf{f}(\mathbf{x})^T \, d\mathbf{x} \, \mathbf{M}^{-1} \mathbf{f}(\mathbf{x})$	$C(\xi)$
q	$q^{-1} \operatorname{tr} \mathbf{M}^{-q}$	$\mathbf{f}(\mathbf{x})^T \mathbf{M}^{-q-1} \mathbf{f}(\mathbf{x})$	$\operatorname{tr} \mathbf{M}^{-q}$

[a] r is the eigenvector for $\lambda_{\min}$.

some measure of the imprecision of estimation, Ψ, let

$$\xi^* = \operatorname*{Arg\,min}_\xi \Psi\{\mathbf{M}(\xi)\}, \qquad (1)$$

that is, ξ^* is the value of the argument ξ for which the minimum occurs. Examples of several important forms for Ψ are given in Table 1. Under very mild assumptions, the most important of which are the compactness of $\mathscr{X}$ and the convexity and differentiability of Ψ, the General Equivalence Theorem states:

(i) ξ^* minimizes $\Psi\{\mathbf{M}(\xi)\}$,

(ii) ξ^* minimizes $\phi(\xi) - \Delta(\xi)$, and

(iii) $\phi(\xi^*) - \Delta(\xi^*) = 0$,

where

$$\phi(\xi) = \max_{\mathbf{x} \in \mathscr{X}} \phi(\mathbf{x}, \xi),$$

$$\phi(\mathbf{x}, \xi) = -\mathbf{f}^T(\mathbf{x}) \frac{\partial \Psi}{\partial \mathbf{M}} \mathbf{f}(\mathbf{x}),$$

and

$$\Delta(x) = -\operatorname{tr} \mathbf{M} \frac{\partial \Psi}{\partial \mathbf{M}}.$$

Table 1 gives a list of $\phi(\xi)$ and $\Delta(\xi)$ for the most widely used optimality criteria. In this table it is assumed that $\mathbf{M}(\xi^*)$ is regular, that is that $\mathbf{M}^{-1}(\xi)$ exists and that, for minimax criteria, the extremal point is unique. Further details are given in Silvey [50] and Fedorov [24]. In practice regularization can be achieved, for instance by the following

perturbation of (1):

$$\xi_\gamma^* = \operatorname*{Arg\,min}_\xi \Psi\{(1 - \gamma)\mathbf{M}(\xi) + \gamma \mathbf{M}(\bar{\xi})\},$$

where $\det \mathbf{M}(\bar{\xi}) = \delta > 0$.

Various further generalizations of the equivalence theorem can be found in Mehra and Lainiotis [37] and Pazman [43, 44]. Pukelsheim and Titterington [46] use results from the differential theory of convex analysis. One consequence is an equivalence theorem for the optimality of singular designs.

NUMERICAL METHODS

The General Equivalence Theorem, in addition to providing a test for the optimality of a proposed design by use of (iii), also provides a general algorithm for the construction of continuous designs. In the case of D-optimality the simplest algorithm can be thought of as the sequential construction of a design by the addition of successive trials at the point where the variance $d(\mathbf{x}, \xi)$ is a maximum. More formally, for an N trial design, let

$$\bar{\mathbf{x}}_N = \operatorname*{Arg\,max}_{\mathbf{x} \in \mathscr{X}} d(\mathbf{x}, \xi_N)$$

and let $\bar{\xi}_N$ be a measure putting unit weight at $\bar{\mathbf{x}}_N$. Then

$$\xi_{N+1} = (N\xi_N + \bar{\xi}_N)/(N + 1).$$

As a simple example consider construction of the D-optimum design for the first-order model with which this entry opened. Suppose that the design region $\mathscr{X}$ is the interval

$[-1, 1]$ and that $N = 3$ with ξ_3 putting weight $\frac{1}{3}$ at x values -1, $\frac{1}{2}$, and 1. Then $d(x, \xi_3) = 3(9 - 4x + 12x^2)/26$ which is a maximum over $\mathcal{X}$ at $x_3 = -1$. The measure ξ_4 given by the algorithm then puts weight $\frac{1}{2}$ at $x = -1$ and $\frac{1}{4}$ at $x = \frac{1}{2}$ and 1. Continuation of this procedure sufficiently far leads arbitrarily close to the design with weight $\frac{1}{2}$ at -1 and $\frac{1}{2}$ at 1.

This first-order, or gradient type, algorithm for the construction of continuous designs can be extended to the general criterion $\Psi\{M(\xi)\}$. The family of algorithms is then to put

$$\xi_{s+1} = (1 - \alpha_s)\xi_s + \alpha_s\bar{\xi}_s. \qquad (2)$$

In regular cases which fall within the framework of the general equivalence theorem given above, $\bar{\xi}_s$ can be chosen from measures concentrated at one point. This leads to a great simplification in numerical procedures. Specifications of (2) can be found in Atwood [7], Fedorov [21, 22, 24], and in Wu and Wynn [56]. Second-order numerical procedures (Atwood [7] and Fedorov [24]) are appreciably more complicated and are not widely used in practice.

Use of an optimum design expressed as a measure ξ^* requires the formation of an exact design ξ_N which specifies the conditions for each of the N trials. If N is relatively large compared with k, a near optimum design may be found from an integer approximation to $N\xi^*$. But for small N such designs may be far in performance from the exact optimum design. Several methods have been proposed for the construction of discrete designs which reduce the problem to a series of searches similar to those for continuous designs. The most widely used for D-optimum designs, due to Mitchell [38, 39], is called DETMAX. Out of a list of n possible sets of conditions, N not necessarily distinct points are to be chosen. Suppose at some stage the trial design contains N' points. A point is added if it causes the determinant of the information matrix to increase by as much as possible, and is deleted if it reduces the determinant by as little as possible, these quantities being determined from $d(\mathbf{x}_{N'}, \xi)$.

This and more complicated algorithms provide powerful methods of finding near optimum exact designs (Mitchell [39] and Mitchell and Bayne [40]). Modifications to, and improved computer implementation of, the method are discussed by Galil and Kiefer [25] who provide a table of designs for quadratic regression with a cuboidal experimental region. Cook and Nachtsheim [17] compare several related algorithms including a two-step exchange algorithm due to Fedorov [21] in which there is simultaneous search for the best point to add and delete. Further developments and comparisons with earlier methods are given by Johnson and Nachtsheim [28].

STOCHASTIC PROCESSES*

So far it has been assumed that the errors of observation are independent. If instead the errors evolve as a stochastic process in time, the design problem becomes appreciably more complicated. We can distinguish three main approaches.

In the first approach, it is required to optimize experiments on a single realization of the stochastic process $Y = Y(\mathbf{x}, t)$. This problem lies outside the scope of the methods described in this article and is traditionally treated under "optimal control theory" (Eykhoff [20] and Åström [1]). Typical results for the second approach, which is related to the optimum storage of information, can be found in Sacks and Ylvisaker [48], who use an asymptotic theory in which the number of points in a fixed time interval increases.

In the third approach, repeated realizations of $Y(\mathbf{x}, t)$ are allowed and optimum designs can be described, as they have been in this article, by a measure on $\mathcal{X}$, where now $\mathcal{X}$ can, for example, belong to an appropriate Hilbert space. The basic ideas are, however, the same as those of traditional convex design theory (Fedorov [23], Pazman [43], Goodwin and Payne [26], and Mehra and Lainiotis [37]).

DESIGNS FOR NONLINEAR MODELS

For a nonlinear regression model the optimum design depends upon the unknown values of the parameters θ. Although there is an element of paradox here, since the purpose of the experiment is to estimate the parameters, the usual procedure is to expand the model in a Taylor series about a preliminary estimate θ_0. D-optimum designs are then sought (Box and Lucas [13] and Klepikov and Sokolov [33]).

Because the design depends upon the parameters, a sequential scheme with allowance for updating the parameters is sensible. If a sequential design is not possible, one obvious strategy is to design the whole experiment using θ_0. Another strategy is the maximin, in which the best design is sought for the worst possible value of θ (Silvey [50, p. 59], Fedorov [23], and Ermakov [19, Chap. 5]). This approach has the disadvantage that the worst value of θ may give rise to such an extreme design that the strategy is unduly defensive.

A less pessimistic view of the world is reflected in the use of prior information about θ. The design criterion can be integrated over the prior distribution of θ (Läuter [35] and Fedorov [24]).

SOME NONSTANDARD APPLICATIONS

Classical procedures for the construction of incomplete block designs* use combinatorial methods. However, for nonstandard cases, designs are often not available or, if they are available, the designs need to be accessed and constructed, perhaps on a computer. Optimum design theory is useful in directing attention toward statistically important features of a design, rather than toward those of combinatorial interest.

An example is the use by John and Mitchell [27] of the DETMAX algorithm to construct optimum block designs. One finding was that the designs had a particularly simple structure when interpreted in terms of graph theory*. Two developments followed from this discovery. One was the construction of optimum designs from regular graphs (Cheng and Wu [16]). The other (Paterson [42]) uses the counts of circuits in the graphs of designs as a surrogate for D-optimality. Candidate designs are generated in a simple manner, for example as special kinds of lattice designs*. The counting of circuits is a fast way of assessing optimality, so that many designs can be compared. The results of this construction are in use as designs for statutory variety trials.

A second application of optimum design theory is to the design of clinical trials in which patients arrive sequentially and are to be assigned to one of several treatments. Information about each patient may include a number of prognostic factors across which the trial should be balanced. With interest in a set of contrasts $A^T\beta$, a sequential D_A-optimum design could be used with the next allocation being that for which $d_A(x, \xi_N)$ was a maximum. Here $\mathscr{X}$ consists of the set of treatments. But, as Efron [18] argued in the development of biased-coin designs, some randomness in the allocation of treatments may be desirable to avoid biases and the suspicion of conscious or unconscious cheating. Atkinson [3] suggests a design of the biased-coin type in which the probability of allocation of the treatments is proportional to $d_A(x, \xi_N)$. The solution provided by optimum design theory to this problem is thus derived from the underlying statistical model.

As a last example of the application of optimum design theory we turn to response surface designs*. Box and Draper [11, 12] stress the importance of designs which allow not only for the variance of the predicted response, as in G-optimality, but also for the bias which arises when the fitted model is viewed as a smooth approximation to an unknown true model. In this case, the approach differs from that of optimum design theory which assumes, in the case of the generalized criterion, that at least one of the models involved is true. A disadvantage of the method of Box and Draper is that it cannot provide designs for arbitrary design regions and arbitrary numbers of trials. Welch [54] extends the original idea of Box and Draper to include a general form of departure from the fitted polynomial. This is subject to an upper bound z_{max} on the magnitude of the departure. The mean squared error of prediction is averaged analytically over all departures to give a criterion to which the procedures of optimum design theory apply. Approximate designs are found by algorithms similar to those for approximate D-optimum designs. For exact designs a variant of Mitchell's DETMAX is employed. Additional results are given in Chap. 6 of Ermakov [19].

LITERATURE

Reference has already been made in this article to several books on the optimum design of experiments. Of these, two are in English, the pioneering book of Fedorov [21] and the succinct account of Silvey [50] which concentrates on the central aspects of the theory covered in Sect. 2 of this article. Several books on the subject have been published in the German Democratic Republic. The introduction provided by Bandemer et al. [9] can be contrasted with the encyclopaedic survey of the theory given by Bandemer et al. [8] and, of applications, by Bandemer and Näther [10]. Other recent books from central Europe include Pilz [45] and Pazman [43, 44]. Of the many books on the subject from the USSR mention has already been made of Ermakov [19]. It is noteworthy that the many books in English on experimental design ignore the development of optimum design. There is thus no equivalent to, for example, Rasch and Herrendörfer [47] in which both classical and optimum experimental design are developed, as they should be, in tandem. Two books, originating from West Germany, which include the optimum design of experiments in the discussion of the linear model are, in English, Bunke and Bunke [15] and, in German, Krafft [34].

References

[1] Åström, K. J. (1970). *Introduction to Stochastic Control Theory*. Academic, New York.

[2] Atkinson, A. C. (1982a). *Int. Statist. Rev.*, **50**, 161–177.

[3] Atkinson, A. C. (1982b). *Biometrika*, **69**, 61–67.

[4] Atkinson, A. C. and Cox, D. R. (1974). *J. R. Statist. Soc. B*, **36**, 321–348.

[5] Atkinson, A. C. and Fedorov, V. V. (1975a). *Biometrika*, **62**, 57–70.

[6] Atkinson, A. C. and Fedorov, V. V. (1975b). *Biometrika*, **62**, 289–303.

[7] Atwood, C. L. (1976). *Proc. 1976 Conf. Inf. Sci. Systems*. Johns Hopkins, Baltimore.

[8] Bandemer, H., Bellmann, A., Jung, W., Le Anh Son, Nagel, S., Nagel, W., Näther, W., Pilz, J., and Richter, K. (1977). *Theorie und Anwendung der Optimalen Versuchsplanung I, Handbuch zur Theorie*. Akademie Verlag, Berlin.

[9] Bandemer, H., Bellmann, A., Jung, W., and Richter, K. (1976). *Optimale Versuchsplanung*, 2nd ed. Akademie Verlag, Berlin.

[10] Bandemer, H. and Näther, W. (1980). *Theorie und Anwendung der Optimalen Versuchsplanung II, Handbuch zur Anwendung*. Akademie Verlag, Berlin.

[11] Box, G. E. P. and Draper, N. R. (1959). *J. Amer. Statist. Ass.*, **54**, 622–654.

[12] Box, G. E. P. and Draper, N. R. (1963). *Biometrika*, **50**, 335–352.

[13] Box, G. E. P. and Lucas, H. L. (1959). *Biometrika*, **46**, 77–90.

[14] Brown, L. D., Olkin, I., Sacks, J., and Wynn, H. P., eds. (1985). *Jack Carl Kiefer, Collected Papers III: Design of Experiments*. Springer, New York.

[15] Bunke, O. and Bunke, H., eds. (1986). *Statistical Inference in Linear Models*, Vol. 1. Wiley, New York.

[16] Cheng, C.-S. and Wu, C.-F. J. (1981). *Biometrika*, **68**, 493–500.

[17] Cook, R. D. and Nachtsheim, C. J. (1980). *Technometrics*, **22**, 315–324.

[18] Efron, B. (1971). *Biometrika*, **58**, 403–417.

[19] Ermakov, S. M., ed. (1983). *Mathematical Theory of Experimental Design*. Nauka, Moscow.

[20] Eykhoff, P. (1976). *System Parameter and State Estimation*. Wiley, London.

[21] Fedorov, V. V. (1972). *Theory of Optimal Experiments*. Academic, New York.

[22] Fedorov, V. V. (1975). *Numerical Aspects of the Least Squares Method in the Analysis and Design of Regression Experiments*. Moscow, State University, Moscow, USSR.

[23] Fedorov, V. V. (1980). *Math. Oper. u. Statist., Ser. Statist.*, **11**, 403–413.

[24] Fedorov, V. V. (1981). *Mathematical Methods in Experimental Design*, V. Penenko, ed. Nauka, Novosibirsk, USSR, pp. 19–73.

[25] Galil, Z. and Kiefer, J. (1980). *Technometrics*, **22**, 301–313.

[26] Goodwin, G. C. and Payne, R. L. (1977). *Dynamic System Identification*. Academic, New York.

[27] John, J. A. and Mitchell, T. J. (1977). *J. R. Statist. Soc. B*, **39**, 39–43.

[28] Johnson, M. E. and Nachtsheim, C. J. (1983). *Technometrics*, **25**, 271–277.

[29] Karlin, S. and Studden, W. J. (1966). *Ann. Math. Statist.*, **37**, 783–815.

[30] Kiefer, J. (1959). *J. R. Statist. Soc. B*, **21**, 272–319.

[31] Kiefer, J. (1975). *Biometrika*, **62**, 277–288.

[32] Kiefer, J. and Wolfowitz, J. (1960). *Canad. J. Math.*, **12**, 363–366.

[33] Klepikov, N. P. and Sokolov, S. N. (1964). *Analysis and Design of Experiments by the Method of Maximum Likelihood*. Fizmatgiz, Moscow, USSR.

[34] Krafft, O. (1978). *Linear Statistical Models and Optimal Experimental Design* (in German). Vandenhoeck and Ruprecht, Göttingen, German Federal Republic.

[35] Läuter, E. (1974). *Math. Oper. u. Statist.*, **5**, 379–398.

[36] Läuter, E. (1976). *Math. Oper. u. Statist.*, **7**, 51–68.

[37] Mehra, R. K. and Lainiotis, D. G. (1976). *System Identification: Advances and Case Studies*. Academic, New York.

[38] Mitchell, T. J. (1974a). *Technometrics*, **16**, 203–210.

[39] Mitchell, T. J. (1974b). *Technometrics*, **16**, 211–220.

[40] Mitchell, T. J. and Bayne, C. K. (1978). *Technometrics*, **20**, 369–383.

[41] Nalimov, V. V., Golikova, T. I., and Granovsky, Y. V. (1985). In *A Celebration of Statistics: the ISI Centenary Volume*, A. C. Atkinson and S. E. Fienberg, eds. Springer, New York, pp. 475–496.

[42] Paterson, L. (1983). *Biometrika*, **70**, 215–225.

[43] Pazman, A. (1980). *Zaklady Optimalizacii Experimentu*. Veda, Bratislava, Czechoslovakia.

[44] Pazman, A. (1986). *Foundations of Optimum Experimental Design*. Reidel, Dordrecht, The Netherlands.

[45] Pilz, J. (1983). *Bayesian Estimation and Experimental Design in Linear Regression Models*. Teubner, Leipzig, German Democratic Republic.

[46] Pukelsheim, F. and Titterington, D. M. (1983). *Ann. Statist.*, **11**, 1060–1068.

[47] Rasch, D. and Herrendörfer, G. (1982). *Statistische Versuchsplanung*. Deutscher Verlag der Wissenschaften, Berlin.

[48] Sacks, J. and Ylvisaker, D. (1966). *Ann. Math. Statist.*, **34**, 66–89.

[49] Sibson, R. (1974). In *Progress in Statistics*, Proc. 9th European Meeting of Statisticians, Budapest, 1972, Vol. 2, J. Gani, K. Sarkadi, and I. Vincze, eds. North-Holland, Amsterdam, Netherlands.

[50] Silvey, S. D. (1980). *Optimal Design: an Introduction to the Theory for Parameter Estimation*. Chapman and Hall, London, England.

[51] Smith, K. (1918). *Biometrika*, **12**, 1–85.

[52] Steinberg, D. M. and Hunter, W. G. (1984). *Technometrics*, **26**, 71–130.

[53] Welch, W. J. (1982). *Technometrics*, **24**, 41–48.

[54] Welch, W. J. (1983). *Biometrika*, **70**, 205–213.

[55] Whittle, P. (1973). *J. R. Statist. Soc. B*, **35**, 123–130.

[56] Wu, C.-F. J. and Wynn, H. P. (1978). *Ann. Statist.*, **6**, 1273–1285.

[57] Wynn, H. P. (1984). *Ann. Statist.*, **12**, 416–423.

(ANALYSIS OF VARIANCE
DESIGN OF EXPERIMENTS
OPTIMAL STOCHASTIC CONTROL
RESPONSE SURFACE DESIGNS
RIDGE REGRESSION
ROBUSTNESS IN EXPERIMENTAL DESIGN
ROTATABLE DESIGNS
WHITTLE EQUIVALENCE THEOREM)

A. C. ATKINSON
V. V. FEDOROV

O-STATISTICS

These are weighted averages of order statistics*. The *O*-statistics of order k from a set of ordered values $X_{1:n} \leqslant X_{2:n} \leqslant \cdots \leqslant X_{n:n}$ are

$$O_{r:k|n} = \sum_{j=r}^{n-k+r} w_{j:r:k|n} X_{j:n},$$

$$r = 1, 2, \ldots, k.$$

The weights

$$w_{j:r:k|n} = \binom{j-1}{r-1}\binom{n-j}{k-r} \Big/ \binom{n}{k}$$

are such that $O_{r:k|n}$ is the arithmetic mean of the rth order statistics in all possible

$\binom{n}{k}$-subsets of size k from $X_{1:n}, X_{2:n}, \ldots,$ $X_{n:n}$. In particular,

$$O_{r:n|n} = X_{r:n}$$

and $O_{1:n|n}$ is the arithmetic mean of the X's.

If the X's are a random sample from a fixed population, the O-statistics are a special class of U-statistics*, and their properties may be studied by methods appropriate to the letter. This was done by Takahasi [4] who considered using O-statistics to estimate parameters associated with the distribution of a particular order statistic $X_{r:n}$. Kaigh and Lachenbruch [3] introduced them as "generalized quantile estimators" and proposed using $O_{[(k+1)p]:k|n}$ as an estimator of the $100p\%$ population quantile, ξ_p, of a continuous random variable X, where

$$\Pr\left[X \leqslant \xi_p \right] = p.$$

([a] denotes "integer part of a.")

They found that although the bias of $O_{[(k+1)p]:k|n}$ was rather greater than that of $X_{[(n+1)p]:n}$ the variance was often so much smaller, that the mean square error, also, was smaller.

Kaigh [1] describes application of the jackknife* to generalized quantile estimators.

Kaigh and Driscoll [2] have described the use of O-statistics in graphical representations, exploiting their smoothing properties. In particular, they are useful in facilitating comparisons among samples of different sizes.

References

[1] Kaigh, W. D. (1983). *Commun. Statist. Theor.-Meth.*, **12**, 2427–2443.

[2] Kaigh, W. D. and Driscoll, M. F. (1987). *Amer. Statist.*, **41**, 25–32.

[3] Kaigh, W. D. and Lachenbruch, P. A. (1982). *Commun. Statist. Theor.-Meth.*, **11**, 2247–2238.

[4] Takahasi, K. (1971). *Ann. Inst. Statist. Math. (Tokyo)*, **22**, 403–412.

(*L*-STATISTICS
ORDER STATISTICS
QUANTILES
RANKED SET SAMPLING
U-STATISTICS)

OSTROGORSKI PARADOX

In a referendum a society decides on a proposal by majority yes/no vote. In a set of referenda it may happen that a majority of the people disagree with the society's majority decision in a majority of the referenda. This phenomenon, which Wagner [6] termed *Anscombe's paradox* (henceforth *A-paradox*) after Anscombe [1], underlies the *Ostrogorski paradox* (henceforth *O-paradox*) of which a simple example goes as follows.

Let the persons $1, \ldots, 5$ and the political parties p, q have the following positions on the issues α, β, γ on which two positions, 1 and -1, are possible:

	p	q	1	2	3	4	5
α	1	-1	1	1	1	-1	-1
β	1	-1	1	1	-1	1	-1
γ	1	-1	1	1	-1	-1	1

$$(1)$$

Let $1, \ldots, 5$ have an election on p and q, and assume that each person chooses the party that agrees with him or her on most issues. Then 1 and 2 vote for p while 3, 4, and 5 vote for q so that q wins by majority rule. Yet, the losing party p shares the views of a majority of $1, \ldots, 5$ on every issue. Indeed, if the election had been organized by referendum, p would have won all three referenda and, hence, the election. This reversal of the society's decision was baptized *O-paradox* by Rae and Daudt [5] as a tribute to Ostrogorski (1854–1919), a Russian political scientist who "...recommended replacing general parties by single-issue ones..." [4].

Since p and q have opposite views in α, β, γ the positions of $1, \ldots, 5$ in (1) constitute person-by-issue choices of either p or q. The above procedures having the opposed outcomes p and q are both two-stage procedures: The votes in (1) are either first amalgamated over *i*ssues and, then, over *p*ersons (procedure IP) or, vice versa, first over *persons* and, then, over *issues* (procedure PI). Writing the right-hand matrix in (1) as choices for either p or q, IP and PI appear as

$$
\begin{array}{c}
\text{IP}\\
\begin{array}{ccccc}
1 & 2 & 3 & 4 & 5\\
p & p & p & q & q\\
p & p & q & p & q\\
p & p & q & q & p\\
\hline
p & p & q & q & q & \Rightarrow q
\end{array}
\end{array}
$$

$$
\begin{array}{c}
\text{PI}\\
\begin{array}{ccccc|c}
1 & 2 & 3 & 4 & 5 & \\
p & p & p & q & q & p\\
p & p & q & p & q & p\\
p & p & q & q & p & p\\
 & & & & & \Downarrow\\
 & & & & & p
\end{array}
\end{array} \qquad (2)
$$

Let x and y be votes with values $0, 1, -1$ representing, respectively, abstention and votes pro and contra a proposal. Let $\mathbf{X} = \{x_1, \ldots, x_n\}$ be a set of votes and let $y = (x_1, \ldots, x_n)$ be the amalgamation of $x_1, \ldots, x_n$ into the vote y by majority rule so that $y = 0$ iff (if and only if) $\Sigma x_i = 0$; $y = 1$ iff $\Sigma x_i > 0$; $y = -1$ iff $\Sigma x_i < 0$.

If we now partition $\mathbf{X}$ into subsets, apply majority rule within subsets, and amalgamate the within subset outcomes by majority rule, the final outcome depends on the partitioning of $\mathbf{X}$. For instance,

$$(1, (1, -1)) = (1, 0) = 1$$

but

$$((1, 1), -1) = (1, -1) = 0.$$

The root of the O-paradox resides in the decision procedure being two-stage, combined with the majority rule not being invariant over partitions of the set of votes to be amalgamated.

The choice under PI in (2) show the A-paradox: $3, 4, 5$—that is, a majority of $1, \ldots, 5$—disagree with the outcomes of 2 of the 3 referenda. Wagner [6] proved the *rule of three-fourths*: The A-paradox cannot occur

if the average proportion of voters comprising the majority within a referendum is at least $3/4$, with the average taken over referenda. Since the A-paradox is necessary for the O-paradox, Wagner's rule gives a sufficient condition for a set of referenda not to show an O-paradox.

Starting from the position that in all person-by-issue choices for either one of two parties the probability of either choice is $1/2$, Bezembinder and Van Acker [2] estimated by computer simulation the probability of an O-paradox in $(n \times m)$-decisions matrices—like the matrices in (2)—for $3 \leq n \leq 6$ issues and $3 \leq m \leq 9$ persons. The estimates so obtained range between zero and 0.114 with mean 0.036, so that an O-paradox seems a rare phenomenon.

The O-paradox has some intricate relations with the voting paradox* and the theory of measurement and scaling, particularly in psychology*. These relations are dealt with by Rae and Daudt [5] and Bezembinder and Van Acker [2].

References

[1] Anscombe, G. E. M. (1976). On frustration of the majority by fulfillment of the majority's will. *Analysis*, **36**, 161–168.

[2] Bezembinder, Th. and Van Acker, P. (1985). The Ostrogorski paradox and its relation to nontransitive choice. *J. Math. Sociol.*, **11**, 131–158.

[3] Gorman, J. L. (1978). A problem in the justification of democracy. *Analysis*, **39**, 46–50.

[4] Lipset, S. M. (1968). Ostrogorski, Moisei Ia. In *International Encyclopedia of the Social Sciences*, D. L. Sills, ed. Macmillan and Free Press, New York, pp. 347–351.

[5] Rae, D. W. and Daudt, H. (1976). The Ostrogorski paradox: A peculiarity of compound majority decision. *European J. Polit. Res.*, **4**, 391–398.

[6] Wagner, C. (1983). Anscombe's paradox and the rule of three-fourths. *Theory and Decision*, **15**, 303–308.

(BALLOT PROBLEMS VOTING PARADOX)

TH. BEZEMBINDER

OXFORD BULLETIN OF ECONOMICS AND STATISTICS

The *Oxford Bulletin of Economics and Statistics* was established in 1939. The editorial board is drawn from members and associates of the Institute of Economics and Statistics at the University of Oxford. The current (1988) editors are D. F. Hendry, J. B. Knight, K. Mayhew, and S. J. Nickell. The editorial address is: Institute of Economics and Statistics, Manor Road, Oxford, OX1 3UL, England. The publisher's address is: Basil Blackwell Publisher Ltd, 108 Cowley Road, Oxford, OX4 1JF, England. The *Bulletin* is published four times a year, each annual volume comprising some 400 pages.

The editorial policy is to publish in all fields of applied economics. The subject areas in which the *Bulletin* has published extensively include labor economics, development economics and international economics—but no subjects are precluded. The editors are willing to publish work on any economy provided that the analysis would be of general interest. Subscribers are distributed throughout the world, as are contributors.

Much of the work published is of an econometric* nature. Although the emphasis is on applications, the editors are willing to publish articles on methodology. Recent (1985) econometric articles include A. Pagan, Time series behaviour and dynamic specification; R. J. Corker and D. K. H. Begg, Rational dummy variables in an intertemporal optimisation framework; F. K. Struth, Modelling expectations formation with parameter-adaptive filters. An empirical application to the Livingston forecasts; and S. Nickell, Error correction, partial adjustment and all that: An expository note.

A recent innovation has been the introduction of the "Practitioners' Corner": The publication of notes intended to introduce applied economists to new developments in statistical and econometric methodology. Recent contributions include P. Kennedy, Logarithmic dependent variables and prediction bias; K. F. Wallis, Comparing time-series and nonlinear model-based forecasts; and D. F. Hendry, Using PC-Give in econometrics teaching.

(ECONOMETRICS)

J. B. KNIGHT

P

PARTIAL CORRELATION

Suppose two variables (say, X and Y) are correlated with each other, but both are also related to a third variable (say, Z). Then one may ask what correlation would remain between X and Y if the relationships with Z were somehow accounted for. This question was first studied by Yule [14] and Pearson [7], who introduced the name *partial correla-* *tion*. The term now refers to various concepts of correlation* between X and Y "given" or "holding constant" or "taking account of" or "controlling for" or "partialing out" Z. Partial correlation will here be symbolized as $C(X, Y|Z)$, where $C(X, Y)$ is any index of total correlation, although the traditional notation due to Yule [15] indicates the index of correlation for a population (sample) by a Greek (Latin) letter, with the variables subscripted: for example $\rho_{XY.Z}$ ($r_{XY.Z}$). Note that, while X and Y must be at least ordinal, Z is unrestricted; it may be only nominal, and it also may be multivariate, although for simplicity of exposition this discussion is limited to the univariate case.

The first basic concept of partial correlation is *correlation between errors* (*or residuals*) *after regression**. Let $\Psi_X(z)$ and $\Psi_Y(z)$ be regression functions of X and Y on Z; then

$$C_1(X, Y|Z = z)$$
$$= C(X - \Psi_X(z), Y - \Psi_Y(z)).$$

The term "regression" is here used very generally: It may refer, for example, to the *mean regression functions*

$$\mu_X(z) = E[X|Z = z],$$
$$\mu_Y(z) = E[Y|Z = z]$$

or to the *least-squares** *lines*

$$\lambda_X(z) = E[X] + \text{cov}(X, Z)\frac{(z - E[Z])}{\text{var}(Z)},$$

$$\lambda_Y(z) = E[Y] + \text{cov}(Y, Z)\frac{(z - E[Z])}{\text{var}(Z)}.$$

With least-squares linear regression, and Pearsonian product–moment correlation (ρ), the well-known *partial correlation formula* holds:

$$\rho_1(X, Y|Z)$$
$$= \frac{\rho(X, Y) - \rho(X, Z)\rho(Y, Z)}{\sqrt{1 - \rho^2(X, Z)}\sqrt{1 - \rho^2(Y, Z)}}.$$

This formula is often used as a definition of partial correlation. It is convenient in that it determines the partial correlation from the total correlations.

The second basic concept of partial correlation is *average conditional correlation*. Let $C(X, Y|Z = z)$ be the correlation between X and Y in their conditional joint distribution given $Z = z$; this is called a *conditional*

correlation. Then

$$C_2(X, Y|Z) = \text{ave}_z C(X, Y|Z = z),$$

where "ave" may indicate any averaging process, though typically a weighted mean. Inherent in this concept is that if $C(X, Y|Z = z)$ has the same value for all z, then the partial correlation C_2 is that common value; and in particular that if X and Y are conditionally independent at every value z, then their partial correlation is zero. If the conditional correlations are not all equal, then X, Y, and Z are said to *interact*, and C_2 is only one possible summary of their relationship.

If X, Y, and Z have a trivariate normal distribution, then the mean regression functions of X and Y on Z are linear, so that the partial correlation formula applies. In addition, the conditional correlations are equal, and their common value is the same as that given by the formula. This coincidence of C_1 and C_2 under normality has led to confusion of the basic concepts. Lawrance [5] showed that it does not hold generally unless the conditional variances and covariance are the same for all z (not just the correlation), in addition to the mean regression functions being linear; see also Lewis and Styan [6].

Another concept of partial correlation is restricted to the situation where the control variable Z is at least ordinal. Consider a pair of observations (x_1, y_1, z_1) and (x_2, y_2, z_2). This pair is concordant, discordant, or tied with respect to X and Y according as

$$(x_1 - x_2)(y_1 - y_2) > , < \text{ or } = 0.$$

Suppose for the moment that tied pairs do not occur. Then the total Kendall correlation (Kendall's tau*) $\tau(X, Y)$ between X and Y equals the probability that the pair will be concordant, less the probability that it will be discordant. The other pairs of variables (X, Z) and (Y, Z) may be treated similarly. It is easily seen that a pair of observations cannot be discordant with respect to all three pairs of variables simultaneously; in fact, there are only four types of nontied pairs, which may be tabulated, with their probabilities, as follows:

		With Respect to X and Z	
		Concordant	Discordant
With respect to Y and Z	Concordant	XYZ-concordant or nondiscordant π_0	YZ-concordant or X-discordant π_X
	Discordant	XZ-concordant or Y discordant π_Y	XY-concordant or Z-discordant π_Z

Kendall [4] proposed a sample measure of partial rank correlation whose population analog is

$$\tau(X, Y|Z)$$
$$= \frac{\pi_0 \pi_Z - \pi_X \pi_Y}{\sqrt{(\pi_0 + \pi_X)(\pi_0 + \pi_Y)(\pi_X + \pi_Z)(\pi_Y + \pi_Z)}},$$

which he showed can also be obtained by substituting the total Kendall correlations into the partial correlation formula. Further development by Somers [12, 13] and Hawkes [2] explains this apparent coincidence in terms of a generalized product–moment system, and provides a clearer logical basis for Kendall's concept of partial correlation. It is also shown that if there are tied pairs, then the variant of Kendall's index known as τ_b should be used.

The primary direct application of partial correlation is in testing conditional independence, and thence indirectly in causal analysis or path analysis*. Actually, one can test only the broader hypothesis that the population value of the partial correlation is zero, except with normal data, for which the two hypotheses are the same. Then, given a sample (X_i, Y_i, Z_i) for $i = 1, \ldots, n$, one may estimate the coefficient ρ_1 by calculating the sample total correlations $r(X, Y)$, $r(X, Z)$, $r(Y, Z)$ and substituting them into the partial correlation formula; this is equivalent to performing least-squares regressions of X and Y on Z and correlating the residuals. If $\rho_1 = 0$ the sampling distribution of r_1 is the same as that of the total correlation r, except for a reduction in the effective sample size: Specifically, the quantity $r_1\sqrt{n - 3} / \sqrt{1 - r_1^2}$ is distributed as Student's t with $(n - 3)$ degrees of freedom. Asymptotic sampling theory for coefficients of the form C_1, not assuming normality, has been developed by Shirahata [11] and Randles [10].

Turning now to the second basic concept of partial correlation, let R be the number from among the $\binom{n}{2}$ pairs of sample observations which are *relevant*, defined, for example, to mean that $|Z_i - Z_j|$ is not greater than some tolerance ϵ, although more complicated definitions may be used: The idea is that such pairs should effectively "hold Z constant." From among the relevant pairs let C be the number concordant with respect to X and Y, and D the number discordant. Then $T = (C - D)/R$ is the measure of partial correlation introduced by Quade [8]; this is a consistent estimator of the difference θ between the conditional probability that a pair will be concordant and the conditional probability that it will be discordant, given that it is relevant. Since the definition of relevance is quite general, there is actually a whole family of such measures; it includes, for example, the Goodman–Kruskal* total correlation (if "relevant" means simply "not tied on X or on Y") and the Goodman–Kruskal partial correlation of Davis [1] (if "relevant" means "not tied on X or on Y, but tied on Z"). Let R_i be the number of relevant pairs which include the observation (X_i, Y_i, Z_i), and W_i the number from among these which are concordant, less the number discordant; then $T = \Sigma W_i / \Sigma R_i$, and $(T - \theta)/S$ is asymptotically standard normal under mild conditions, where

$$S =$$

$$2\sqrt{\frac{\sum W_i^2\left(\sum R_i\right)^2 - 2\sum W_i\sum R_i\sum W_iR_i + }{\left(\sum W_i\right)^2\sum R_i^2/\left(\sum R_i\right)^2}}.$$

Thus asymptotic tests and confidence intervals concerning θ can be constructed.

An estimate $t(X, Y|Z)$ of Kendall's partial correlation $\tau(X, Y|Z)$ can be obtained by substituting the sample total Kendall correlations into the partial correlation formula. Hoeffding [3] showed that the sampling distribution of this statistic is asymptotically normal, and gave an intractably complicated expression for its variance; as a special case, if X and Y are independent of Z, then the asymptotic distribution of the partial is the same as that of the total correlation coefficient. All subsequent work on the distribution seems limited to the case where X, Y, and Z are all mutually independent, rather than where X and Y are only conditionally independent given Z, so no tests of conditional independence or of $\tau(X, Y|Z) = 0$ are at present available.

As a final remark, it is interesting to note a relationship between partial correlation and two-group analysis of covariance*. Let Y_{ij} be the response and X_{ij} the covariate value for the jth observation in the ith group, and define a dummy variable Z which only distinguishes the groups. Then the standard analysis of covariance test for equality of the mean responses after adjustment for the covariable is equivalent to a test that $\rho_1(Y, Z|X) = 0$. Similarly, the covariance analysis of matched differences between the groups, as given by Quade [9], is equivalent to the test that $T(Y, Z|X) = 0$.

References

[1] Davis, J. A. (1967). *J. Amer. Statist. Ass.*, **62**, 189–193.

[2] Hawkes, R. K. (1971). *Amer. J. Sociol.*, **76**, 908–926.

[3] Hoeffding, W. (1948). *Ann. Math. Statist.*, **19**, 293–325.

[4] Kendall, M. G. (1942). *Biometrika*, **32**, 277–283.

[5] Lawrance, A. J. (1976). *Amer. Statist.*, **30**, 146–149.

[6] Lewis, M. C. and Styan, G. H. (1981). In *Statistics and Related Topics*, M. Csörgő, D. A. Dawson, J. N. K. Rao, and A. K. Md. E. Saleh, eds. North-Holland, Amsterdam, pp. 57–65.

[7] Pearson, K. (1902). *Philos. Trans. R. Soc., Lond., A*, **200**, 1–66.

[8] Quade, D. (1974). In *Measurement in the Social Sciences: Theories and Strategies*, H. M. Blalock, Jr., ed. Aldine, Chicago, Chap. 13.

[9] Quade, D. (1982). *Biometrics*, **38**, 597–611.

[10] Randles, R. (1984). *J. Amer. Statist. Ass.*, **79**, 349–354.

[11] Shirahata, S. (1977). *Biometrika*, **64**, 162–164.

[12] Somers, R. H. (1959). *Biometrika*, **46**, 241–246.

[13] Somers, R. H. (1974). *Social Forces*, **53**, 229–246.

[14] Yule, G. U. (1897). *Proc. R. Soc., Lond., A*, **60**, 477–489.

[15] Yule, G. U. (1907). *Proc. R. Soc., Lond., A*, **79**, 182–193.

(ASSOCIATION, MEASURES OF
CORRELATION
DEPENDENCE, CONCEPTS OF
DEPENDENCE, MEASURES AND INDICES OF
GOODMAN–KRUSKAL TAU AND GAMMA
KENDALL'S TAU
PARTIAL REGRESSION
REGRESSION (various entries))

D. QUADE

PERCENTILE POINTS

Any value, x, such that

$$\Pr[X \leqslant x] = \alpha$$

is called a *$100\alpha\%$ point* of the distribution of X. (For absolutely continuous distributions with positive PDFs the $100\alpha\%$ point is unique. For discrete distributions, the point will not exist for some values of α; for other values it may not be unique.)

These points are called, collectively, the *percentage points*, or *percentile points*, or more briefly, the *percentiles* or *centiles* of X—more precisely, of the distribution of X. A $100(1 - \alpha)\%$ point is sometimes called an *upper* $100\alpha\%$ point, and a $100\alpha\%$ point a *lower* $100\alpha\%$ point.

For some specific values of α, special names are used thus:

$\alpha = 1/2,$ *median,*

$\alpha = 1/4$ and $3/4$, *lower* and *upper quartiles,*

$\alpha = j/5, \; j = 1, 2, 3, 4,$ *quintiles,*

$\alpha = j/6, \; j = 1, \ldots, 5,$ *sextiles,*

$\alpha = j/8, \; j = 1, \ldots, 7,$ *octiles,*

$\alpha = j/10, \; j = 1, \ldots, 9,$ *deciles,*

and so on.

(QUANTILES)

PERMISSIBLE ESTIMATORS

An alternative name for *range-preserving* estimators*. The term *permissible estimators* was introduced by Verdooren [1] (see also ref. 2).

References

[1] Verdooren, L. R. (1980). *Statist. Neerlandica*, **34**, 83–106.

[2] Verdooren, L. R. (1988). *Commun. Statist. Theor. Meth.*, **17**, 1027–1051.

(RANGE-PRESERVING ESTIMATORS
VARIANCE COMPONENTS)

PERMUTATION MATRIX—II

A *permutation matrix* (p.m.) is a square matrix with a single unit element in each row and each column, the other elements being zero. Alternatively, a p.m. results from permuting the rows or the columns of an identity matrix. If $\mathbf{x}$ is a k-vector and $\mathbf{P}(k \times k)$ is some p.m., then $\mathbf{Px}$ is the k-vector with elements of $\mathbf{x}$ permuted in the same way as the rows of $\mathbf{I}_k$ were permuted to obtain $\mathbf{P}$. Some properties of $\mathbf{P}$ are $\mathbf{P'P} = \mathbf{PP'} = \mathbf{I}_k$, $\mathbf{P'} = \mathbf{P}^{-1}$, and $\mathbf{P}^r$ is also a p.m. for any positive integer r. There are $k!$ different p.m.'s of order k, and if $\mathbf{P}$ is randomly drawn from these, $E[\mathbf{P}] = \mathbf{J}/k$ and $D(\mathbf{P}) = (\mathbf{E} \otimes \mathbf{E})/k$, with $\mathbf{J}$ a $k \times k$-matrix of ones,

$\mathbf{E} \equiv \mathbf{I}_k - \mathbf{J}/k$, $D(\cdot)$ the variance of a matrix (*see* VEC-OPERATOR), and $\otimes$ the (right) Kronecker product*.

A particular type of p.m. with many applications in statistics is the *commutation matrix* (alternatively called the *Kronecker product permutation matrix*, the *permuted identity* matrix, the *vec-permutation* matrix, the *universal flip* matrix, etc.). An *implicit* or *operational* definition of the commutation matrix $\mathbf{P}_{n,m}$ of order $mn \times mn$ is $\mathbf{P}_{n,m}\text{vec}\,\mathbf{A} = \text{vec}\,\mathbf{A}'$ (with $\mathbf{A}'$ the transpose of $\mathbf{A}$) for any $m \times n$-matrix $\mathbf{A}$; "vec" denotes the vec-operator*. So $\mathbf{P}$ changes the running order of a vector of double-indexed variables.

An *explicit* or *descriptive* definition of $\mathbf{P}_{n,m}$ is as follows. It consists of an array of $m \times n$ blocks each of order $n \times m$. The (i, j)th block has a unit element in the (j, i)th position and zeros elsewhere. Then

$$\mathbf{P}_{n,m} = \sum_{i=1}^{m} \sum_{j=1}^{n} \mathbf{e}_i^m (\mathbf{e}_j^n)' \otimes \mathbf{e}_j^n (\mathbf{e}_i^m)',$$

where $\mathbf{e}_i^m$ is the ith unit column vector of order m, and $\otimes$ denotes the Kronecker product. For example,

$$\mathbf{P}_{3,2} = \begin{pmatrix} 1 & 0 & 0 & 0 & 0 & 0 \\ 0 & 0 & 1 & 0 & 0 & 0 \\ 0 & 0 & 0 & 0 & 1 & 0 \\ 0 & 1 & 0 & 0 & 0 & 0 \\ 0 & 0 & 0 & 1 & 0 & 0 \\ 0 & 0 & 0 & 0 & 0 & 1 \end{pmatrix}.$$

Some useful properties are:

$$\mathbf{P}'_{n,m} = \mathbf{P}_{m,n},$$

$$\mathbf{P}_{n,m}\mathbf{P}_{m,n} = \mathbf{I}_{mn},$$

$$\mathbf{P}_{n,1} = \mathbf{P}_{1,n} = \mathbf{I}_n,$$

$$\mathbf{P}_{m,p}(\mathbf{A} \otimes \mathbf{B})\mathbf{P}_{q,n} = \mathbf{B} \otimes \mathbf{A},$$

$$\mathbf{P}_{m,p}(\mathbf{A} \otimes \mathbf{B}) = (\mathbf{B} \otimes \mathbf{A})\mathbf{P}_{n,q},$$

$$\text{vec}(\mathbf{A} \otimes \mathbf{B}) = (\mathbf{I}_n \otimes \mathbf{P}_{m,q} \otimes \mathbf{I}_p)$$
$$\times (\text{vec}\,\mathbf{A} \otimes \text{vec}\,\mathbf{B}),$$

with $\mathbf{A}$: $m \times n$ and $\mathbf{B}$: $p \times q$; and with $\mathbf{A}$: $m \times n$ and $\mathbf{B}$: $n \times m$,

$$\text{tr}\,\mathbf{AB} = \text{tr}(\mathbf{A} \otimes \mathbf{B})\mathbf{P}_{m,n}.$$

A useful compendium is ref. 1. An overview

is given in ref. 3. The characteristic polynomial is derived in ref. 2. Applications in statistics are given in refs. 4–6. Some of these are: (i) If **x**: $p \times 1$ and **y**: $q \times 1$ are mutually independent random vectors with $E[\mathbf{xx}'] = \mathbf{V}_1$ and $E[\mathbf{yy}'] = \mathbf{V}_2$, then

$$E[\mathbf{xy}' \otimes \mathbf{yx}'] = (\mathbf{V}_1 \otimes \mathbf{V}_2)\mathbf{P}_{q,\,p}.$$

(ii) If **X**: $n \times k$ is random with rows independent and identically distributed $N(\mathbf{0}, \Sigma)$, then

$$D(\mathbf{X'X}) = n(\mathbf{I}_{k^2} + \mathbf{P}_{k,\,k})(\Sigma \otimes \Sigma).$$

(iii) If $\mathbf{X} \sim N_k(\mathbf{0}, \mathbf{V})$, then

$$E[\mathbf{X} \otimes \mathbf{X} \otimes \mathbf{X} \otimes \mathbf{X}]$$
$$= \operatorname{vec}\mathbf{V} \otimes \operatorname{vec}\mathbf{V} + \operatorname{vec}(\mathbf{V} \otimes \mathbf{V})$$
$$+ ((\mathbf{V} \otimes \mathbf{V})\mathbf{P}_{kk}).$$

References

[1] Balestra, P. (1976). *La Dérivation Matricielle*. Sirey, Paris.

[2] Don, F. J. H. and Van der Plas, A. P. (1981). *Linear Algebra Appl.*, **37**, 135–142.

[3] Henderson, H. V. and Searle, S. R. (1981). *Linear Multilinear Algebra*, **9**, 271–288.

[4] Kapteyn, A., Neudecker, H., and Wansbeek, T. J. (1986). *Psychometrika*, **51**, 269–275.

[5] Magnus, J. R. and Neudecker, H. (1979). *Ann. Statist.*, **7**, 381–394.

[6] Neudecker, H. and Wansbeek, T. J. (1983). *Canad. J. Statist.*, **11**, 221–231.

(VEC OPERATOR)

TOM WANSBEEK

PROBABILITY AND MATHEMATICAL STATISTICS

This journal is published by the Institute of Mathematics, Polish Academy of Sciences, Wrocław University, Poland. The Editor-in-Chief is Kazimierz Urbanik. The main purpose of the journal is to foster development in fields of probability theory of special relevance to statistics. These cover a very wide range.

Volume 7 appeared in 1986. Papers may be in English, French, German, or Russian. Most papers actually appearing are in English. The editorial address is Pl. Grunwaldzki 2/4, 50-384 Wrocław, Poland.

PROBABILITY IN THE ENGINEERING AND INFORMATIONAL SCIENCES

This journal commenced publication in 1987; the Editor is Sheldon M. Ross and the Editorial Office is Department of Industrial Engineering and Operations Research, University of California, Berkeley, CA 94720. All papers must be in English.

"This journal will be concerned with articles detailing innovative uses of probability in a variety of areas. Among others, we will be especially interested in applications to computer and information science, the engineering sciences, applied statistics, and reliability theory." [From the Introductory Editorial, Vol. 1, No. 1 (1987).]

The journal is published by Cambridge University Press, New York.

PROBABILITY MOMENTS AND FREQUENCY MOMENTS

For a random variable X having a discrete distribution, with

$$\Pr[X = x_j] = p_j, \qquad j = \dots, -1, 0, 1, \dots,$$

the rth *probability moment* is

$$\omega_r = \sum_{j=-\infty}^{\infty} p_j^r. \tag{1}$$

If the distribution is continuous with probability density function $f(x)$, it is defined as

$$\Omega_r = \int_{-\infty}^{\infty} \{f(x)\}^r \, dx.$$

(Note that $\omega_1 = \Omega_1 = 1$.)

These quantities were introduced by Sichel [1, 2], who used sample analogs of these parameters in fitting distributions. He also defined ratios

$$\alpha_r = \begin{cases} \Omega_{r+2}/\Omega_2^{r+1}, & r = 1, 2, 3, \ldots, \\ \Omega_{(2r+5)/2}/\Omega_{3/2}^{(2r+1)/2}, & r = \frac{1}{2}, \frac{3}{2}, \frac{5}{2}, \ldots, \end{cases}$$

as "measures of kurtosis*." For continuous distributions he suggested using ω_r, as defined in (1) with a "discretized" (grouped) form of the distribution, as an approximation to Ω_r, and termed the sample analogs "*working*" probability moments.

The rth working *frequency moment* for a sample of size n, is n^r times the rth working probability moment.

Sichel [2] gives an account of accuracy to be expected using probability moments to estimate parameters of a distribution.

References

[1] Sichel, H. S. (1947). *J. R. Statist. Soc. Ser. A*, **110**, 337–347.

[2] Sichel, H. S. (1949). *Biometrika*, **36**, 404–425.

(METHOD OF MOMENTS
PROBABILITY-WEIGHTED MOMENTS)

PROBABILITY-WEIGHTED MOMENTS

The (r, s, t) probability-weighted moment of a random variable X, having a cumulative distribution function (CDF) $F(x)$ is

$$M_{r,s,t} = E\left[X^r \{ F(X) \}^s \{ 1 - F(X) \}^t \right]$$

$$= \int_0^1 \{ x(F) \}^r F^s (1 - F)^t \, dF,$$

where $x(F)$ is the inverse of the CDF. Equivalently, $M_{r,s,t}$ is $(s!t!)/(s + t + 1)!$ times the expected value of the rth power of the $(s + 1)$th order statistic* in a random sample of size $(s + t + 1)$. The quantities $\{ M_{r,0,0} \}$ are the moments about zero of X.

The quantities $M_{r,s,t}$ were introduced by Greenwood et al. [1]. Equating sample to population values may be used for estimation of parameters. [Note, however, that the sample size (n) need not be equal to $(s + t + 1)$.] Hosking et al. [2] consider use of $\{ M_{1,s,0} \}$ in this connection. Hosking [3] gives detailed theory.

References

[1] Greenwood, J. A., Landwehr, J. M., Matalar, N. C., and Wallis, J. R. (1979). *Water Resources Res.*, **15**, 1049–1054.

[2] Hosking, J. R. M., Wallis, J. R., and Wood, E. F. (1985). *Technometrics*, **27**, 251–261.

[3] Hosking, J. R. M. (1986). The Theory of Probability-Weighted Moments, *IBM Res. Rep. RC 12210 (#54860)*.

(ESTIMATION, POINT
METHOD OF MOMENTS
ORDER STATISTICS
PROBABILITY MOMENTS AND
 FREQUENCY MOMENTS
U-STATISTICS)

PROPENSITY SCORE

This is a parameter that describes one aspect of the organization of a clinical trial*. It is an estimate of the conditional probability of assignment to a particular treatment given a vector of values of concomitant variables*.

Rosenbaum [1] and Rosenbaum and Rubin [2, 3] discuss the use of propensity scores in correcting the analysis of observational data* to allow for possible biases in the assignment of individuals to treatments as indicated by the propensity scores.

References

[1] Rosenbaum, P. R. (1984). *J. Amer. Statist. Ass.*, **79**, 565–574.

[2] Rosenbaum, P. R. and Rubin, H. (1983). *Biometrika*, **70**, 41–55.

[3] Rosenbaum, P. R. and Rubin, H. (1984). *J. Amer. Statist. Ass.*, **79**, 516–524.

(CLINICAL TRIALS
OBSERVATIONAL DATA)

Q

QUASILINEAR MEAN *See* WEIGHTED QUASILINEAR MEAN

QÜESTIIÓ

This journal was founded by a number of Departments of the Universidad Politécnica de Catalunya in 1977. From the outset it has appeared four times a year, the four issues constituting a volume. The goal of the journal was to publish research work in any of the languages spoken in Spain, in French, or in English. The editors of the journal have been Xavier Berenguer (1977–1979), Narcis Nabona (1980–1983), Albert Prat (1983–1985), and Josep M². Oller (1986–). The current address is: Centre de Calcul de la U.P.C., Av. Dr. Gregorio Marañón s/n, Barcelona 08028, Spain.

The editorial policy of the journal has changed little since its foundation. It is aimed to achieve a balance between theoretical and applied articles in the following areas:

STATISTICS

E-0 Methods. E-1 Statistical inference. E-2 Stochastic processes. E-3 Statistical methods for estimate and decision making. E-4 Data analysis. E-5 Planning of experiments. E-6 Sampling of finite populations. E-7 Nonparametric statistics. E-15 Applications.

SYSTEMS

S-0 Methods. S-1 Modeling, estimation, and simulation. S-2 Adaptive and learning systems. S-3 Form recognition. S-4 Biosystems. S-5 Control theory (linear, nonlinear, and stochastic systems, optimum control...). S-15 Specific applications of control (generative systems and transmission of energy flows—industrial applications, communications, agriculture, and other natural systems —socio-economic systems).

COMPUTER SCIENCE

I-0 Methods. I-1 Supervisory systems (monitors, operative systems, data bases...). I-2 Processors (compilers, generators, interpreters...). I-3 Programming languages. I-4 Software evaluation. I-5 Logic design of hardware. I-6 Components and circuits. I-7 Logic and formal systems. I-8 Automata. I-9 Program analysis and effectiveness. I-15 Applications (engineering, scientific calculation, artificial intelligence, research, real time...).

OPERATIONS RESEARCH

IO-0 Methods. IO-1 Linear programming (continuous, complete and mixed, stochastic). IO-2 Nonlinear programming. IO-3 Directed exploration. IO-4 Network analysis. IO-5 Dynamic programming (deterministic, probabilistic). IO-6 Decisions and games. IO-7 Queueing. IO-9 Research. IO-10 Simulation. IO-15 Applications.

No ranking is implied in this list.

All submissions are refereed by at least two referees, some of them from outside Spain.

The journal also accepts case studies that are of potential interest to people working in industry.

The contents of recent issues include:

E. J. Hannan, Multivariable ARMA systems and practicable calculations.

Tunnicliffe Wilson, Problems in scientific time series analysis.

Arthur B. Treadway, On the properties typical of economic time series.

William Kruskal, Concepts of relative importance.

G. Arnaiz Tovar and C. Ruiz Rivas, Outliers in circular data: A Bayesian approach.

M. Pilar García-Carrasco, Distribuciones mínimo informativas, caso de espacio paramétrico finito.

M. P. Galindo Villardón, Una alternativa de representación simultánea: HJ-BIPLOT.

J. Dopazo, A. Barberá, and A. Moya, Generación de selección dependiente de frecuencias en sistemas de dos competidores: 1 modelo del espacio biológico unidad aplicado a competidores móviles.

Andrés Moya, Problemas de metrización y medición en biología de poblaciones.

Finally, the journal invites well-known specialists to publish surveys on their subject matter.

ALBERT PRAT

R

RAI AND VAN RYZIN DOSE–RESPONSE MODEL (for teratological experiments involving quantal responses)

Rai and Van Ryzin [1] proposed a dose–response model specifically for teratological experiments involving quantal responses* which they characterize as a mechanistic model because it can represent two stages in the development of a defect in a fetus whose parent has been exposed to a toxicant just before or during gestation. This model deals with the relatedness within litters (intralitter correlation) by accounting for the variability among litters in the same parental dose group by using litter size as a covariate, structured on the assumption that larger litter sizes represent better maternal health and hence lower fetal risk. The model has the form

$$P(d, s) = \lambda(d) \times P[d, s; \lambda(d)],$$

where $P(d, s)$ is the probability of an adverse response in a fetus from a litter of size s with parent exposed to dose level d, $\lambda(d)$ is the probability of a predisposing disturbance in the litter environment, and $P[d, s; \lambda(d)]$ is the conditional probability of an adverse response in a fetus with parental exposure dose d, in a litter of size s, given a predisposing disturbance in its litter environment.

In their article the first stage, $\lambda(d)$, is represented by a one-hit model, but other models could be used. The second stage is expressed as an exponential function in dose containing litter size s. $P(d, s)$ is thus formulated as

$$P(d, s) = \{1 - \exp[-(\alpha + \beta d)]\}$$
$$\times \exp[-s(\theta_1 + \theta_2 d)].$$

The marginal probability of fetal response, $P(d)$, is obtained by assuming a distribution for litter size, such as the Poisson*, and taking the expectation of $P(d, s)$ over all litter sizes.

The likelihood function* for parameter estimation and hypothesis testing uses the binomial distribution* with parameter $P(d, s)$. Since the log-likelihood for $P(d)$ divides into two discrete sections, the dose–response parameters and the litter-size distribution parameters are estimated separately. Standard errors estimation and a goodness-of-fit* test are based on the information matrix and the assumption of joint asymptotic normality of the parameter estimates.

Two methods are proposed for estimating doses that produce very low specified additional risks and their confidence intervals. Method 1 uses the assumption of normality of the dose estimate and a first-order Taylor expansion to obtain an approximation of its standard error. Method 2 provides a conservative method of low-dose interpolation by using the upper bound on $P(d, s)$, i.e., $P(d, 0)$, for estimation of the "virtually safe dose" associated with a specific risk.

References

[1] Rai, K. and Van Ryzin, J. (1985). *Biometrics*, **41**, 1–10

[2] Van Ryzin, J. (1985). *Toxicology and Environmental Health*, **1** 299–310. (This is a version of ref. 1 for toxicologists.)

(BIOASSAY, STATISTICAL METHODS IN DOSAGE–RESPONSE CURVE QUANTAL RESPONSE ANALYSIS)

D. G. WELLINGTON

RAO AND SCOTT (TYPE) TESTS

Chi-square tests* provide a basis for the evaluation and selection of log-linear models and other parametric models for contingency tables* or other categorical data* under simple sampling situations. The properties of the *Pearson* and *likelihood-ratio** chi-square tests are well known when the observed data arise from multinomial* or Poisson* distributions. Generally, critical regions* for these tests are given approximately by the upper tail of the chi-square distribution* on an appropriate number of degrees of freedom*.

When the categorical data arise from a complex sample, such as those typically used in sample surveys*, the sampling distribution of the two test statistics under the null hypothesis may be substantially different from the corresponding behavior under the multinomial distribution. Often, the effect of the complex sample is to induce the chi-square tests to reject at a much higher rate than the nominal level, if the usual critical values are employed. These effects may be quite extreme when the sample estimates are based upon weighted data.

An alternative approach to Pearson or likelihood-ratio chi-square tests in the context of complex samples is to perform a Wald test based upon an estimated variance–covariance matrix for the complex sample design. (*See* CHI-SQUARE TESTS.) A disadvantage of this method, however, is that users of published categorical data frequently do not have access to detailed information on covariances required by the Wald test. The Wald test in some situations has also been shown [8] to be less stable than the tests to be described here.

Scott and Rao proposed an alternative method based upon reinterpreting the Pearson or likelihood-ratio chi-square tests computed directly from the sample estimates by compensating for the effect of the complex sample design upon the distribution of the test statistics under the null hypothesis. They developed two variants of this basic approach. The simpler variant divides the original value of the test statistic by a correction factor. The advantage of this choice is that estimates of covariances, which rarely accompany published cross-classifications, are not required for many models. The more complex variant, which incorporates a Satterthwaite [7] approximation* to the distribution of the test statistic under the null hypothesis, represents a further improvement but requires estimates of the covariances.

Both variants of their basic approach stem from the same theoretical results. Rao and Scott [5, 6] showed that, when maximum likelihood* estimates appropriate for the multinomial distribution are computed from the estimated frequencies from the complex sample, the Pearson X^2 or likelihood-ratio G^2 tests have the approximate distribution

$$X^2 \sim \sum_{i=1}^{u} \delta_i W_i, \tag{1}$$

where the W_i are independent χ_1^2 variables, the δ_i are strictly positive constants, and u is the number of degrees of freedom for the test under multinomial sampling. More precisely, let the $T \times 1$ vector π represent the true proportions in the population, and $\mathbf{p}$, a sample estimate of π. Suppose $n^{1/2}(\mathbf{p} - \pi) \rightarrow N_T(\mathbf{0}, \mathbf{V})$, as the sample size $n \rightarrow \infty$. Let $\mathbf{P}$ be a $T \times T$ matrix with elements $P_{ii} = \pi_i(1 - \pi_i)$ and $P_{ij} = -\pi_i\pi_j$, $i \neq j$. Suppose π satisfies a log-linear model of the form

$$\mu = \tilde{u}(\theta)\mathbf{1} + \mathbf{X}\theta,$$

where $\mu_i = \ln \pi_i$, $\mathbf{X} = (\mathbf{X}_1, \mathbf{X}_2)$, for the $T \times s$ matrix $\mathbf{X}_1$ and the $T \times u$ matrix $\mathbf{X}_2$; $\theta = (\theta_1', \theta_2')'$, with $s \times 1$ θ_1 and $u \times 1$ θ_2; and $\tilde{u}(\theta)$ is a function of θ. Let X^2 or G^2 denote the chi-square test of $\theta_2 = \mathbf{0}$. Let $\tilde{\mathbf{X}}_2$ denote the projection of $\mathbf{X}_2$ onto the orthogonal complement of the space spanned by the columns of $\mathbf{X}_1$ with respect to $\mathbf{P}$, i.e.,

$$\tilde{\mathbf{X}}_2 = \left(\mathbf{I} - \mathbf{X}_1(\mathbf{X}_1'\mathbf{P}\mathbf{X}_1)^{-1}\mathbf{X}_1'\mathbf{P}\right)\mathbf{X}_2.$$

Then, (1) gives the asymptotic distribution of X^2 or G^2 under the null hypothesis, where the δ_i are the eigenvalues of $(\tilde{\mathbf{X}}_2'\mathbf{P}\tilde{\mathbf{X}}_2)^{-1}(\tilde{\mathbf{X}}_2'\mathbf{V}\tilde{\mathbf{X}}_2)$.

Rao and Scott [5] called the δ_i *generalized design effects* (d_{eff}) (*see* DESIGN EFFECT) be-

cause the δ_i reflect components of the ratio of design-based variance to multinomial variance for $\mathbf{p}$. When $\mathbf{V} = \mathbf{P}$, the δ_i are identically one; more generally, if $\mathbf{V}$ is a scalar multiple of $\mathbf{P}$, i.e., $\mathbf{V} = c\mathbf{P}$, the δ_i all assume the value c.

The simple variant proposed by Rao and Scott is to compare X^2/δ. to the chi-square distribution on u degrees of freedom, where δ. is an estimate of $\Sigma\delta_i/u$. In ref. 6, Rao and Scott discussed the calculation of δ. from the variances of the cell estimates and of specific marginal tables, when the maximum likelihood estimates of the log-linear model(s) are given in closed form. Bedrick [1] published similar results for this problem.

For the important special case of testing independence in an $I \times J$ table, δ. is given by

$$u\delta. = \sum_i \sum_j \left(1 - (\pi_{i+})(\pi_{j+})\right)d_{ij}$$
$$- \sum_i (1 - \pi_{i+})d_i(r)$$
$$- \sum_j (1 - \pi_{+j})d_j(c),$$

where the d_{ij} are the design effects for the cells, $d_i(r)$ the design effects for the row totals, and $d_j(c)$ the design effects for the columns [6].

The second version of the test employs the Satterthwaite approximation to (1) by treating $X^2/(\delta.(1 + a^2))$ as χ_ν^2, where $\nu = u/(1 + a^2)$, and $a = \{\Sigma(\delta_i - \delta.)^2/(u\delta.^2)\}^{1/2}$ is the coefficient of variation of the δ_i's. Direct evaluation of the eigenvalues of an estimate of $(\tilde{\mathbf{X}}_2'\mathbf{P}\tilde{\mathbf{X}}_2)^{-1}(\tilde{\mathbf{X}}_2'\mathbf{V}\tilde{\mathbf{X}}_2)$ gives an estimate of a^2.

In addition to the examples in refs. 5 and 6, Binder et al. [2] illustrated the application of these methods.

Fay [4] developed an alternative "jackknifed chi-squared test" based upon replication methods. Comparisons of this statistic with those of Rao and Scott may be found in refs. 4 and 8. A general computer program implementing the jackknifed test was described in ref. 3.

References

[1] Bedrick, E. J. (1983). *Biometrika*, **70**, 591–595.

[2] Binder, D. A., Gratton, M., Hidiriglou, M. A., Kumar, S., and Rao, J. N. K. (1985). *Surv. Methodology*, **10**, 141–156.

[3] Fay, R. E. (1982). *Proc. Sect. Surv. Res. Meth. Amer. Statist. Ass.*, pp. 44–53.

[4] Fay, R. E. (1985). *J. Amer. Statist. Ass.*, **80**, 148–157.

[5] Rao, J. N. K. and Scott, A. J. (1981). *J. Amer. Statist. Ass.*, **76**, 221–230.

[6] Rao, J. N. K. and Scott, A. J. (1984). *Ann. Statist.*, **12**, 46–60.

[7] Satterthwaite, F. E. (1946). *Biometrics*, **2**, 110–114.

[8] Thomas, D. R. and Rao, J. N. K. (1984). *Proc. Sect. Surv. Res. Meth.* Amer. Statist. Ass., pp. 207–211.

(CATEGORICAL DATA
CHI-SQUARE TESTS
CONTINGENCY TABLES
JACKKNIFE)

Robert E. Fay

RAO DISTANCE

The *Rao distance* is defined as the geodesic distance of the Riemannian metric induced from the Fisher information matrix* over the parameter space of a parametric family of probability distributions. This distance was first introduced in 1945 by Rao [4] as a measure of dissimilarity between two probability distributions and it was derived by using heuristic considerations based on the information matrix (*see* PROBABILITY SPACES, METRICS AND DISTANCES ON). A more abstract approach to these concepts can be achieved by representing the space of distributions by means of an embedding in a Hilbert space (see also refs. 2 and 3).

HILBERT SPACE EMBEDDING

Let μ be a σ-finite measure, defined on a σ-algebra of the subsets of a measurable set $\mathcal{X}$, and let $\mathcal{M} = \mathcal{M}(\mathcal{X}: \mu)$ be the space of all μ-measurable functions on $\mathcal{X}$. We let $\mathcal{M}_+$ denote the set of all $p \in \mathcal{M}$ such that $p(x) > 0$ for μ-almost all $x \in \mathcal{X}$, while for a real $\alpha \neq 0$, $\mathcal{L}^\alpha$ designates the space of all $p \in \mathcal{M}$ so that

$$\|p\|_\alpha \equiv \left\{ \int_{\mathcal{X}} |p|^\alpha \, d\mu \right\}^{1/\alpha} < \infty,$$

and we define $\mathcal{L}_+^\alpha = \mathcal{L}^\alpha \cap \mathcal{M}_+$. Moreover, for $0 < r < \infty$ we also let $\mathcal{L}^\alpha(r) = \{ p \in \mathcal{L}^\alpha : \|p\|_\alpha = r \}$ and $\mathcal{L}_+^\alpha(r) = \mathcal{L}^\alpha(r) \cap \mathcal{M}_+$, and we write $\mathcal{P}^\alpha$ for $\mathcal{L}_+^\alpha(1)$. In this notation, $\mathcal{P} = \mathcal{P}^1$ is the space of probability distributions and $\mathcal{L}^2$ is a Hilbert space with the inner product and norm

$$(p, q) = \int pq \, d\mu, \qquad \|p\| = \|p\|_2,$$

$$p, q \in \mathcal{L}^2.$$

For $\alpha \neq 0$ and $p \in \mathcal{M}_+$ we define

$$T_\alpha(p) = \frac{2}{|\alpha|} p^{\alpha/2}. \tag{1}$$

Then T_α is a bijection of $\mathcal{M}_+$ onto $\mathcal{M}_+$ and T_α embeds $\mathcal{L}_+^\alpha$ into $\mathcal{L}^2$ with

$$T_\alpha(\mathcal{L}_+^\alpha) = \mathcal{L}_+^2, \qquad T_\alpha(\mathcal{P}^\alpha) = \mathcal{L}_+^2\left(\frac{2}{|\alpha|}\right).$$

The induced distance on $\mathcal{L}_+^\alpha$ is

$$\rho_\alpha(p_1, p_2) = \frac{2}{|\alpha|} \|p_1^{\alpha/2} - p_2^{\alpha/2}\|,$$

$$p_1, p_2 \in \mathcal{L}_+^\alpha, \tag{2}$$

and is called the "*α-order Hellinger distance*" on $\mathcal{L}_+^\alpha$. We consider a parametric family $\mathcal{F}_\Theta^\alpha = \{ p(\cdot|\theta) \in \mathcal{L}_+^\alpha : \theta \in \Theta \}$ of positive distributions $p(\cdot|\theta)$, $\theta \in \Theta$, having suitable regularity properties and where Θ is a parameter space embedded in $\mathcal{L}_+^\alpha$. We also consider the subfamily $\mathcal{P}_\Theta^\alpha = \mathcal{F}_\Theta^\alpha \cap \mathcal{P}^\alpha$ of $\mathcal{F}_\Theta^\alpha$.

On $\mathcal{F}_\Theta^\alpha$ we have

$$\rho_\alpha^2(p(\cdot|\theta), p(\cdot|\theta + d\theta)) = ds_\alpha^2(\theta),$$

to the second-order infinitesimal displacements. Here $ds_\alpha^2(\theta)$ is the "*α-information metric*"

$$ds_\alpha^2(\theta) = \int_{\mathcal{X}} p^\alpha (d \log p)^2 \, d\mu,$$

where the dependence on $x \in \mathcal{X}$ and $\theta \in \Theta$ in the integrand has been suppressed. This may also be written as

$$ds_\alpha^2(\theta) = d\theta' \, \mathscr{I}_\alpha(\theta) \, d\theta$$

where

$$\mathscr{I}_\alpha(\theta) = \int_{\mathcal{X}} p_\alpha (\partial_\theta \log p)^2 \, d\mu$$

is the "*α-order information matrix.*" In particular, ds_1^2 and $\mathscr{I}_1$ are the ordinary "*information metric*" and the "*information matrix,*" respectively (*see* PROBABILITY SPACES, METRICS AND DISTANCES ON for additional details).

GEOMETRY OF SPACES OF DISTRIBUTIONS

The geometries of $\mathcal{L}_+^\alpha$ and $\mathcal{P}^\alpha$, $\alpha \neq 0$, under the α-order ds_α^2 may be read off from the embedding T_α in (1) of $\mathcal{L}_+^\alpha$ into $\mathcal{L}^2$,

$$q = T_\alpha(p), \qquad p \in \mathcal{L}_+^\alpha.$$

We then have

$$ds_\alpha^2(p) = ds_2^2(q) = \|dq\|^2, \qquad q \in \mathcal{L}^2,$$

where the parameter space Θ may now be taken as a subset of $\mathcal{L}^2$.

The geometry of $\mathcal{L}_+^\alpha$ under ds_α^2 is induced by the representation $T_\alpha(\mathcal{L}_+^\alpha) = \mathcal{L}_+^2$ with the Euclidean metric ds_α^2. It follows that this geometry is essentially Euclidean with the Riemann–Christoffel tensor* of the first kind being identically zero. The geodesic curves $p[s] \equiv p(\cdot|s) \in \mathcal{L}_+^\alpha$ are given by

$$p[s] = (as + b)^{2/\alpha}, \qquad 0 \leqslant s < \infty,$$

where a and b are functions in $\mathcal{L}^\alpha$ which are independent of the parameters. Moreover, the geodesic distance of ds_α^2 on $\mathcal{L}_+^\alpha$ is identical with the α-order Hellinger distance* $\rho_\alpha(p_1, p_2)$ in (2).

The geometry of $\mathcal{P}^\alpha$ under ds_α^2, on the other hand, is induced by the spherical representation $T_\alpha(\mathcal{P}^\alpha) = \mathcal{L}_+^2(2/|\alpha|)$ with the Euclidean metric ds_α^2. This is essentially the spherical geometry with the Riemann–

Christoffel tensor of the first kind

$$R_\alpha(x, y: u, v)$$
$$= \tfrac{1}{4}\{(x, u)(y, v) - (x, v)(y, u)\},$$

where $x, y, u, v \in \mathcal{L}^2$. The mean Gaussian curvature is then

$$\kappa_\alpha(x, y) = \frac{R_\alpha(x, y: x, y)}{\|X\|^2 \|y\|^2 - [(x, y)]^2} \equiv \frac{1}{4},$$

$$x, y \in \mathcal{L}^2.$$

When $\alpha = 1$, the above quantities give the "*first information curvature tensor*" and the "*information curvature*" on $\mathcal{P} = \mathcal{P}^1$, respectively (*see* PROBABILITY SPACES, METRICS AND DISTANCES ON for details). The geodesic curves $p[s] \equiv p(\cdot|s) \in \mathcal{P}^\alpha$, $0 \leqslant s \leqslant L$, where s is the arc-length parameter, are given by

$$p[s] = \left\{ a \cos \frac{|\alpha|}{2} s + b \sin \frac{|\alpha|}{2} s \right\}^{2/\alpha},$$

$$0 \leqslant s \leqslant L,$$

where a and b are parameter-independent orthonormal functions in $\mathcal{P}^2$, i.e.,

$$\|a\| = \|b\| = 1, \qquad (a, b) = 0,$$

$$a, b \in \mathcal{M}_+.$$

It is also assumed that

$$a \cos \frac{|\alpha|}{2} s + b \sin \frac{|\alpha|}{2} s \in \mathcal{M}_+,$$

$$0 \leqslant s \leqslant L.$$

The geodesic distance on $\mathcal{P}^\alpha$ is then

$$S_\alpha(p_1, p_2) = \frac{2}{|\alpha|} \cos^{-1}\left(p_1^{\alpha/2}, p_2^{\alpha/2} \right),$$

$$p_1, p_2 \in \mathcal{P}^\alpha,$$

or

$$S_\alpha(p_1, p_2) = \frac{2}{|\alpha|} \cos^{-1} \int_{\mathcal{X}} (p_1 p_2)^{\alpha/2} \, d\mu,$$

$$p_1, p_2 \in \mathcal{P}^\alpha.$$

Moreover, the geodesic curve $p[s] \in \mathcal{P}^\alpha$ connecting p_1 and p_2 of $\mathcal{P}^\alpha$ has the representation

$$p[s] = A^{1/\alpha} \cos^{2/\alpha}\left(B - \frac{|\alpha|}{2} s \right),$$

$$0 \leqslant s \leqslant L,$$

where

$$A = \frac{p_1^\alpha + p_2^\alpha - 2(p_1 p_2)^{\alpha/2} \cos \frac{1}{2}|\alpha| L}{\sin^2 \frac{1}{2}|\alpha| L}$$

$$B = \tan^{-1}\left\{ \left[\left(\frac{p_2}{p_1} \right)^{\alpha/2} - \cos \frac{|\alpha|}{2} L \right] \sin \frac{|\alpha|}{2} L \right\},$$

and

$$L = S_\alpha(p_1, p_2), \qquad p_1, p_2 \in \mathscr{P}^\alpha.$$

When $\alpha = 1$, we find that the Rao distance on $\mathscr{P} = \mathscr{P}^1$ is

$$S(p_1, p_2) = S_1(p_1, p_2)$$

$$= 2 \cos^{-1} \int_{\mathscr{X}} (p_1 p_2)^{1/2} d\mu,$$

$$p_1, p_2 \in \mathscr{P},$$

which is effectively the *Hellinger–Bhattacharyya distance* [1]. This distance was obtained previously in Rao [4] by using concrete methods, and later in Dawid [3] and Burbea [2] by using abstract methods.

References

[1] Bhattacharyya, A. (1943). On a measure of divergence between two statistical populations. *Bull. Calcutta Math. Soc.*, 35, 99–109.

[2] Burbea, J. (1986). Informative geometry of probability spaces. *Expo. Math.*, 4, 347–378.

[3] Dawid, A. P. (1977). Further comments on some comments on a paper by Bradley Efron. *Ann. Statist.*, 5, 1249.

[4] Rao, C. R. (1945). Information and accuracy attainable in the estimation of statistical parameters. *Bull. Calcutta Math. Soc.*, 37, 81–91.

(FISHER INFORMATION
HELLINGER DISTANCE
INFORMATION THEORY AND CODING
   THEORY
J-DIVERGENCES AND RELATED
   CONCEPTS
PROBABILITY SPACES, METRICS AND
   DISTANCES ON
STATISTICAL CURVATURE
TENSORS)

JACOB BURBEA

RATES

DEFINITIONS

Let $y = y(x)$ be a continuous increasing function of x, $\Delta x > 0$ an increment of x, and $\Delta y = y(x + \Delta x) - y(x) > 0$ an increment of y. Then

$$\alpha(x) = \lim_{\Delta x \to 0} \frac{\Delta y}{\Delta x} = \frac{dy}{dx} \qquad (1)$$

is called the *instantaneous absolute rate at point x*. It measures (conceptually) the change in y per unit change of independent variable at point x. The quantity $\Delta y / \Delta x$ may be considered as an *average* absolute rate over the interval x to $x + \Delta x$. In most situations, x represents time (or age). It will be so used in the remaining text of this article.

In describing biological or chemical processes, the *relative rate*,

$$\mu(x) = \frac{1}{y(x)} \lim_{\Delta x \to 0} \frac{\Delta y}{\Delta x} = \frac{1}{y(x)} \frac{dy}{dx}$$

$$= \frac{d[\log y(x)]}{dx} \qquad (2)$$

is more appropriate. (The word "relative" is usually omitted.) For example, $y(x)$ may represent a growth curve*, and $\mu(x)$ the growth rate.

It is customary to express rates as positive quantities, so that for decreasing functions, the right-hand side in (1) and (2) is given a negative sign.

In lifetime analysis, if $y(x)$ represents the survival distribution function, then $\alpha(x) = -dy/dx$ is called (in actuarial science*) the "curve of deaths" (it is simply the probability density function) while

$$\mu(x) = -\frac{1}{y(x)} \frac{dy}{dx} = -\frac{d \log[y(x)]}{dx}$$

$$\qquad (2a)$$

is called the *hazard rate* or *intensity rate* (reliability*), or *force of mortality** (actuarial science).

In chemical reactions $y(x)$ may represent a "mass" which is subject to decay in time, and (2a) is often called the *reaction velocity*.

The average (relative) rate, called in life-table analysis the *central rate*, is

$$_{\Delta x}m_x = -\frac{\Delta y}{\int_x^{x+\Delta x} y(t)\, dt} = -\frac{1}{y(x')}\frac{\Delta y}{\Delta x},\tag{3}$$

where $x \leqslant x' < x + \Delta x$. Commonly, the two following approximations to $y(x)$ are used

$$y(x') \doteq y\left(x + \tfrac{1}{2}\Delta x\right),\tag{4a}$$

$$y(x') \doteq y(x) + \tfrac{1}{2}\Delta y.\tag{4b}$$

In special situations

$$y(x') \doteq y(x)\tag{4c}$$

is also used.

In practice, the mathematical form of $y(x)$ is usually unknown and the right-hand sides in (4) have to be estimated from the data.

VITAL RATES

Central rates play important roles in vital statistics*, especially in life-table* analysis.

Example 1. Consider population mortality data grouped in age intervals $[x_i, x_{i+1})$ with width $x_{i+1} - x_i = h_i$. If D_i is the observed number of deaths in age interval $[x_i, x_{i+1})$, then the central rate, m_i (more precise notation would be $_{h_i}m_{x_i}$), can be approximated by the *age-specific death rate*,

$$m_i \doteq M_i = D_i/P_i,\tag{5}$$

where P_i is the midyear (as July 1) population in this age group for a given calendar year. [Note: Here the approximation (4a) is used.] The overall rate for the whole population,

$$M = \left(\sum_i D_i\right)\Big/\left(\sum_i P_i\right) = D/P\tag{6}$$

is called the "crude" death rate.

Example 2. Consider follow-up* data of a cohort by age. Let t be the follow-up time, and suppose that the mortality data are grouped in intervals $[t_i, t_{i+1})$, with $t_{i+1} - t_i = h_i$. Let N_i be the number of survivors at t_i, and D_i the number of deaths in $[t_i, t_{i+1})$. Then

$$M_i = D_i/\left\{h_i\left(N_i - \tfrac{1}{2}D_i\right)\right\}\tag{7}$$

can be used as an estimate of central rate. [Note: Here the approximation (4b) is used.]

Example 3. Consider a study population, with mortality data grouped in fixed age intervals as in Example 1. Let $\mathscr{S}_i$ be the *risk set*, that is, the set of individuals who were observed for at least part of the interval $[x_i, x_{i+1})$, and let h_{ij} be length of the period of observation for the jth individual in this interval. Then the age-specific rate is

$$M_i = D_i\Big/\left(\sum_{j \in \mathscr{S}_i} h_{ij}\right),\tag{8}$$

where $\sum_{j \in \mathscr{S}_i} h_{ij}$ represents the amount of person-years exposed to risk* in age interval $[x_i, x_{i+1})$. Similar formulas can be used for follow-up data.

OTHER RATES

1. If $y(t)$ denotes height at time t, and $y(t + h)$ the height at time $(t + h)$, then

$$_h g_t \doteq \frac{1}{y(t)}\frac{y(t + h) - y(t)}{h}\tag{9}$$

is an average *growth rate* over the interval $(t, t + h)$. [Note: In this situation, the approximation (4c) is used.]

2. If B is the number of live births in a given year, and P is the size of midyear population, then

$$b = B/P\tag{10}$$

is the *birth rate*. Here, not every member of the population is "exposed to risk"; birth rate, in a certain sense, can be considered as a "growth rate," of a population.

3. In *discrete point processes* rates are expressed by the numbers of events occurring in a time unit. For example, number of customers arriving to a store in an hour, number of telephone calls per month for a given household, number of hurricanes per year in a specified area, etc.

MISUSE OF THE TERM "RATE"

There is a tendency to use the term "rate" for indices which represent a quotient, or more particularly, a proportion. In studying disease occurrence, the *incidence rate* refers (correctly) to the number of *new* cases of a disease occurring in a unit time per 1000, say, individuals. On the other hand, the number of cases already *existing* in a given population at the specified time point per 1000 individuals is called (incorrectly) the "prevalence rate"; this is clearly a proportion, not a rate, and might be simply termed "prevalence." (*See also* MORBIDITY.)

Similarly, the proportion of deaths within a year among individuals alive at exact age x (denoted in life-table analysis by q_x) is clearly a (conditional) death probability. However, it is sometimes called a "mortality rate" as opposed to the central "death rate," m_x. This confusion arises because in both situations, the event (occurrence of a disease, death) is a phenomenon related to time.

Further examples on use and misuse of "rate" can be found in medical, epidemiological*, demographic, biochemical, and many other related journals. See also, refs. 1–3.

References

[1] Elandt-Johnson, R. C. (1975). Definition of rates: Some remarks on their use and misuse. *Amer. J. Epid.*, **102**, 267–271.

[2] Elandt-Johnson, R. C. and Johnson, N. L. (1980). *Survival Models and Data Analysis*. Wiley, New York, Chap. 2.

[3] Shryock, H., Siegel, J. S., and Associates (1973). *The Methods and Materials in Demography*. U.S. Department of Commerce, Washington, D.C., Chap. 14.

(GROWTH CURVES
HAZARD RATE AND OTHER
    CLASSIFICATIONS OF DISTRIBUTIONS
RATES, STANDARDIZED)

REGINA C. ELANDT-JOHNSON

RAYLEIGH QUOTIENT

The *Rayleigh quotient* of the symmetric $n \times n$ matrix $\mathbf{A}$ with respect to the $n \times 1$ vector $\mathbf{X}$ is $\mathbf{X}'\mathbf{AX}/\mathbf{X}'\mathbf{X}$.

This function arises in the conditional maximization of a quadratic form $\mathbf{y}'\mathbf{Ay}$ subject to the constraint $\mathbf{y}'\mathbf{y} = 1$ (or some other positive constant). See ref. 1 for more details.

Reference

[1] Basilevsky, A. (1983). *Applied Matrix Algebra in the Statistical Sciences*. North-Holland, Amsterdam, Netherlands.

(QUADRATIC FORMS)

RAYLEIGH'S TEST

This is one of the most popular circular single sample tests*, which seek to test the null hypothesis that an angular direction ϕ has a uniform distribution—that is, probability density function

$$f_\phi(t) = (2\pi)^{-1}, \qquad 0 \leqslant t \leqslant 2\pi,$$

—based on a random sample of n observed values $t_1, t_2, \ldots, t_n$.

The test statistic is

$$r^2 = n^{-1}\left\{\left(\sum_{i=1}^{n} \cos t_i\right)^2 + \left(\sum_{i=1}^{n} \sin t_i\right)^2\right\}.$$

Sufficiently large values of r^2 indicate "one-sidedness" of the parent population. Significance limits are available in refs. 1–4.

The test is uniformly most powerful* against circular normal* (von Mises distributions) alternatives. For unimodal alternatives the Rayleigh test is sensitive not only to "one-sidedness" but also to concentration around a mean direction—that is, there is a *preferred* direction.

The *V* test* is a competitor to Rayleigh's test.

References

[1] Batschelet, E. (1981). *Circular Statistics in Biology.* Academic, New York.

[2] Greenwood, J. A.. and Durand, D. (1955). *Ann. Math. Statist.*, **26**, 233–246.

[3] Stephens, M. A. (1969). *J. Amer. Statist. Ass.*, **64**, 280–289.

[4] Zar, J. H. (1974). *Biostatistical Analysis.* Prentice-Hall, Englewood Cliffs, NJ.

(CIRCULAR NORMAL DISTRIBUTION
DIRECTIONAL DATA ANALYSIS
PERIODOGRAM ANALYSIS
V TEST)

RECIPROCAL MODEL

The regression* model

$$E[Y|x] = (\alpha + \beta x)^{-1}.$$

Bibliography

Mead, R. (1979). *Biometrics*, **35**, 41–54.

Shinozaki, K. and Kira, T. (1956). *J. Inst. Poly., Osaka City Univ., Ser. D*, **7**, 35–72.

RECURSIVE PARTITIONING

A name for a collection of methods of classification*, the essential features of which are: (i) Successive dichotomies of a population according to the value(s) of some function(s) of observed characters (variables). These are set out in the form of a "tree," the subgroups forming "nodes." At each stage the aim is to subdivide nodes into two new nodes as distinct ("diverse") from each other as possible in respect of members of specified classes in the population. (ii) The process is continued until a set of "terminal nodes" is obtained, such that the contents of each node are sufficiently homogeneous. (iii) Balancing the requirement of "sufficient homogeneity" against excessive subdivision.

Application of the procedures must be based on several, more or less arbitrary, optimality criteria. These include maximizing an index of "distinctiveness" (or diversity*) or minimizing (for each node) an index of "impurity" (containing individuals from more than one class in the population). These are counterbalanced by indices depending on the number of terminal nodes. It is often recommended to start by continuing dichotomization excessively, resulting in a large number of terminal nodes, and then reducing them by a process of recombination, termed "pruning."

Many useful practical details are contained in ref. 1, which is a standard text on the subject [although the term "recursive partitioning" (RP) does not appear in it explicitly]. A computer program CART, developed in association with ref. 1 provides a means of implementing the procedures. Some workers regard the term "CART" as a replacement for "recursive partitioning."

It is important to remember that in constructing a "tree," a large number of adjustments are made, which tend to result in an over optimistically good fit to the data actually used. It is essential, therefore, that some form of cross-validation* be applied, e.g., using the tree to classify fresh data, to obtain a just appreciation of the goodness of fit*. One way of doing this is to split the data arbitrarily into a "training" and a "test" sample, and use the latter to test a fit obtained from the former.

RP differs from cluster analysis in that in the former the existence and identity of the relevant classes in the population is supposed to be known, while in the latter it is the *existence* of classes differing systematically in regard to the observed values which is under investigation. Discriminant analysis*, on the other hand, starts from the same premise as RP partitioning. As competitors, RP provides for a wider range of possibilities, though discriminant analysis gives a speedier result when the discriminant function (e.g., linear or quadratic) happens to be approximately relevant. Both techniques, of course, aim at providing an accurate method for assigning individuals to classes on the basis of the values of the observed variables.

Reference

[1] Breiman, L., Friedman, J. H., Olshen, R. A., and Stone, C. J. (1984). *Classification and Regression Trees*. Wadsworth, Belmont, CA.

(CLASSIFICATION
DISCRIMINANT ANALYSIS
DISTANCE FUNCTIONS
DIVERSITY INDICES
HIERARCHICAL CLUSTER ANALYSIS
PARTIAL-ORDER SCALOGRAM ANALYSIS
STEPWISE REGRESSION
TWOING INDEX)

REDESCENDING *M*-ESTIMATORS

A fundamental statistical problem is to estimate the "center" of a distribution function F, a concept which is well defined if, as we will assume, the distribution is symmetric (see Bickel and Lehmann [3]). It is of special interest to define robust estimators; that is, estimators which are reasonably efficient* for an assumed model and relatively insensitive to slight departures from the model. Of course, an important case concerns data which have a nearly Gaussian (i.e., normal) distribution* but not exactly so either because of outliers* or the actual physical model; Hampel [8] quotes evidence suggesting that such deviations from the Gaussian model are common and are usually in the direction of "heavy-tailed" distributions*.

Two standard estimators of the center are the sample mean and median*, the former being the maximum likelihood estimator* for the Gaussian model, the latter maximum likelihood for the double exponential* model. The sample mean is not robust, because it is sensitive to outliers and inefficient when the distribution has heavy tails. The sample me-

dian, while robust, is inefficient at the normal model (Andrews et al. [2] and ROBUST ESTIMATION).

In his seminal work, Huber [10] showed that *M*-estimators are robust and efficient; they can be defined either by

$$\sum_{i=1}^{n} \rho\big((Y_i - T_n)/s_n\big) = \text{minimum}, \quad (1)$$

or

$$\sum_{i=1}^{n} \psi\big((Y_i - T_n)/s_n\big) = 0, \quad (2)$$

where $Y_1, \ldots, Y_n$ are the sample from a population with center θ, T_n is the *M*-estimator of θ, s_n is an estimator of scale (*see* *M*-ESTIMATORS), and ψ is the derivative of ρ. When $\rho(x) = x^2$ we obtain the least-squares* estimator (the sample mean), while if $\rho(x) = |x|$ we obtain the sample median. Huber assumes that ρ is convex (and ψ is hence monotone) and found that a minimax estimator can be defined by

$$\begin{aligned} \psi(x) &= -k, & x &< -k, \\ &= x, & -k &\leqslant x \leqslant k, \\ &= k, & x &> k. \end{aligned} \quad (3)$$

He showed that the estimator was relatively efficient at the normal model, and robust.

In his thesis and related papers, Hampel [7–9] cast the estimation problem in a functional framework and developed the idea of an influence curve*. For general classes of estimators, he showed that if θ is the center, then

$$T_n \doteq \theta + n^{-1} \sum_{i=1}^{n} \text{IC}(Y_i - \theta, F). \quad (4)$$

For *M*-estimators defined by (possibly monotone) ψ, he showed that

$$\text{IC}(Z, F) = \psi(Z)/B[\psi'(Z)]. \quad (5)$$

He thus showed that the influence function $\text{IC}(\cdot, F)$ expresses in (4) the influence of an observation on the estimator. This influence is unbounded for the sample mean [since $\psi(x) = x$] and is bounded but nonzero for Huber's *M*-estimator.

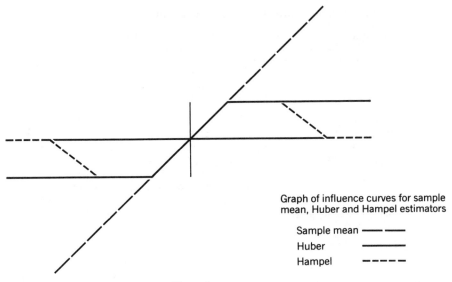

Graph of influence curves for sample mean, Huber and Hampel estimators

Sample mean —— ——
Huber   ————————
Hampel   — — — —

Figure 1

Ideally, one wants the influence of a gross outlier to be zero. This idea led Hampel (in Andrews et al. [2]) to suggest the class of *redescending M-estimators* in which ψ [and hence the influence curve IC$(\cdot, F)$] eventually redescends to zero. Here are three such choices of ψ in common use (see Gross [5, 6]):

Hampel

$$\psi(x) = -\psi(-x)$$

$$\begin{aligned}
&= x, & 0 \leqslant x \leqslant a, \\
&= a, & a \leqslant x \leqslant b, \\
&= a(c - x)/(c - b), & b \leqslant x \leqslant c, \\
&= 0, & x > c;
\end{aligned}$$

Andrews

$$\psi(x) = -\psi(-x)$$

$$\begin{aligned}
&= \sin(x), & 0 \leqslant x \leqslant \pi, \\
&= 0, & x > \pi;
\end{aligned}$$

Turkey's biweight

$$\psi(x) = -\psi(-x)$$

$$\begin{aligned}
&= x(1 - x^2)^2, & 0 \leqslant x < 1, \\
&= 0, & x \geqslant 1.
\end{aligned}$$

Note that the Huber *M*-estimators are not within the class of redescending *M*-estimators. In Fig. 1 we graph the influence curves for the sample mean and the Huber and Hampel estimators. One can see that these overlap for a good portion of the range, indicating from (2) that they will be virtually identical for clean Gaussian data. Note that for Hampel and Huber, the effect of a moderate observation is also the same. Finally, we see that Hampel gives gross outliers no influence, while Huber gives positive and bounded, and the sample mean gives unbounded influence to outliers.

In practice, (2) may not have a unique solution. This poses conceptual difficulties if one approaches the definition of redescending estimates from a pseudo-likelihood* context (see Huber [11, p. 103]). To get around this difficulty, one can either take the solution for (2) closest to a Huber estimate or take one or two steps of a Newton–Raphson* iteration from a Huber estimate toward solving (2); the latter is easier to apply in regression.

As regards the distributional robustness, we reproduce in Table 1 the results of the Princeton Monte Carlo* study (Andrews et al. [2]) for the following estimators: least squares, Huber ($k = 1.5$, $s_n = $ MAD/ 0.6745), Hampel ($a = 2.5$, $b = 4.5$, $c = 9.5$, $s_n = $ MAD), and Andrews ($s_n = 2.1$ MAD), where MAD is the median of the absolute deviations from the median. The table shows

Table 1   Monte Carlo Variances Multiplied by the Sample Size *N* = 20

Distribution	Mean	Huber	Hampel	Andrews
Normal	1.00	1.05	1.05	1.07
5% 3N[a]	1.42	1.17	1.16	1.16
10% 3N[a]	1.88	1.33	1.32	1.31
5% 10N[a]	6.49	1.22	1.13	1.13
Double Exponential	2.10	1.55	1.58	1.54
t − 3 d.f.	3.14	1.67	1.67	1.64
Cauchy	∞	4.50	3.70	3.50

[a] The convention $\alpha\%$ mN means that the data are standard Gaussian with probability $1 - \alpha$, but are Gaussian with zero expected value and standard deviation m with probability α.

that for the normal model the *M*-estimators are slightly less efficient than the sample mean, but as we move to heavy-tailed distributions the *M*-estimators clearly dominate, as expected. Note also that the redescending *M*-estimators are virtually identical to the Huber estimators for the most part, while they are significantly better at the Cauchy* and 5% 10*N*. This suggests, as in Fig. 1, that using redescending *M*-estimators can be very useful when there are large outliers.

Andrews [1] and Carroll [4] give numerical examples which illustrate the use of redescending *M*-estimators in regression*. Huber [1, pp. 100–103] is less than enthusiastic about redescending *M*-estimates. Besides noting the possible problem with multiple solutions of (2), he also emphasizes that one must not let ψ redescend to zero too quickly. The role of the idea of a redescending influence function for bounded influence regression is unclear and needs to be explored.

References

[1] Andrews, D. F. (1974). *Technometrics*, **16**, 523–532. (Gives a nice practical application of redescending *M*-estimates in a messy data context.)

[2] Andrews, D. F., Bickel, P. J., Hampel, F. R., Huber, P. J., Rogers, W. H., and Tukey, J. W. (1972). *Robust Estimation of Location: Survey and Advances*. Princeton University Press, Princeton,

NJ. (This book is an extensive Monte Carlo study written by some of the leading figures in the area. It compares many estimators and introduces the redescending *M*-estimators.)

[3] Bickel, P. J. and Lehmann, E. L. (1975). *Ann. Statist.*, **3**, 1045–1069.

[4] Carroll, R. J. (1980). *Appl. Statist.*, **29**, 246–251.

[5] Gross, A. M. (1976). *J. Amer. Statist. Ass.*, **71**, 409–416.

[6] Gross, A. M. (1977). *J. Amer. Statist. Ass.*, **72**, 341–354.

[7] Hampel, F. R. (1971). *Ann. Math. Statist.*, **42**, 1887–1896. (This paper introduces the functional approach which at present dominates much of the theoretical robustness literature.)

[8] Hampel, F. R. (1973). *Z. Wahrscheinlicheitsth. Verw. Geb.*, **27**, 87–104. (A subjective, pointed, and entertaining survey.)

[9] Hampel, F. R. (1974). *J. Amer. Statist. Ass.*, **69**, 383–393. (Gives a nice review of the influence curve.)

[10] Huber, P. J. (1964). *Ann. Math. Statist.*, **35**, 73–101. (The seminal paper and a remarkable tour de force.)

[11] Huber, P. J. (1981). *Robust Statistics*. Wiley, New York. (The theoretical guide to robustness, sprinkled with valuable advice on practical aspects.)

Bibliography

Bickel, P. J. (1975). One-step Huber estimates in the linear model. *J. Amer. Statist. Ass.*, **70**, 428–434.

Bickel, P. J. (1976). Another look at robustness. *Scand. J. Statist.*, **3**, 145–168. (A review of robustness which also has new ideas for robust regression. A lively, skeptical but useful discussion of robustness follows the paper.)

Boos, D. D. and Serfling, R. J. (1980). On differentials and the CLT and LIL for statistical functions, with applications to *M*-estimates. *Ann. Statist.*, **8**, 197–204.

Carroll, R. J. (1978). On almost sure expansions for *M*-estimates. *Ann. Statist.*, **6**, 314–318. (This, and the paper by Boos and Serfling, present almost sure approximations for *M*-estimators by an average of bounded random variables.)

Collins, J. R. (1976). Robust estimation of a location parameter in the presence of asymmetry. *Ann. Statist.*, **4**, 68–85. (Optimality results in the class of redescending *M*-estimators.)

Hampel, F. R., Rousseeuw, P. J., and Ronchetti, E. (1981). The change-of-variance curve and optimal redescending *M*-estimators. *J. Amer. Statist. Ass.*, **76**, 643–648.

Huber, P. J. (1972). Robust statistics: A review. *Ann. Math. Statist.*, **43**, 1041–1067.

Huber, P. J. (1973). Robust regression: Asymptotics, conjectures and Monte-Carlo. *Ann. Statist.*, **1**, 799–821.

Huber, P. J. (1977). *Robust Statistical Procedures*. SIAM, Philadelphia. (A readable monograph which summarizes the work of Huber and Hampel.)

(ESTIMATION, POINT
INFLUENCE FUNCTION
LOCATION–SCALE FAMILIES
OUTLIERS
M-ESTIMATORS
ROBUSTNESS)

R. J. CARROLL

REFINEMENT

The concept of *refinement* was introduced by DeGroot and Fienberg [2, 3] in the context of comparing forecasters whose predictions are presented as their subjective probabilities that various events will occur. As a specific example, consider a weather forecaster who each day must specify his or her probability that it will rain in a particular location, where it is assumed that the event "rain" has been carefully defined.

We will refer to the probability x specified by a forecaster on a given day as his or her *prediction* on that day, and for convenience of exposition we will assume that x is restricted to a finite set of values $0 = x_0 < x_1 < \cdots < x_k = 1$. We shall let $X = \{x_0, x_1, \ldots, x_k\}$ denote the set of allowable predictions. The behavior of any forecaster is then characterized by the following two functions: (i) the probability function $\nu(x)$ which specifies the probability that the forecaster's prediction on a particular day will be x, where $\sum_{x \in X} \nu(x) = 1$, and (ii) the conditional probability $\rho(x)$ of rain on a particular day given that the forecaster's prediction was x. (*See also* PROBABILITY FORECASTING.)

The functions ν and ρ can be interpreted from either the frequentist* or the Bayesian* view of probability. From the frequentist view, $\nu(x)$ is the relative frequency with which the forecaster has made the prediction x in an appropriate reference sequence of days, and $\rho(x)$ is the relative frequency of rain among those days on which the prediction was x. From the Bayesian view, $\nu(x)$ is an observer's subjective probability that the forecaster's prediction will be x based on whatever past data and other information the observer might have about the forecaster, and $\rho(x)$ is the observer's subjective probability of rain after learning that the forecaster's prediction is x.

A forecaster is said to be *well-calibrated** if $\rho(x) = x$ for every $x \in X$. The concept of refinement pertains to the comparison of well-calibrated forecasters. Consider two well-calibrated forecasters A and B characterized by the probability functions ν_A and ν_B. Roughly speaking, forecaster A is at least as refined as forecaster B if, from A's prediction on each day and an auxiliary randomization*, we can simulate a prediction that has the same stochastic properties as B's prediction. The precise definition of this concept is as follows:

A function $h(x|y)$ defined on $X \times X$ is called a *stochastic transformation* if

$$h(x|y) \geqslant 0 \quad \text{for every } x \in X \text{ and } y \in X,$$

and

$$\sum_{x \in X} h(x|y) = 1 \quad \text{for every } y \in X.$$

Then forecaster A is at least as refined as forecaster B if the following two relations are satisfied:

$$\sum_{y \in X} h(x|y) \nu_A(y) = \nu_B(x)$$

for every $x \in X$,

$$\sum_{y \in X} h(x|y) y \nu_A(y) = x \nu_B(x)$$

for every $x \in X$.

The first of these relations guarantees that the predictions resulting from the stochastic transformation, or auxiliary *randomization**, will have the same distribution ν_B as B's predictions; and the second relation guarantees that the predictions will still be well calibrated.

If A is at least as refined as B, then any decision maker who is given a choice between learning the prediction of A or the prediction of B, will prefer to learn the prediction of A regardless of the decision problem in which that information will be used. For this reason, the relation that A is at least as refined as B is very strong. Given any two well-calibrated forecasters, it is not necessarily true that one of them is at least as refined as the other. Thus this concept introduces a *partial ordering* among all well-calibrated forecasters.

Since the mean $\mu = \sum_{x \in X} x \nu(x)$ must be the same for all well-calibrated forecasters, the concept of refinement can be regarded as introducing a partial ordering* in the class of all distributions on X with mean μ. This partial ordering is studied by DeGroot and Eriksson [1] and is shown to be the same as that obtained in other types of problems from the concept of second-degree stochastic dominance, the comparison of statistical experiments, the comparison of Lorenz curves*, mean-preserving spreads, and the theory of majorization*.

The extension of the concept of refinement to problems in which a forecaster's prediction is his or her subjective probability* distribution of a random variable that can take more than two values is given in DeGroot and Fienberg [4].

The term "refinement" is also used, in a different sense, in the theory of belief functions* (Shafer [5], Chapter 6).

References

[1] DeGroot, M. H. and Eriksson, E. A. (1985). In *Bayesian Statistics 2*, J. M. Bernardo, M. H. DeGroot, D. V. Lindley, and A. F. M. Smith, eds. Elsevier–North-Holland, Amsterdam, Netherlands, pp. 99–118.

[2] DeGroot, M. H. and Fienberg, S. E. (1982). In *Statistical Decision Theory and Related Topics III*, Vol. 1, S. S. Gupta and J. O. Berger, eds. Academic, New York, pp. 291–314.

[3] DeGroot, M. H. and Fienberg, S. E. (1983). *The Statistician*, London, **32**, 12–22.

[4] DeGroot, M. H. and Fienberg, S. E. (1986). In *Bayesian Inference and Decision Techniques: Essays in Honor of Bruno de Finetti*, P. K. Goel and A. Zellner, eds. North-Holland, Amsterdam, Netherlands, pp. 247–264.

[5] Shafer, G. (1976). *A Mathematical Theory of Evidence*. Princeton University Press, Princeton, NJ.

(BAYESIAN INFERENCE
DECISION THEORY
FORECASTING
PROBABILITY FORECASTING
WELL-CALIBRATED FORECASTS)

MORRIS H. DEGROOT

REGRESSION, ITERATIVE

An iterative regression algorithm was developed by Bilenas and Gibson in ref. 1 as a method for calculating standardized regression coefficients* in a multiple linear regression* model without having to invert the correlation or variance–covariance matrix*. The method—which is based on Hotelling's [2] iteration procedure—was initially used as an education tool to provide students with a procedure for calculating regression coefficients by hand. In ref. 1 this algorithm is compared with ridge regression* in regard to multicollinearity* problems. (Bilenas and Gibson [1] contains many further details.)

References

[1] Bilenas, J. V. and Gibson, W. A. (1984). *Proc. 1984 Annual Meeting Amer. Statist. Ass., Statist. Comput. Sect.*, pp. 123–125.

[2] Hotelling, H. (1933). *Psychometrika*, **7**, 27–35.

(LINEAR REGRESSION
MULTICOLLINEARITY
MULTIPLE LINEAR REGRESSION
RIDGE REGRESSION)

J. V. BILENAS
W. GIBSON

REGRESSION QUANTILE

This is a statistic developed by Koenker and Bassett [1] as a robust tool in fitting linear regression* (simple or multiple). It is an extension of the concept of a sample quantile*. For sample values $y_1, y_2, \ldots, y_n$ a θth quantile is any number b such that the proportion of y's less than b is no greater than θ and the proportion greater than b is no greater than $(1 - \theta)$. Stated in this way, the concept does not generalize naturally to situations in which

$$\Pr[Y_j < y] = F\left(y - \sum_{i=1}^{p} x_{ji}\beta_i\right)$$
$$= F(y - \mathbf{x}_j'\boldsymbol{\beta}),$$

where $F(\cdot)$ is a CDF of unknown form and $\boldsymbol{\beta}$ is a $p \times 1$ vector of unknown constants.

Koenker and Bassett [1] point out that the ordinary θth quantile can be equivalently defined as any value of b minimizing

$$\theta \sum_{y_j \leqslant b} |y_j - b| + (1 - \theta) \sum_{y_j > b} |y_j - b|.$$

This naturally generalizes to the definition of θth *regression quantile* as a value of the vector $\boldsymbol{\beta}$ which minimizes

$$\theta \sum_{y_j \leqslant \mathbf{x}_j\boldsymbol{\beta}'} |y_j - \mathbf{x}_j\boldsymbol{\beta}'| + (1 - \theta) \sum_{-y_j > \mathbf{x}_j\boldsymbol{\beta}'} |y_j - \mathbf{x}_j\boldsymbol{\beta}'|.$$

In particular, a *regression median* is obtained by taking $\theta = \frac{1}{2}$. Regression quantiles are used in robust methods of fitting regressions.

("Regression quantile" should not be confused with "quantile regression," which is the conditional quantile of the distribution of Y, given $\mathbf{x}$, regarded as a function of $\mathbf{x}$.)

Reference

[1] Koenker, R. and Bassett, G. (1978). *Econometrics*, **16**, 33–50. (Defines quantile regression, and gives examples of application.)

(QUANTILES
REGRESSION, MEDIAN
ROBUSTNESS)

RETRACTED DISTRIBUTIONS

A term used by Nogami [1] for families of distributions obtained by truncating* a given distribution to intervals $[\theta, \theta + 1)$. The parameter θ indexes the family.

Reference

[1] Nogami, Y. (1985). In *Statistical Theory and Data Analysis*, K. Matusita, ed. Elsevier, Amsterdam, Netherlands, pp. 499–515.

(TRUNCATION
WEIGHTED DISTRIBUTIONS)

RETRODICTION

The prediction of missing data or unrecoverable observables from the present data is referred to as *retrodiction*. The terminology was introduced by Geisser [2] in the context of parametric models in which there are missing or unobserved responses in the data. Inference about these unobserved responses is called retrodiction.

The term "imputation" has often been inappropriately used in place of retrodiction as indicated by Geisser [2]. The inferential aim of the statistical analysis underlies this terminological distinction in addition to the inferential procedure that is most appropriately used in achieving the aim. If our interest centers on the values of the unobserved data, then we retrodict the missing values; when parametric estimation is our aim and it is necessary to "fill in" values for the missing data so the complete data (data + "filled in" values) may be used to estimate parameters, then the "filled in" values are imputations. Simplicity and convenience of estimation is often the motivation for imputing the missing data. Butler [1] points out, however, that the best retrodictors are not always in agreement with the imputations suggested by an estimative inference. A simple example helps in clarifying these ideas. Suppose the data are a sequence of 10 multinomial trials with three categories. Let the data result in frequency counts 2, 3, and 4 and suppose one observation is "missing" but known not to lie in category 2. Methods of retrodiction, such as the method of predictive likelihood (Butler [1], and Hinkley [3]) or the Bayesian approach with a uniform prior on the parameters, lead to odds of $\frac{2}{3}$ in favor of the missing datum deriving from category 1 rather than category 3. On the other hand, the missing datum imputation which leads to the maximum likelihood estimates of the multinomial parameters from the complete data (via the EM algorithm) assigns weight $\frac{1}{3}$ to category 1 and $\frac{2}{3}$ to category 2. The complete data are $2\frac{1}{2}$, 3, $4\frac{1}{2}$, and the imputative odds in favor of category 1 are $\frac{1}{3} \div \frac{2}{3} = \frac{1}{2}$, not $\frac{2}{3}$. For this example the retrodictive odds differ from the imputative odds. In a more general context, the best point retrodictor does not necessarily agree with the imputation value associated with maximum likelihood estimation*. For many normal theory examples, however, they do agree (see Butler [1], rejoinder).

References

[1] Butler, R. W. (1986). Predictive likelihood inference with applications (with Discussion). *J. R. Statist. Soc. B*, **48**, 1–38.

[2] Geisser, S. (1986). In Discussion of Butler [1].

[3] Hinkley, D. V. (1979). Predictive likelihood. *Ann. Statist.*, **7**, 718–728 (corrig., **8**, 694).

(INCOMPLETE DATA
MISSING INFORMATION PRINCIPLE
PREDICTIVE ANALYSIS)

R. W. BUTLER

REVIEW OF ECONOMICS AND STATISTICS, THE

This journal was founded, as the *Review of Economic Statistics*, in 1919; the present name has been used since 1948. It is published quarterly by Elsevier (Amsterdam), under the auspices of the Department of Economics, Harvard University.

It is a scholarly journal which publishes both theoretical and empirical articles in economics and statistics. The majority of articles are in applied economics. Virtually all fields of economics are dealt with, but a preponderance of articles appears in the areas of industrial organization, consumer demand, labor economics, and public finance. Many articles emphasize the quantitative methodology employed rather than the economic relationships that are discussed. Methods include input–output analysis, hedonic approaches, state–space methods, logit models, and many other statistical and econometric estimating methodologies.

A typical issue contains 15 regular articles plus about 10 shorter articles in a "Notes" section. Many articles include an appendix. A brief abstract precedes each article.

ROTATABILITY INDEX

A rotatable design* is one for which the variance of the predicted response at a general point $(x_1, x_2, \ldots, x_k)$ in the coded predictor variable space depends only on the distance of that point from the design origin [namely $(0, 0, \ldots, 0)$ in coded units]. More formally, $V\{\hat{y}(\mathbf{x})\} = f(r^2 = x_1^2 + x_2^2 + \cdots + x_k^2)$. Rotatability is a desirable feature but not an essential one; it often pays to give up exact rotatability in order to achieve some other design feature such as, for example, orthogonal blocking. Nevertheless a near-rotatable design would still be a good choice. How do we know when a design is near-rotatable? The only sure way is to draw and inspect the variance contours, a procedure so tedious, especially when the number of factors, k, exceeds 2, that it is rarely done. Two measures of rotatability have been suggested. One, due to Khuri [2], leads to a percentage value, which essentially provides a comparison of how closely the design moments of a specific design match up to the moments of a rotatable design. A value of 100% means the design is rotatable. A second measure of rotatability for symmetric designs, due to Draper and Guttman [1], approximates a variance contour near the edges of the design by a contour of the type $|x_1|^m + |x_2|^m + \cdots + |x_k|^m = 1$. With a value of m in hand, one refers to a reference diagram showing the shapes of contours $|x_1|^m + |x_2|^m = 1$ for several preselected benchmark values of m, to gain an immediate grasp of the shape. Values of m for certain standard composite designs* were provided. Both measures are useful, both have valuable features and drawbacks, and they are complementary rather then competitive.

References

[1] Draper, N. R. and Guttman I. (1988). An index of rotatability. *Technometrics*, **30**, 105–111.

[2] Khuri, A. I. (1988). A measure of rotatability for response-surface designs. *Technometrics*, **30**, 95–104.

(RESPONSE SURFACE DESIGNS
ROTATABLE DESIGNS)

NORMAN R. DRAPER
IRWIN GUTTMAN

RUN LENGTHS, TESTS OF

The runs* in a sequence of elements of two types can be tested for randomness* in different ways, one of which is the Wald–Wolfowitz test (*see* RUNS), which tests whether the *number* of runs is abnormally small or large. However, other possible tests focus on the *lengths* of runs.

One test based on run lengths simply uses the length of the *longest* run of either type of element as the test statistic. Alternatively, the length of the longest run of just one type of element can be used. For some details, including tables, see Mosteller [5], Bateman [1], and Takashima [9]. A related type of possible test, using the number of runs with length greater than a given value as the test statistic, receives brief mention by Lehmann [4, p. 314].

Let the two types of elements be **0**'s and **1**'s. Let n_0 and n_1 denote the respective numbers of **0**'s and **1**'s in the sequence, and let r_0 and r_1 denote the respective numbers of *runs* of **0**'s and **1**'s. The different tests already mentioned and the ones covered below all have the same null hypothesis. It specifies simply that the elements of the sequence are randomly arranged, i.e., that all $(n_0 + n_1)!/(n_0!n_1!)$ arrangements are equally likely. The competing tests differ, however, in their effectiveness against different types of alternatives. In particular, the tests of refs. 6, 7, and 8 described below are directed against alternatives under which the *variability* of run lengths is greater (or conceivably, less) than usual.

Aiming mainly at situations where one type of element is far more frequent than the other, O'Brien [6] utilized a test of Dixon [3] (see also Blum and Weiss [2] for further theoretical details) to try to detect whether the variability in the run lengths of the more frequent element is consistent with randomness. The test is a conditional test, conditional on n_0 and n_1. Let the **1**'s be the more frequent type of element. The test statistic is based on the sample variance* of the numbers of **1**'s before the first **0**, between each pair of successive **0**'s, and after the last **0**. Equivalently, this is the variance of the $(n_0 + 1)$ integers consisting of $(n_0 + 1 - r_1)$ zeros along with the lengths of the r_1 runs of **1**'s. Under the null hypothesis of random arrangement of the n_0 **0**'s and n_1 **1**'s, this variance, times

$$c = \frac{n_0(n_0 + 2)(n_0 + 3)(n_0 + 4)}{2(n_0 + 1)(n_1 - 1)(n_0 + n_1 + 2)},$$

follows approximately the chi-square distribution* with

$$\nu = \frac{cn_1(n_0 + n_1 + 1)}{(n_0 + 1)(n_0 + 2)}$$

degrees of freedom. (Generally, ν is not an integer.) The critical region ordinarily consists of large values of the test statistic.

Example 1. The rather short sequence **0111100111111111011** will serve to illustrate the calculations. Here $n_0 = 4$, $n_1 = 15$, $r_0 = 3$, and $r_1 = 3$. The five integers whose variance is to be found are $0, 4, 0, 9, 2$. The variance is thus $(0^2 + 4^2 + 0^2 + 9^2 + 2^2 - 15^2/5)/(5 - 1)$, or 14. Then $c = 16/35$, and c times 14 is 6.4, which is referred to the chi-square distribution with $\nu = 4.57$ degrees of freedom.

O'Brien and Dyck [7] recently proposed a test that treats the **0**'s and **1**'s symmetrically. It uses a linear combination of s_0^2 and s_1^2, where s_0^2 denotes the sample variance of the r_0 run lengths of **0**'s and s_1^2 is defined similarly for the **1**'s. It is a conditional test, conditional not only on n_0 and n_1 but also on r_0 and r_1. Its theoretical underpinning is again provided by the test of Dixon [3], but less directly so this time, as the derivation begins with a step that effectively reduces the length of each run by one. This time there cannot be any zeros among the integers whose variances are calculated, since none of the runs can be of length zero. For ref. 7 as well as ref. 6, the motivation was a medical application, involving Schwann cell disease.

The test statistic may be written as

$$u = k\left(a_0 s_0^2 + a_1 s_1^2\right).$$

The first two moments* of the null distribution of u can be obtained with the aid of ref. 3 and the fact that s_0^2 and s_1^2 are independent for fixed n_0, n_1, r_0, r_1. For any choice of a_0 and a_1, one can choose k so that these first two moments are the same as those of a

chi-square distribution, which, in turn, can be used to approximate the null distribution of u. The best choice for the ratio of a_0 to a_1, however, is open to question. O'Brien and Dyck [7], arguing from considerations of an accurate approximating distribution and also of power*, proposed a weighting with $k = 1$ and a_i, $i = 0, 1$, equal to

$$a_i^* = \frac{(r_i - 1)(r_i + 1)(r_i + 2)(r_i + 3)}{2r_i(n_i + 1)(n_i - r_i - 1)},$$

for which the degrees of freedom of the approximating chi-square distribution are $\nu^* = \nu_0 + \nu_1$, where

$$\nu_i = \frac{a_i^* n_i(n_i - r_i)}{r_i(r_i + 1)}.$$

Potthoff [8], using a different power criterion that he thought could provide better power, suggested a weighting with a_i equal to

$$a_i' = \frac{r_i^2(r_i + 1)(n_i - r_i - 1)}{n_i^2(n_i - r_i)}$$

and k equal to

$$k' = 2(g_0 + g_1)/(h_0 + h_1),$$

where

$$g_i = r_i(n_i - r_i - 1)/n_i$$

and

$$h_i = \frac{4r_i^4(n_i - r_i - 1)^3(n_i + 1)}{n_i^3(r_i - 1)(r_i + 2)(r_i + 3)(n_i - r_i)};$$

the approximating chi-square distribution has $\nu' = k'(g_0 + g_1)$ degrees of freedom.

Example 2. Males (**0**'s) and females (**1**'s) in a sequence of 30 persons leaving a cafeteria line appeared as follows:

010110000000011111101101010100.

Here $n_0 = 16$, $n_1 = 14$, $r_0 = 8$, $r_1 = 7$, $s_0^2 = 1^2 + 1^2 + 8^2 + \cdots + 2^2 - 16^2/8)/(8 - 1) = 6$, and $s_1^2 = 10/3$. Thus $a_0^* = 495/136$, $a_1^* = 24/7$, $\nu_0 = 110/17$, and $\nu_1 = 6$, so that $a_0^* s_0^2 + a_1^* s_1^2 = 33.27$ is referred to the chi-square distribution with $\nu^* = 12.47$ degrees of freedom. Alternatively, one obtains $g_0 = 7/2$, $g_1 = 3$, $h_0 = 833/220$, $h_1 = 3$, $k' = 2860/1493$, $a_0' = 63/32$, and $a_1' = 12/7$, for which $k'(a_0's_0^2 + a_1's_1^2) = 33.57$ is referred to the chi-square distribution with $v' = 12.45$ degrees of freedom. In either case u is easily large enough to be significant at the 0.01 level.

References

[1] Bateman, G. (1948). *Biometrika*, **35**, 97–112. (Generalizes ref. 5 to cover the longest-run test when n_0 and n_1 may differ.)

[2] Blum, J. R. and Weiss, L. (1957). *Ann. Math. Statist.*, **28**, 242–246. (Establishes some desirable properties of the test of ref. 3.)

[3] Dixon, W. J. (1940). *Ann. Math. Statist.*, **11**, 199–204. (Developed a basic test that can be adapted to provide tests of variability of run lengths.)

[4] Lehmann, E. L. (1975). *Nonparametrics: Statistical Methods Based on Ranks*. Holden-Day, San Francisco.

[5] Mosteller, F. (1941). *Ann. Math. Statist.*, **12**, 228–232. (Covers the longest-run test for the case where $n_0 = n_1$.)

[6] O'Brien, P. C. (1976). *Biometrics*, **32**, 391–401.

[7] O'Brien, P. C. and Dyck, P. J. (1985). *Biometrics*, **41**, 237–244.

[8] Potthoff, R. F. (1985). *Biometrics*, **41**, 1071–1072.

[9] Takashima, M. (1955). *Bull. Math. Statist.*, **6**, 17–23. (Has tables for situations not covered by tables of refs. 1 and 5.)

(HYPOTHESIS TESTING
RANDOMNESS, TESTS OF
RUNS
TWO-SAMPLE TESTS)

RICHARD F. POTTHOFF

S

SELF-CONSISTENT ESTIMATORS

Suppose $T_1, \ldots, T_n$ are a sample of independent, right-censored observations, where for each i, $T_i = \min(X_i, C_i)$, in which X_i is the failure time and C_i is the censoring time. Let the censoring indicator for the ith observation be δ_i, so that $\delta_i = 1$ if $T_i = X_i$ and $\delta_i = 0$ otherwise. Define

$$\pi_S(t) = n^{-1}\left\{ \#(T_i > t) \right.$$

$$\left. + \sum_{T_i \leqslant t} (1 - \delta_i) S(t)/S(T_i) \right\},$$

where $S(\cdot)$ denotes the survival function* for the failure times: $S(t) = P(X_i > t)$. Note that $(1 - \delta_i) S(t)/S(T_i) = P(X_i > t | T_i, \delta_i)$. Then $\hat{S}(\cdot)$ is called a self-consistent estima-

tor [2] of $S(\cdot)$ if

$$\hat{S}(t) \equiv \pi_{\hat{S}}(t). \tag{1}$$

[Note that (1) is intuitively reasonable as a source of estimators of $S(\cdot)$.] Equation (1) is uniquely solved by the Kaplan–Meier* estimator and (1) motivates a convergent algorithm based on $\hat{S}^{(j+1)}(t) \equiv \pi_{\hat{S}^{(j)}}(t)$, $j = 0, 1, \ldots$. The algorithm is an example of an *EM algorithm* [1] and self-consistent estimators can be discovered for more complicated censoring structures [3, 4] as well as other incomplete-data* problems.

References

[1] Dempster, A. P., Laird, E. M., and Rubin, D. B. (1977). *J. R. Statist. Soc. B*, **39**, 1–38.

[2] Efron, B. (1967). *Proc. 5th Berkeley Symp. Math. Statist. Prob.*, Vol. 4. University of California Press, Berkeley, CA, pp. 831–853.

[3] Turnbull, B. W. (1974). *J. Amer. Statist. Ass.*, **69**, 169–173.

[4] Turnbull, B. W. (1976). *J. R. Statist. Soc. B*, **38**, 290–295.

(CENSORED DATA
SURVIVAL ANALYSIS)

D. M. TITTERINGTON

SEPARABLE STATISTIC

If $X_1, X_2, \ldots, X_n$ are mutually independent random variables with

$$\Pr[X_i = j] = p_j, \quad \text{for } j = 1, 2, \ldots, N$$

and all $i = 1, 2, \ldots, n$,

and

$$H_j = H_j(n)$$

is the number of X_i's equal to j, Medvedev [5, 6] terms any function of the form

$$L(\mathbf{H}) = L_N(\mathbf{H}(n)) = \sum_{j=1}^{N} f_j(H_j(n))$$

a *separable statistic*. (A better name might be *linearly* separable statistic.)

Many important statistics, such as chi-squared*, log (likelihood ratio)*, and empty cells statistics are of this form.

Separable statistics are of value in the construction of test criteria for hypotheses relating to probability distributions associated with random discrete sequences (as $N \to \infty$) especially for tests based on grouped data*. Ivchenko and Medvedev [3, 4] give detailed results on asymptotic properties of separable statistics in the multinomial* model when $n/N \leqslant \alpha$.

Recently Ivanov [1, 2] has extended the concept of separable statistics and applied it to problems relating to random allocation of balls to cells. (*See also* URN MODELS.)

References

[1] Ivanov, V. A. (1983). In *Economic–Mathematical Models and Numerical Solution of Applied Problems*. Akad. Nauk, Kiev, Ukr SSR, pp. 43–47.

[2] Ivanov, V. A. (1983). *Trudy Steklov. Mat. Inst.*, **177**, 47–59.

[3] Ivchenko, G. I. and Medvedev, Yu. I. (1978). *Theor. Prob. Appl.*, **23**, 798–806.

[4] Ivchenko, G. I. and Medvedev, Yu. I. (1980). *Theor. Prob. Appl.*, **25**, 540–551.

[5] Medvedev, Yu. I. (1970). *Dokl. Akad. Nauk SSSR*, **192**, 987–989.

[6] Medvedev, Yu. I. (1977). *Theor. Prob. Appl.*, **22**, 1–15 and 607–614.

(*U*-STATISTICS)

SEQUENTIAL REJECTIVE BONFERRONI PROCEDURE *See* SI-MULTANEOUS TESTING

SIAM JOURNALS

The Society for Industrial and Applied Mathematics (SIAM) publishes several mathematical and statistical journals including:

SIAM Review (4) (mostly expository and survey articles) and *SIAM Journals* on:
Applied Mathematics (6)
Computing (6)
**Control and Optimization* (6)
Discrete Mathematics (4)
Mathematical Analysis (6)
**Matrix Analysis and Applications* (4)
Numerical Analysis (6)
**Scientific and Statistical Computing* (6).

(Numbers of issues per year in parentheses.)

Those journals of relatively greater statistical interest are marked with an asterisk.

The address of the Society is Suite 1400, 117 South 17th Street, Philadelphia, PA 19013-5052.

SIMULATED ANNEALING

A common mathematical problem arising in many scientific disciplines is that of finding the minimum value of a function of several variables, $f(\mathbf{x}) = f(x_1, x_2, \ldots, x_n)$. One practical difficulty in finding this minimum is that the function $f(\mathbf{x})$ may have more than a single minimum, and classical numerical methods for solving such problems tend to locate a local rather than a global minimum. The method of simulated annealing is a probabilistic method for overcoming this difficulty. The original idea for this method comes from the subject of statistical mechanics*, but it has since been generalized and requires no knowledge of physics for its application.

The basic idea behind the method is a simple one which was first proposed in the context of statistical mechanics by Metropolis et al. [9]. Let $p(x)$ be a function taking values between 0 and 1, which will be interpreted as a probability. The form of this function which is most natural in the context of statistical mechanics is

$$p(x) = \exp(-\lambda x^2), \qquad (1)$$

where λ is a control parameter that we discuss later. Consider a numerical algorithm that generates two successive evaluations of the function to be minimized, $f(\mathbf{x}_1)$ and $f(\mathbf{x}_2)$. If $f(\mathbf{x}_1) > f(\mathbf{x}_2)$, then $f(\mathbf{x}_2)$ is always taken as the estimated minimum. In the contrary case, even though $f(\mathbf{x}_2)$ has the larger value, it is nevertheless chosen as the estimate of the minimum with a probability equal to $p[\{ f(\mathbf{x}_2) - f(\mathbf{x}_1)\}]$. It is this probabilistic step that allows the numerical algorithm to escape from local minima.

In practice the minimization is run a number of times. In each succeeding run the value of the control parameter λ is increased (in physical applications λ is inversely proportional to a temperature, the reduction in temperature being analogous to the process of annealing in metallurgy). A very small value of λ allows the algorithm to explore large parts of the underlying space while a very large λ tends to localize the search. An unanswered question at the time of writing is what constitutes an optimal program for changing the control parameter. Some scattered results are known about the convergence of the method [1, 5], but much more

remains to be done in this area. A second problem of current interest is that of finding an "optimal" $p(x)$. The probability in (1) is the choice suggested by the physical annealing model, but the method is by no means restricted to that particular choice of a probability function. Szu and Hartley [10] have recently shown that replacing (1) by

$$p(x) = \left\{ 1 + (x/T)^2 \right\}^{-1} \qquad (2)$$

leads to a considerable improvement in convergence time in the context of a particular problem.

Thus far, there have been a large number of applications of the method to practical problems with encouraging success for many further applications. Original applications outside statistical mechanics included finding an optimal or nearly optimal solution of the traveling-salesman problem* and computer design [6]. Many further applications have been made to circuit-board design [3], to Bayesian restoration of images [4], and even to the calculation of efficient garbage collection routes in Grenoble [4]. At the time of writing the full potential of the method of simulated annealing is unknown, and the methodology for determining optimal variation of parameters and determining convergence rates is still in its infancy. It should be noted that a number of other techniques based on the concepts of neural networks and other ideas about biological models are being explored that might also lead to considerably more efficient calculation of extrema in multidimensional problems [2]. Klein [7] describes applications in exploratory data analysis*.

References

[1] Aarts, E. H. L., Korst, J. H. M., and van Laarhoven, P. J. M. (1988). *J. Statist. Phys.*, **50**, 187–206.

[2] Bounds, D. G. (1987). *Nature*, **329**, 215–219.

[3] Darema, S., Kirkpatrick, S., and Norton, V. A. (1987). *IBM J. Res. Dev.*, **31**, 391–402.

[4] Geman, S. and Geman, D. (1984). *IEEE Trans. Patt. Anal. Mach. Intell.*, **17**, 721–724.

[5] Gidas, B. (1985). *J. Statist. Phys.*, **39**, 73–132.

[6] Kirkpatrick, S., Gelatt, C. D., Jr., and Vecchi, M. P. (1983). *Science*, **220**, 671–680.

[7] Klein, R. A. (1987). Projection and clustering by simulated annealing. M.S. thesis, Dept. Computer Science, Michigan State University, East Lansing, MI.

[8] van Laarhoven, P. J. M. and Aarts, E. H. L. (1987). *Simulated Annealing: Theory and Applications*. Kluwer, Dordrecht, Netherlands. (A good general introduction to the subject and a number of applications.)

[9] Metropolis, M., Rosenbluth, A., Rosenbluth, M., Teller, A., and Teller, E. (1953). *J. Chem. Phys.*, **21**, 1087–1092.

[10] Szu, H. H., and Hartley, R. L. (1987). *Phys. Lett. A*, **122**, 157–162.

(OPTIMIZATION IN STATISTICS
OPTIMIZATION, STATISTICS IN
QUANTUM MECHANICS
STATISTICAL PHYSICS)

GEORGE H. WEISS

SKEWNESS, MEASURES OF

Skewness is usually defined, broadly, as lack of symmetry*. *Measures of skewness* necessarily reflect specific types of symmetry. All measures take the value zero for symmetric distributions, but a value of zero does not imply that the distributions must be symmetrical.

Generally, distributions with relatively long right-hand (left-hand) tails are called positively (negatively) skew. The measure

$$\sqrt{\beta_1} = \alpha_3 = \mu_3/\mu_2^{3/2}$$

(where μ_r is the rth central moment*) was used by K. Pearson*. The measure

$$\frac{[(\text{Upper quartile*}) - (\text{Median})] + [(\text{Lower quartile}) - (\text{Median})]}{(\text{Upper quartile}) - (\text{Lower quartile})}$$

is less affected by tail behavior.

Sample analogs of the population indices can be constructed in a natural way.

See also MEAN, MEDIAN, MODE, AND SKEWNESS.

SMITH–BAIN LIFETIME DISTRIBUTION

This distribution is obtained by supposing that $(T/\alpha)^\beta$ has a truncated type III extreme-value distribution*. The hazard function is

$$\lambda_T(t) = \beta\alpha^{-\beta}t^{\beta-1}\exp\{-(t/\alpha)^\beta\},$$
$$t > 0, \alpha, \beta > 0.$$

For $0 < \beta < 1$ the graph of $\lambda_T(t)$ is bathtub*-shaped.

Methods of fitting the parameters α and β are described by Smith and Bain [1].

Reference

[1] Smith, R. M. and Bain, L. J. (1975). *Commun. Statist.*, **4**, 469–481.

(SURVIVAL ANALYSIS)

SPEARMAN'S FOOTRULE

A somewhat neglected statistic proposed by Spearman in 1906 [3] as an alternative to Spearman's ρ. It is defined as

$$D(\mathbf{p}, \mathbf{q}) = \sum_{i=1}^{n} |p_i - q_i|,$$

where **p** and **q** are two rankings of n objects. [Compare with $\sum_{i=1}^{n}(p_i - q_i)^2$—the basic constituent of Spearman's ρ.] Spearman's footrule was recently "revived" by Diaconis and Graham [1] who studied its relations to other association measures*. Extensive tabulation of the distribution of this statistic was carried out by Franklin [2]. Due to the current (1988) popularity of L_1-procedures, the Spearman footrule is likely to receive wider attention.

References

[1] Diaconis, P. and Graham, R. L. (1977). *J. R. Statist. Soc. B*, **39**, 262–268.

[2] Franklin, L. A. (1988). *Statist. Prob. Lett.*, **6**, 399–406.

[3] Spearman, C. (1906). *Brit. J. Psychol.*, **2**, 89–108.

(ASSOCIATION, MEASURES OF
SPEARMAN RANK CORRELATION
   COEFFICIENT)

SPITZER–ROSÉN THEOREM

Let $X, X_1, X_2, \ldots$ be independent and identically distributed random variables with zero mean and unit variance, and set $S_n = \sum_{j=1}^{n} X_j$. It follows from the central limit theorem* that the probability $\Pr[S_n \leqslant 0]$ tends to $\frac{1}{2}$ as $n \to \infty$. The Spitzer–Rosén theorem provides a description of the rate of convergence in this result. Spitzer (1960) proved that the infinite series,

$$s = \sum_{n=1}^{\infty} n^{-1} \{ \Pr[S_n \leqslant 0] - \tfrac{1}{2} \}, \quad (1)$$

converges. Rosén (1961) showed that the series converges absolutely.

This simple theorem has implications which extend well beyond the rather abstract formulation given above. Spitzer used the result to derive several properties of the time of first entry of the parallel sums $S_1, S_2, \ldots$ into the positive half-plane. For example, let $Z = S_N$ denote the first positive sum. Then the mean of Z is given by $E[Z] = 2^{-1/2} e^s$, and $n^{1/2} P[N > n] \to \pi^{-1/2} e^s$ as $n \to \infty$, where s denotes the series in (1). Baum and Katz (1963) proved that if $E[|X|^{2+\alpha}] < \infty$ for $0 \leqslant \alpha < 1$, then the convergence in (1) can be sharpened to

$$\sum_{n=1}^{\infty} n^{-1+\alpha/2} |\Pr[S_n \leqslant 0] - \tfrac{1}{2}| < \infty. \quad (2)$$

Heyde (1966) and Koopmans [4] considered the case of nonidentically distributed summands.

The results described above provide only *sufficient* conditions for the convergence of series. The problem of deriving necessary conditions is of a more difficult nature. To see this, observe that if the summands $X_1, X_2, \ldots, X_n$ are continuous and symmetrically distributed, then $\Pr[S_n = 0] = \frac{1}{2}$ for all n. Therefore the series in (2) converges, without regard to any moment conditions. In a sense, this awkward behavior is due to the Spitzer–Rosén problem being posed in too restrictive a context. Recall that the central limit theorem guarantees the convergence of $\Pr[S_n \leqslant n^{1/2} x]$ to the standard normal distribution function, $\Phi(x)$, for all values of x, and that the convergence is uniform in x. Therefore we might ask for conditions under which

$$\sum_{n=1}^{\infty} n^{-1+\alpha/2} \sup_{-\infty < x < \infty} \left| \Pr[S_n \leqslant n^{1/2} x] - \Phi(x) \right| < \infty. \quad (3)$$

Clearly, the convergence of this series implies convergence in (2). It follows from results of Friedman, Katz, and Koopmans (1966) and Heyde (1967), that if $0 < \alpha < 1$, condition (3) is *equivalent* to the moment condition, $E[|X|^{2+\alpha}] < \infty$. The case $\alpha = 0$ is a little more complicated, and is discussed in detail in Friedman et al. (1966), Heyde (1967), Egorov [1], and Heyde [3]. Hall [2] showed that under appropriate conditions, the series (2) and (3) converge or diverge together.

The references cited above which are not listed below explicitly are listed together in [4].

References

[1] Egorov, V. A. (1973). *Theor. Prob. Appl.*, **18**, 175–180. (Describes rates of convergence in CLT which are equivalent to existence of finite variance.)

[2] Hall, P. (1981). *Ann. Prob.*, **9**, 633–641. (Gives necessary and sufficient conditions for the Spitzer–Rosén theorem.)

[3] Heyde, C. C. (1973). *Zeit. Wahrscheinlichkeitsth. Verw. Geb.*, **25**, 83–95. (Describes rates of convergence in CLT which are equivalent to existence of finite variance.)

[4] Koopmans, L. H. (1968). *Ann. Math. Statist.*, **39**, 897–904. (Considers the Spitzer–Rosén theorem in non-i.i.d. case, and reviews earlier work.)

(ASYMPTOTIC NORMALITY
CONVERGENCE OF
   SEQUENCES OF RANDOM VARIABLES
LIMIT THEOREM, CENTRAL)

PETER HALL

SPLINE FUNCTIONS

UNIVARIATE POLYNOMIAL SPLINES

The name "spline function" was given by I. J. Schoenberg to the piecewise polynomial functions now known as *univariate polynomial splines*, because of their resemblance to the curves obtained by draftsmen using a mechanical spline—a thin flexible rod with weights or "ducks," used to position the rod at points through which it was desired to draw a smooth interpolating curve. See Schoenberg [42]. A univariate natural polynomial (unp) spline, f, is a function on $[0, 1]$ (any interval will do, of course), with the following properties: Given the positive integer m, and $n \geq m$ points $0 < t_1 < t_2 < \cdots < t_n < 1$, called "*knots*"

$$f \in \pi^{m-1}, \qquad t \in [0, t_1], \, t \in [t_n, 1],$$
$$f \in \pi^{2m-1}, \qquad t \in [t_i, t_{i+1}],$$
$$\qquad\qquad\qquad i = 1, \ldots, n - 1,$$
$$f \in C^{2n-2}, \qquad t \in [0, 1],$$

where π^k is the class of polynomials of at most degree k and C^k is the class of functions with k continuous derivatives. Thus f is a piecewise polynomial of degree $2m - 1$ with the pieces joined at the knots so that f has $2m - 2$ continuous derivatives, satisfying the $2m$ boundary conditions $f^{(k)}(0) = f^{(k)}(1) = 0$ for $k = m, m + 1, \ldots, 2m - 1$. "Natural" was the term given by Schoenberg to functions satisfying these (Neumann) boundary conditions which arise "naturally" from the solution to a variational problem, to be described below.

If f is represented by its polynomial coefficients, it is seen that it requires $2m$ coefficients to describe f in $[0, t_1]$ and f in $[t_n, 1]$, $(n - 1)2m$ coefficients to describe f in the $n - 1$ intervals $[t_i, t_{i+1}]$, $i = 1, \ldots, n - 1$, for a total of $2mn$ unknowns. The continuity conditions provide $(2m - 1)n$ conditions, which can be shown to be linearly independent, leaving n conditions to specify f completely. These conditions can be provided by specifying the values of f at $t_1, \ldots, t_n$.

Interpolating (unp) splines have been of interest to numerical analysts at least since Schoenberg's 1964 work. Suppose that f is some function which possesses a Taylor expansion* with remainder to order $m - 1$, and let f_n be the unp spline of interpolation to f at the points $t_1, \ldots, t_n$. Then f_n and its first $m - 1$ derivatives tend pointwise to f and its first $m - 1$ derivatives as n gets large, provided the t_i's are distributed "nicely," and if f possesses $2m$ continuous derivatives and satisfies the Neumann boundary conditions, then all $2m$ of the derivatives of f_n will converge to those of f. Integrals of f_n also converge to integrals of f, and this fact can be used to generate quadrature* formulas; see Schoenberg [43]. These favorable approximation theoretic properties, as well as the fact that splines are easy to compute, have lead to their popularity among numerical analysts. Interpolating splines are frequently the functions of choice when it is desired to represent everywhere a function whose values are given exactly only on a finite set of points. Piecewise polynomial functions satisfying other boundary and continuity conditions are also called splines. The scholarly work of Schumaker [44] provides a history of interpolating splines from a numerical analyst's point of view. Prenter [39] describes their role in the numerical solution of differential equations and De-Boor [12] is the standard reference on algorithms for generating (univariate) splines.

Splines are of interest to statisticians for the same reasons that they are of interest to numerical analysts, as well as because of their favorable properties in smoothing noisy data. The two major types of splines of interest for smoothing noisy data as a function of one variable are regression splines and smoothing splines. We will discuss these in turn, and then go on to some other splines.

REGRESSION SPLINES

To discuss regression splines, we first want to describe the B-splines (B stands for "basis"). *B*-splines can be conveniently defined in terms of truncated power functions and divided difference operators. Given a function $f(\cdot)$ and "knots" $t_i, \ldots, t_{i+k}$ define the divided difference operator $[t_i, \ldots, t_{i+k}]f(\cdot)$ as

$$[t_i, t_{i+1}]f(\cdot) = \frac{f(t_{i+1}) - f(t_i)}{t_{i+1} - t_i},$$

$$[t_i, t_{i+1}, t_{i+2}]f(\cdot) =$$

$$\left[\frac{f(t_{i+2}) - f(t_{i+1})}{t_{i+2} - t_{i+1}} - \frac{f(t_{i+1}) - f(t_i)}{t_{i+1} - t_i} \right] \bigg/ (t_{i+2} - t_i),$$

and so forth. For fixed x, we will let $f(\cdot) = (\cdot - x)_+^{k-1}$ be the truncated power function, where $(u)_+ = u$ if $u \geqslant 0$ and 0 otherwise. Then the *B*-spline of degree $k - 1$ for the knots $t_i, \ldots, t_{i+k}$ is defined as

$$B_i(x) = (t_{i+k} - t_i)[t_i, \ldots, t_{i+k}](\cdot - x)_+^{k-1}.$$

For example, for $k = 2$,

$$B_2(x) = \frac{(t_{i+2} - x)_+ - (t_{i+1} - x)_+}{t_{i+2} - t_{i+1}}$$
$$- \frac{(t_{i+1} - x)_+ - (t_i - x)_+}{t_{i+1} - t_i},$$

which is a tent function on $[t_i, t_{i+2}]$. $B_i(x)$ is a piecewise polynomial of degree $k - 1$ and can be shown to possess $k - 2$ continuous derivatives, be zero outside $[t_i, \ldots, t_{i+k}]$, and positive in the interior. The *B*-splines of degree $k - 1$ can also be obtained as projections on the real line of the volumes of convex polyhedra in k dimensions. Figure 1 illustrates this for $k = 2$. *B*-splines with coalescing knots are allowed; the effect is to reduce the continuity conditions at the multiple knots. If the t_i's are equally spaced, then one can also get the *B*-splines by shifting and rescaling the convolution of k uniform densities, and then the *B*-splines for $k \geqslant 1$ will be hill functions. Given k and a set of knots $t_1, \ldots, t_{N+k}$, one can define a set of N *B*-splines $B_1, \ldots, B_N$ where B_j is the *B*-spline of degree $k - 1$ with knots $t_j, \ldots, t_{j+k}$. These N functions provide a basis for a space of smooth functions with $k - 2$ continuous derivatives and may be used as regression functions when one wants

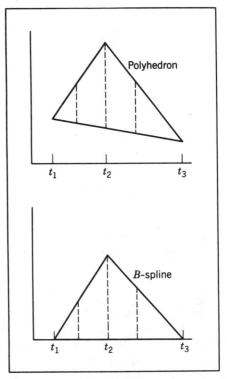

Figure 1   A convex polyhedron and its *B*-spline projection.

to fit a smooth function without otherwise specifying its form. Specifically, suppose we are given data $\mathbf{y} = (y_1, \ldots, y_n)$, from the model

$$y_i = f(x_i) + \epsilon_i, \quad i = 1, \ldots, n, \quad (1)$$

where f is known to be "smooth" and $\epsilon = (\epsilon_1, \ldots, \epsilon_n) \sim N(0, \sigma^2 I)$. Given knots $t_1, \ldots, t_{N+k}$ and the corresponding *B*-splines of degree $k - 1$, one may estimate f as f_N, where

$$f_N \sim \sum_{l=1}^{N} c_l B_l$$

by doing ordinary least-squares regression*, that is, choosing $\mathbf{c} = (c_1, \ldots, c_N)$ to minimize

$$\sum_{i=1}^{n} \left(y_i - \sum_{l=1}^{N} c_l B_l(x_i) \right)^2 .$$

If $N = n$, then f_N will interpolate the data, and as N becomes much smaller than n, f_N

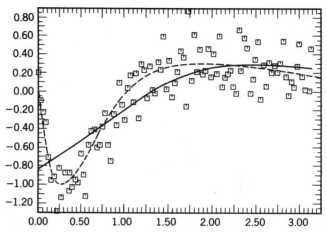

Figure 2 Model function (dashed curve), simulated data (open squares), and smoothing spline with too large value of λ (solid curve).

will have an increasingly smooth appearance, and the residuals will tend to increase. In principle, the knots $t_1, \ldots, t_N$ can be left as unknowns and chosen along with the coefficients to minimize the above sum of squares, but in practice, with noisy data, the determination of more than just a few knots this way is difficult and complicated by multiple local minima. Choosing the knots when interpolating a smooth function which is given exactly seems to be easier. See DeBoor [12]. The "eyeball" or trial and error method is also frequently used to choose the knots. Agarwal and Studden [1] give theoretical asymptotic results on the optimal number and location of knots for approximation to f in the model (1). Loosely speaking, if f has two derivatives then the optimal number of knots is of the order of $n^{1/5}$, so that there will be many fewer B-splines than data points. Regression splines are easy to compute using standard regression programs and the B-spline programs given in DeBoor [12], and if the true f is in the span of the B-splines chosen, then the estimate of f shares all of the usual properties of least-squares regression estimates. In general, however, the estimates of f may be biased, with the order of the bias similar to the order of the variance if N is chosen to minimize mean square error. See Buse and

Lim [8], Poirier [38], and Winsberg and Ramsay [72] for applications of regression splines.

SMOOTHING SPLINES

The other popular spline method for fitting the model (1) is to find f in an appropriate space of functions to minimize

$$\frac{1}{n} \sum_{i=1}^{n} (y_i - f(x_i))^2 + \lambda J_m(f), \quad (2)$$

where

$$J_m = \int_0^1 f^{(m)}(t) \, dt,$$

for some integer m. The minimizer f_λ to this problem was obtained by Schoenberg [42]. If there are at least m distinct x_i's, the solution, which is known as a *smoothing spline*, is unique and is a unp spline of degree $2m - 1$, with knots at the data points $x_1, \ldots, x_n$. The smoothing parameter λ controls the trade-off between the fit to the data as measured by the residual sum of squares and the smoothness, as measured by J_m. For $m = 2$, then $f^{(2)}$ is curvature, and a small J_2 corresponds to visual, or psychological, smoothness. As $\lambda \to \infty$, the solution tends to the polynomial of degree m best fitting

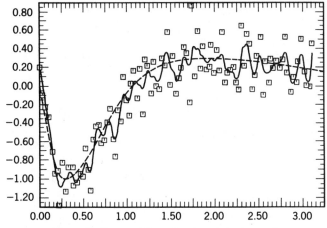

Figure 3   Same as Figure 2, except λ is too small.

the data in a least-squares* sense, and as λ → 0, f_λ tends to the unp spline which interpolates to the data.

Figure 2 from Wahba and Wold [69] shows a model function f (dotted line), data generated according to the model (1), and a smoothing spline fit to the data (solid line) with a value of λ which is too large. Figure 3 shows the same model f and data, and a smoothing spline with λ too small. Figure 4 shows the same f and data, and the fitted smoothing spline with λ chosen by ordinary cross-validation* (OCV). OCV involves deleting a data point and solving the opti-

mization problem with a trial value of λ, computing the difference between the predicted value and the deleted observation with this trial value of λ, accumulating the sums of squares of these differences as one runs through each of the data points in turn, and finally choosing the λ for which the accumulated sum is smallest. Generalized cross-validation (GCV) was developed later by Craven and Wahba [10]. and Golub et al. [19], and is an improvement over OCV both on asymptotic theoretical grounds and computational ease, although numerical results is an example like the one given can be ex-

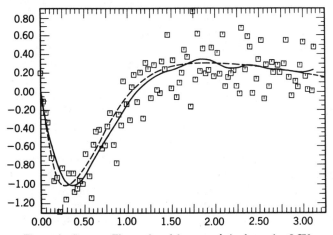

Figure 4   Same as Figures 2 and 3, except λ is chosen by OCV.

pected to be quite similar. Although originally these cross-validation methods for choosing λ were computationally expensive, fast $O(n)$ transportable code is now readily available for the smoothing spline with GCV, see the "Algorithms" section below.

CHOOSING BETWEEN REGRESSION AND SMOOTHING SPLINES

We make a few remarks on the choice of regression vs. smoothing splines for smoothing data from the model (1). Asymptotic theoretical convergence rates for the two methods are the same under the same assumptions (compare Agarwal and Studden [1] and Wahba [55]), provided the smoothing parameters N and λ are both chosen optimally. For very large data sets (say $n >$ 1000), the results for data from the model (1) are likely to be practically nearly the same, that is, indistinguishable on an $8\frac{1}{2}'' \times 11''$ plot, and the regression spline will require less storage to manipulate the results. Similar remarks concerning comparison of the results hold for medium-to-small sample sizes if the underlying f is close to being in the span of a small number of B-splines. (Recall that the optimal number of B-splines is $n^{1/5}$ under typical circumstances.) However it is likely that for examples like the one shown, and for example, the multimodal cases in Craven and Wahba [10], the regression spline with the optimal number of B-splines does not have the resolution to follow local features that can, in fact, be followed by a smoothing spline. A hybrid approach for very large data sets has been suggested by Nychka et al. [34], where the variational problem of (2) is solved in a space spanned by enough B-splines to avoid losing resolution. For estimation of derivatives or other local features such as maxima, the cross-validated smoothing spline is probably the method of choice, and in fact good results have been obtained with biomechanical data, which frequently satisfy the assumptions of model (1) well. See Woltring [73]. Some authors have found the use of regression splines

to be useful as a smoother in such applications as projection pursuit where a smoothing operation needs to be repeatedly carried out and sharp detail in the function is not expected.

SPLINES AS BAYES ESTIMATES

Smoothing splines can be interpreted as Bayes estimates if one views f as a sample function from a zero mean Gaussian stochastic process*. Loosely speaking, the stochastic process can be described as $f^{(m)}$ = $b \times$ white noise*. Then it can be shown that for each t the conditional expectation of $f(t)$ given the data $y_1, \ldots, y_n$ is $f_\lambda(t)$ with $\lambda = \sigma^2/nb$. See Kimeldorf and Wahba [27] and Wahba [57]. It can be seen from these references (and originally from the work of Parzen [37]) that there is a duality between Bayes estimates given discrete data on continuous-time stochastic processes and the solution to variational problems like (2), which extends to a very general class of penalty functionals. In particular, the splines in the remainder of this article which can be obtained as the solution to a variational problem also have interpretations as Bayes estimates.

MULTIVARIATE SPLINES

There are several generalizations of univariate splines to several variables. The multivariate B-splines are generalizations of the univariate B-splines, which are piecewise polynomials, satisfy certain continuity conditions, and have compact support. The thin-plate smoothing splines generalize the univariate polynomial splines as the solution to a variational problem, and are popular in meteorology*, computational vision, and other applications for smoothing two- and three-dimensional noisy data. They are not piecewise polynomials, however. The tensor* product smoothing splines also generalize the univariate splines as the solution to a variational problem, and are the foundation for the newly developing theory of interaction splines. We will discuss each of these separately.

MULTIVARIATE B-SPLINES

The multivariate B-splines in d variables are piecewise polynomials of degree $k - 1$, which are 0 outside a convex polyhedron and positive inside. They can be obtained as projections of the volume of convex polyhedra in $k + d$ dimensions onto Euclidean d-space. See DeBoor [11] and Hollig [22]. Tensor products of B-splines are special cases of multivariate B-splines. In two dimensions, by a tensor product B-spline we mean a function of two variables, say x_1 and x_2 of the form $f(x_1, x_2) = B_i(x_1)B_j(x_2)$. The multivariate B-splines have found applications in computer-aided design and other fields where it is desired to model a smooth surface in two or three dimensions given exact values of it at a finite number of points. Tensor products of B-splines have also been used as a basis for bivariate regression where the data are given on a regular grid.

THIN-PLATE SPLINES

The *thin-plate* splines are a popular generalization of the unp splines as solutions to a variational problem. In two dimensions ($d = 2$), with $m = 2$, the variational problem leading to the thin-plate spline is: Find $f \in X$ to minimize

$$\frac{1}{n} \sum_{i=1}^{n} \left(y_i - f(x_1(i), x_2(i)) \right)^2 + \lambda J_2^2(f),$$

where

$$J_2^2(f) = \int_{-\infty}^{\infty} \int_{-\infty}^{\infty} \left(f_{x_1 x_1}^2 + 2 f_{x_1 x_2}^2 + f_{x_2 x_2}^2 \right) dx_1 \, dx_2. \quad (3)$$

X is an abstract function space of functions of two variables for which (3) is finite, defined in Meinguet [32]. In d dimensions with general m, the variational problem becomes:

Find f in X (a space of functions of d variables) to minimize

$$\frac{1}{n} \sum_{i=1}^{n} \left(y_i - f(x_1(i), \ldots, x_d(i)) \right)^2 + \lambda J_m^d(f), \quad (4)$$

where

$$J_m^d(f) = \sum_{\alpha_1 + \cdots + \alpha_d = m} \frac{m!}{\alpha_1! \cdots \alpha_d!}$$
$$\times \int_{-\infty}^{\infty} \cdots \int_{-\infty}^{\infty} \left(\frac{\partial^m f}{\partial x_1^{\alpha_1} \cdots \partial x_d^{\alpha_d}} \right)^2 dx_1 \cdots dx_d. \quad (5)$$

It is necessary that $2m - d > 0$. An explicit representation for f_λ, the solution to this variational problem, was given by Duchon [13] and Meinguet [32], and is discussed further in a smoothing context by Wahba and Wendelberger [68]. The solution is known to lie in the span of a certain set of $n + m$ easily generated functions; see the "Algorithms" section for transportable software. Let $t_i = (x_1(i), \ldots, x_d(i))$, and let Δ be the Laplacian operator, that is, $\Delta = \partial^2/\partial x_1^2 + \cdots + \partial^2/\partial x_d^2$. Then f_λ has the property that $\Delta^m f_\lambda(t) = 0$ for all $t \neq t_1, \ldots, t_n$. This is a generalization of the analogous property of the unp spline in one variable, namely, $\Delta^m f_\lambda \equiv f_\lambda^{(2m)} = 0$ for $t \neq t_1, \ldots, t_n$, since f_λ is a polynomial of degree $2m - 1$ in the intervals between the knots. In several dimensions f_λ is a linear combination of $\binom{d + m - 1}{d}$ monomials in the d variables $x_1, \ldots, x_d$ of total degree less than m, and n other functions each of which is a Green's function for the mth iterated Laplacian. The thin-plate spline is also a Bayes estimate, and loosely speaking, one can think of the prior as f satisfying $\Delta^{m/2} f = b \times$ white noise. The thin-plate spline appears to be a particularly useful tool for smoothing data from diffusion processes* and other phenomena that may be thought of as representing the solution to an elliptic partial differential equation driven by white noise. The thin-plate spline is also a special case of an estimate of one of the "intrinsic random functions" of Matheron

[30]; see Duchon [13]. The integral over d-space in (5) can be replaced by an integral over a bounded region Ω containing the data points (see Dyn and Wahba [14]) and then the minimizer will satisfy Neumann boundary conditions on the boundary of Ω. For $d = 1$, the solution inside Ω coincides with the unp spline previously described, but for $d > 1$, the minimizer of (4) satisfies the Neumann boundary conditions only at ∞, and the two multivariate thin-plate splines will be different.

TENSOR PRODUCT SMOOTHING SPLINES

The space of functions of one variable referred to in connection with the univariate smoothing spline is known in the approximation theory literature as W_2^m and is a Hilbert space of functions with square integrable mth derivative. To smooth functions of two variables, one can define a tensor* product space $X = W_2^m \otimes W_2^m$ which consists of sums and limits of sums of functions of the form $f(x_1, x_2) = f_1(x_1)f_2(x_2)$ with f_1 and f_2 in W_2^m and find $f \in X$ to minimize

$$\frac{1}{n} \sum_{i=1}^{n} \left(y_i - f(x_1(i), x_2(i)) \right)^2 + \lambda J(f),$$

where now

$$J(f) = \int_0^1 \int_0^1 \left(\frac{\partial^{2m} f}{\partial x_1^m \, \partial x_2^m} \right)^2 dx_1 \, dx_2$$

$$+ \text{other terms.}$$

The other terms involve lower-order derivatives and guarantee that the solution will be unique under general conditions. Generalizations to d variables can be made. See Mansfield [29], Wahba [56], and Wahba [67]. These splines are piecewise polynomials in d variables, where the boundaries of the pieces are horizontal and vertical lines drawn through the data points. These splines have recently interested statisticians because of their role in the development of the interaction splines which are described below.

SPLINES ON THE CIRCLE AND THE SPHERE

Splines on the circle can be obtained by supposing that f is periodic with a representation of the form:

$$f(t) = a_0 + \sum_{\nu=1}^{\infty} a_\nu \cos 2\pi\nu t + \sum_{\nu=1}^{\infty} b_\nu \sin 2\pi\nu t,$$

$$\sum_{\nu=1}^{\infty} \left(a_\nu^2 + b_\nu^2 \right)(2\pi\nu)^{2m} < \infty,$$

and finding f to minimize

$$\frac{1}{n} \sum_{\nu=1}^{n} \left(y_i - f(t_i) \right)^2 + \int_0^1 \left(f^{(m)}(u) \right)^2 du.$$

A closed-form expression for f_λ as a piecewise polynomial which satisfies m periodic boundary conditions may be obtained by using the fact that

$$\sum_{\nu=1}^{\infty} \left(\cos 2\pi\nu s \cos 2\pi\nu t \right.$$

$$+ \sin 2\pi\nu s \sin 2\pi\nu t \left. \right)(2\pi\nu)^{-2m}$$

$$= \sum_{\nu=1}^{\infty} \cos 2\pi\nu(s - t)(2\pi\nu)^{-2m}$$

and this latter infinite series has a closed-form expression in terms of the $2m$th Bernoulli polynomial*; see Craven and Wahba [10].

The Bayes model here is

$$f(t) = a_0 + \sum_{\nu=1}^{\infty} \left(\alpha_\nu \cos 2\pi\nu t + \beta_\nu \sin 2\pi\nu t \right),$$

where the α_ν, β_ν are independent, zero mean normal random variables with $E\alpha_\nu^2 = E\beta_\nu^2 = b(2\pi\nu)^{-2m}$. This Bayes model also can be thought of as satisfying $f^{(m)} = b \times$ white noise, along with the periodic boundary conditions. A periodic smoothing spline with equally spaced data points can also be shown to be a kernel estimate. With unequally spaced data points in the general case there is an approximately equivalent variable kernel estimate*; see Silverman [48].

The spherical harmonics Y_{ls}, $s = -l, \ldots,$ l, $l = 0, 1, \ldots,$ play the same role on the sphere as sines and cosines on the circle. See

Sansone [41] for more on spherical harmonics. The spherical harmonics are the eigenfunctions of the surface Laplacian Δ on the sphere, with

$$\Delta Y_{ls} = -l(l+1)Y_{ls},$$

which is analogous to

$$\frac{d^2}{dt^2}\cos 2\pi\nu t = -(2\pi\nu)^2 \cos 2\pi\nu t.$$

Splines on the sphere are defined as the solution to the variational problem: find $f \in X$ (an appropriate space) to minimize

$$\frac{1}{n}\sum_{i=1}^{n}(y_i - f(P_i))^2 + \lambda \int_S (\Delta^{m/2}f)^2\,dP,$$

where S is the sphere, and P is a point on the sphere. A closed-form expression is available for f_λ in the cases $m = 2, 3$ (Wendelberger [71]). Approximate closed-form expressions may be found in Wahba [59, 60]. The corresponding Bayes model is

$$f(P_i) = \theta + \sum_{l=0}^{\infty}\sum_{s=-l}^{l} f_{ls}Y_{ls}(P_i),$$

where $E[f_{ls}^2] = b[l(l+1)]^{-m}$. This can also be viewed as $\Delta^{m/2}f = b \times$ white noise. Splines on the sphere have a number of interesting applications in geophysics and meteorology; see for example Shure et al. [46]. Vector smoothing splines can also be defined on the sphere and are useful in estimating horizontal vector fields from discrete, noisy measurements on, for example, the horizontal wind field, the magnetic field, etc.; see Wahba [62].

CHOOSING THE SMOOTHING PARAMETER, CONFIDENCE INTERVALS, DIAGNOSTICS

GCV appears to be the most popular method for choosing the smoothing parameter λ from the data in the context of smoothing splines, for various theoretical and practical reasons; see Craven and Wahba [10], Li [28], Speckman [49, 50], Utreras [52], and Wahba [55, 66]. GCV can be obtained from OCV by an invariance argument by rotating the system to a standard coordinate system, doing OCV, and rotating back. (Ordinary leaving out one is not invariant under rotations of the observation coordinate system.)

Let $A(\lambda)$ be the influence* matrix associated with f_λ, that is, $A(\lambda)$ satisfies

$$\begin{pmatrix} f_\lambda(t_1) \\ \vdots \\ f_\lambda(t_n) \end{pmatrix} = A(\lambda)\begin{pmatrix} y_1 \\ \vdots \\ y_n \end{pmatrix}.$$

An explicit representation of A in the unp case can be found in Craven and Wahba [10], and in general in Wahba [66]. The GCV estimate $\hat{\lambda}$ of λ is obtained as the minimizer of

$$V(\lambda) = \frac{(1/n)\|(I - A(\lambda))y\|^2}{((1/n)\mathrm{Tr}(I - A(\lambda)))^2}.$$

$A(\lambda)$ has many of the properties of the influence or hat matrix* in ordinary least-squares* regression, and this can be used to build a theory of confidence intervals and spline regression diagnostics*.

Trace $A(\hat{\lambda})$ can be viewed as the "degrees of freedom for signal." The posterior covariance matrix of $(f_\lambda(t_1), \ldots, f_\lambda(t_n))'$ is $\sigma^2 A(\lambda)$ and this fact has been used to propose posterior "confidence intervals" based on $\hat{\sigma}^2 A(\hat{\lambda})$, where $\hat{\sigma}^2 = RSS(\hat{\lambda})/\mathrm{Tr}(I - A(\hat{\lambda}))$. See Wahba [63]. These "confidence intervals" appear to have useful frequentist properties if interpreted "across the function," rather than pointwise; see Hall and Titterington [20], Nychka [33], Silverman [48], and Wahba [61].

Eubank [18] has proposed methods for detecting outliers* and influential observations*, by exploiting the influence matrix analogy. See also Silverman [48]. For example, letting $\epsilon_j(\hat{\lambda})$ be the jth residual, it is suggested that the quantities

$$T_j = \epsilon_j(\hat{\lambda})/\{\hat{\sigma}(1 - a_{jj}(\hat{\lambda}))\}^{1/2}$$

be called "studentized residuals" and an observation be considered an outlier if $|T_j|$ exceeds an appropriate critical value from a Student's t-distribution* with approximate degrees of freedom $\mathrm{Tr}(I - A(\hat{\lambda}))$.

PARTIAL SPLINES

Consider the model

$$y_i = f(t_i) + \sum_{j=1}^{p} \theta_j \Phi_j(t_i, z_i) + \epsilon_i.$$

Here $t_i \in E^d$, $z_j = (z_1(i), \ldots, z_q(i))'$, f is assumed to be "smooth" and the Φ_j's are known functions of $t \in E^d$ and the concomitant variables z. An estimate of θ and f may be obtained by finding $f \in X$ and $\theta \in E^p$ to minimize

$$\frac{1}{n} \sum_{i=1}^{n} \left(y_i - f(t_i) - \sum_{i=1}^{p} \theta_j \Phi_j(t_i, z_i) \right)^2$$

$$+ J_m^d(f).$$

Let $S_{n \times p}$ be the matrix with i, jth entry $\Phi_j(t_i, z_i)$. If the $n \times (M + p)$ matrix $(T : S)$ is of full column rank, then there will be a unique minimizer $(f_\lambda, \hat{\theta})$, and f_λ will be a thin-plate spline (in the case $d = 1$, a unp spline). Such models are extremely flexible and are attractive in a variety of applications; see Engle et al. [17], Ansley and Wecker [2], Shiau et al. [45], and Wahba [64] and references cited therein. Properties of the estimate of θ are an area of active research; see, for example, Heckman [21] and Rice [40].

SPLINES AS PENALIZED LIKELIHOOD ESTIMATES

Let

$$y_i \sim \text{Binomial}(1, p(t_i)),$$

where the logit $f(t) = \log p(t)/(1 - p(t))$, $t \in E^d$, is assumed to be a smooth function of t. The likelihood of y is

$$p^y (1 - p)^{1-y} = \exp y(p/(1 - p))$$

$$- \ln(1/(1 - p))$$

$$= \exp(yf - \ln(1 + e^f)).$$

The log likelihood of $y_1, \ldots, y_n$ then becomes

$$\log L = Q(y, f)$$

$$= \sum_i y_i f(t_i) - \ln(1 + e^{f(t_i)}).$$

A penalized* log-likelihood estimate of f is then the minimizer in X of

$$Q(y, f) + \lambda J_m^d(f). \tag{6}$$

The minimizer f_λ can be shown to be a thin-plate spline. If $t \in S$, then the spline penalty functional on the sphere can be used, and the solution is a spline on the sphere, and so forth. The binomial distribution above can be replaced by the Poisson distribution* $\Lambda(t)$ or, any member of the exponential family. These penalized likelihood estimates generalize the GLIM* (generalized linear models) described in McCullagh and Nelder [31] and reduce to the parametric GLIM model in the null space of the penalty functional in the case $\lambda = \infty$. Equation (6) generally has to be solved by numerical methods, typically in a sufficiently large but finite-dimensional space of convenient basis functions. GCV estimates of λ are defined for these models as follows: For a trial value of λ, (6) is minimized by a Gauss–Newton iteration*, equivalently, by minimizing an approximating sequence of quadratic optimization problems, At convergence, one evaluates the GCV function $V(\lambda)$ for the final quadratic optimization problem, repeats the calculation for another trial value of λ, until a minimizer is found; see O'Sullivan et al. [36]. A discussion of thin-plate basis functions which can be used in a hybrid numerical method for minimizing (6) is in Wahba [58]. Silverman [47] has proposed a density estimate in the context of penalized likelihood with the spline penalty functional. *See also* PENALIZED MAXIMUM LIKELIHOOD ESTIMATION.

INTERACTION SPLINES

Let $t = (x_1, \ldots, x_d)$ be in the unit cube in E^d. Stone [51] and others have suggested modeling f as

$$f(x_1, \ldots, x_d) = f_0 + \sum_{\alpha=1}^{d} f_\alpha(x_\alpha),$$

where $\int_0^1 f_\alpha(x_\alpha) \, dx_\alpha = 0$, $\alpha = 1, \ldots, d$, and the f_α are "smooth" functions. Both regression and smoothing splines have been proposed for the f_α. This idea has been extended to model f as

$$\begin{aligned} f(x_1, \ldots, x_d) \\ = f_0 + \sum_{\alpha=1}^{d} f_\alpha(x_\alpha) + \sum_{\alpha < \beta} f_{\alpha\beta}(x_\alpha, x_\beta) \\ + \sum_{\alpha < \beta < \gamma} f_{\alpha\beta\gamma}(x_\alpha, x_\beta, x_\gamma) + \cdots, \end{aligned}$$

where appropriate marginal integrals are 0 to guarantee identifiability, and the number of terms to be included is to be determined. Such splines are known as *interaction splines*, by analogy with analysis of variance*. The individual component functions $f_{\alpha\beta}$, $f_{\alpha\beta\gamma}$, etc., are *tensor product splines* satisfying certain conditions on their integrals analogous to conditions found in analysis of variance. See Barry [3, 4], Wahba [67], and Chen [9].

SPLINES WITH LINEAR INEQUALITY CONSTRAINTS

Splines satisfying a family of linear inequality constraints can be found as the solution to the problem: Find $f \in X$ to minimize

$$\frac{1}{n} \sum (y_i - f(t_i))^2 + \lambda J(f)$$

subject to

$$a_i \leqslant L_i f \leqslant b_i,$$

where L_i is a bounded linear functional. Included are discretized positivity and monotonicity constraints; see Utreras [53] and Villalobos and Wahba [54].

HISTOSPLINES

Histosplines is the name given to splines which are constructed to have a volume matching or volume smoothing property. They were introduced into the statistical literature in the context of density estimation* by Boneva et al. [7]. They constructed a univariate spline which had the volume matching property

$$\int_{x_i}^{x_{i+1}} f(x) \, dx = n_i/n,$$

where n_i is the number of observations from a random sample of size n which fell in the bin with boundaries x_i and x_{i+1}. Volume smoothing histosplines arise when one observes

$$y_i = \int_{\Omega_i} f \, dt + \epsilon_i$$

and chooses f as the minimizer of

$$\frac{1}{n} \sum_{i=1}^{n} \left(y_i - \int_{\Omega_i} f(t) \, dt \right)^2 + \lambda J_m^d(f). \quad (7)$$

See Wahba [60] and Dyn et al. [15] and references cited therein.

INDIRECT SENSING PROBLEMS

These occur when one observes

$$y_i = \int K(t_i, s) f(s) \, dt + \epsilon_i$$

and are of major practical importance. f can be estimated by solving a variational problem analogous to (7). See O'Sullivan [35] and references cited therein.

ALGORITHMS AND SOFTWARE

This is an area of active research and we only briefly mention a few results. In one dimension the unp spline has special structure which allows fast algorithms for computing both the spline and the GCV estimate of λ. Transportable code CUBGCV based on the fast algorithm proposed by Hutchinson and DeHoog [25] may be found in Hutchinson [24]; this algorithm with some additions is incorporated in GCVSPL of Woltring [74]. Literature connecting the

Markov properties of the unp spline and its relationship to Kalman filtering* has suggested fast algorithms; an early reference is Weinert and Kailath [40]. Older code (ICSSCV) for the unp spline with the GCV estimate of λ can be found in the IMSL library [6]. In more than one variable, the special structure of the one-dimensional case does not appear to exist and more general methods are required. The bidiagonalization approach of Elden [16] and the truncated singular value decomposition, Bates and Wahba [5] may be used to speed the calculation. Transportable code for thin-plate splines using thin-plate basis functions is available in Hutchinson [23], and for partial thin-plate splines and general problems using the truncated singular value decomposition, in GCVPACK (Bates et al. [26]). GCVSPL, GCVPACK, code for generating B-splines based on DeBoor [12] and other spline code may be obtained via an electronic mail daemon on the arpanet by writing netlib @ anl-mcs.arpa. The message "send index" will cause instructions for the use of the system to be returned to the sender.

References

[1] Agarwal, G. and Studden, W. J. (1980). *Ann. Statist.*, **8**, 1307–1325.

[2] Ansley, C. F. and Wecker, W. E. (1981). Extensions and examples of the signal extraction approach to regression. *Proceedings of ASA-CENSUS-NBER Conference on Applied Time Series Analysis of Economic Data*. Washington, D.C., pp. 181–192.

[3] Barry, D. (1983). Nonparametric Bayesian Regression. Thesis, Yale University, New Haven, CT.

[4] Barry, D. (1986). *Ann. Statist.*, **14**, 934–953.

[5] Bates, D. M. and Wahba, G. (1982). Computational methods for generalized cross-validation with large data sets. In *Treatment of Integral Equations by Numerical Methods*, C. T. Baker and G. F. Miller, eds. Academic, London, pp. 283–296.

[6] Bates, D. M., Lindstrom, M. J., Wahba, G., and Yandell, B. (1987). *Commun. Statist. Simul. Comp.*, **16**, 263–297.

[7] Boneva, K., Kendall, D., and Stefanov, I. (1971). *J. R. Statist. Soc. B*, **33**, 1–70.

[8] Buse, A. and Lim, L. (1977). *J. Amer. Statist. Ass.*, **72**, 64–68.

[9] Chen, Z. (1987). *Commun. Statist. Theor. Meth.*, **16**, 877–895.

[10] Craven, P. and Wahba, G. (1979). *Numer. Math.*, **31**, 377–403.

[11] DeBoor, C. (1976). Splines as linear combinations of *B*-splines: A survey. In *Approximation Theory II*, G. G. Lorenz, C. K. Chui, and L. L. Schumaker, eds. Academic, New York, pp. 1–48.

[12] DeBoor, C. (1978). *A Practical Guide to Splines*. Springer, New York.

[13] Duchon, J. (1977). Splines minimizing rotation-invariant semi-norms in Sobolev spaces. In *Constructive Theory of Functions of Several Variables*, W. Schempp and K. Zeller, eds. Springer, Berlin, pp. 85–100.

[14] Dyn, N. and Wahba, G. (1982). *SIAM J. Math. Anal.*, **13**, 134–152.

[15] Dyn, N., Wahba, G., and Wong, W. (1979). *J. Amer. Statist. Ass.*, **74**, 530–535.

[16] Elden, L. (1984). *BIT*, **24**, 467–472.

[17] Engle, R., Granger, C., Rice, J., and Weiss, A. (1986). *J. Amer. Statist. Ass.*, **81**, 310–320.

[18] Eubank, R. L. (1985). *J. R. Statist. Soc. B*, **47**, 332–341.

[19] Golub, G. H., Heath, M., and Wahba, G. (1979). *Technometrics*, **21**, 215–224.

[20] Hall, P. and Titterington, D. M. (1987). *J. R. Statist. Soc. B*, **49**, 184–198.

[21] Heckman, N. E. (1986). *J. R. Statist. Soc. B*, **48**, 244–248.

[22] Hollig, K. (1986). Multivariate splines. In *Approximation Theory*, C. deBoor, ed. American Mathematical Society, Providence, RI, pp. 103–127.

[23] Hutchinson, M. F. (1984). A Summary of Some Surface Fitting and Contouring Programs for Noisy Data. *Consulting Report No. ACT 84/6*, CSIRO Division of Mathematics and Statistics, Canberra, Australia.

[24] Hutchinson, M. F. (1985). *ACM Trans. Math. Software*, **12**, 150–153.

[25] Hutchinson, M. F. and DeHoog, F. R. (1985). *Numer. Math.*, **47**, 99–106.

[26] IMSL Library (1986).

[27] Kimeldorf, G. and Wahba, G. (1971). *J. Math. Anal. Appl.*, **33**, 82–95.

[28] Li, K. (1986). *Ann. Statist.*, **14**, 1101–1112.

[29] Mansfield, L. (1972). *Numer. Math.*, **20**, 99–114.

[30] Matheron, G. (1973). *Adv. Appl. Prob.*, **5**, 439–468.

[31] McCullagh, P. and Nelder, J. A. (1983). *Generalized Linear Models*. Chapman and Hall, London, England.

[32] Meinguet, J. (1979). *J. Appl. Math. Phys. (ZAMP)*, **30**, 292–304.

[33] Nychka, D. (1986). Frequency Interpretation of Bayesian "Confidence" Intervals for Smoothing Splines. *Mimeo Series 1699*, Institute of Statistics, North Carolina State University, Raleigh, NC.

[34] Nychka, D., Wahba, G., Goldfarb, S., and Pugh, T. (1984). *J. Amer. Statist. Ass.*, **79**, 832–846.

[35] O'Sullivan, F. (1986). *Statist. Sci.*, **1**, 502–527.

[36] O'Sullivan, F., Yandell, B., and Raynor, W. (1986). *J. Amer. Statist. Ass.*, **81**, 96–103.

[37] Parzen, E. (1970). Statistical inference on time series by rkhs methods. In *12th Biennial Seminar Canadian Mathematical Congress Proc.* R. Pyke, ed. Canadian Mathematical Congress, Montreal, pp. 1–37.

[38] Poirier, D. (1973). *J. Amer. Statist. Ass.*, **68**, 515–524.

[39] Prenter, P. (1975). In *Splines and Variational Methods*. Wiley, New York.

[40] Rice, J. (1986). *Statist. Prob. Lett.*, **4**, 203–208.

[41] Sansone, G. (1959). *Orthogonal Functions*. Wiley, New York.

[42] Schoenberg, I. (1964). *Proc. Natl. Acad. Sci. U.S.A.*, **52**, 947–950.

[43] Schoenberg, I. (1968). Monosplines and quadrature formulae. In *Theory and Application of Spline Functions*, T. N. E. Grenville, ed. University of Wisconsin Press, Madison, WI.

[44] Schumaker, L. (1981). *Spline Functions*. Wiley, New York.

[45] Shiau, J., Wahba, G., and Johnson, D. R. (1986). *Atmos. Ocean Tech.*, **3**, 714–725.

[46] Shure, L., Parker, R. L., and Backus, G. E. (1982). *J. Phys. Earth Planet Interiors*, **28**, 215–229.

[47] Silverman, B. W. (1982). *Ann. Statist.*, **10**, 795–810.

[48] Silverman, B. W. (1985). *J. R. Statist. Soc. B*, **46**, 1–52.

[49] Speckman, P. (1982). Efficient Nonparametric Regression with Cross-Validated Smoothing Splines. Dept. of Statistics, University of Missouri, Columbia, MO.

[50] Speckman, P. (1985). *Ann. Statist.*, **13**, 970–983.

[51] Stone, C. J. (1985). *Ann. Statist.*, **13**, 689–705.

[52] Utreras, F. (1978). Quelques Resultats D'Optimalité Pour La Methode De Validation Croissée. *Seminaire d'Analyse Nuumerique*, *301*, Grenoble, France.

[53] Utteras, F. (1985). *Numer. Math.*, **47**, 611–625.

[54] Villalobos, M. and Wahba, G. (1987). *J. Amer. Statist. Ass.*, **82**, 239–248.

[55] Wahba, G. (1975). *Numer. Math.*, **24**, 383–393.

[56] Wahba, G. (1978a). Interpolating Surfaces: High Order Convergence Rates and Their Associated Designs, with Applications to X-Ray Image Reconstruction. *Technical Report No. 523*, Statistics Dept., University of Wisconsin, Madison, WI.

[57] Wahba, G. (1978b). *J. R. Statist. Soc. B*, **40**, 364–372.

[58] Wahba, G. (1980). Spline bases, regularization, and generalized cross validation for solving approximation problems with large quantities of noisy data. In *Approximation Theory III*, W. Cheney, ed. Academic, New York, pp. 905–912.

[59] Wahba, G. (1981a). *SIAM J. Sci. Statist. Comp.*, **2**, 5–16.

[60] Wahba, G. (1981b). *Commun. Statist. Theor. Meth.*, **10**, 2475–2514.

[61] Wahba, G. (1982a). *SIAM J. Sci. Statist. Comp.*, **3**, 385–386.

[62] Wahba, G. (1982b). Vector splines on the sphere, with application to the estimation of vorticity and divergence from discrete, noisy data. In *Multivariate Approximation Theory*, Vol. 2, W. S. K. Zeller, ed. Verlag, Birkhäuser, pp. 407–429.

[63] Wahba, G. (1983). *J. R. Statist. Soc. B*, **45**, 133–150.

[64] Wahba, G. (1984). Cross validated spline methods for the estimation of multivariate functions from data on functionals. In *Statistics: An Appraisal, Proceedings 50th Anniversary Conference Iowa State Statistical Laboratory*, H. A. David and H. T. David, eds. Iowa State University Press, Ames, IA, pp. 205–235.

[65] Wahba, G. (1985a). *J. R. Statist. Soc. B*, **46**, 44.

[66] Wahba, G. (1985b). *Ann. Statist.*, **13**, 1378–1402.

[67] Wahba, G. (1986). Partial and interaction splines for the semiparametric estimation of functions of several variables. In *Computer Science and Statistics: Proceedings of the 18th Symposium on the Interface*, T. J. Boardman, ed. American Statistical Association, Washington, D.C., pp. 75–80.

[68] Wahba, G. and Wendelberger, J. (1980). *Monthly Weather Rev.*, **108**, 1122–1145.

[69] Wahba, G. and Wold, S. (1975). *Commun. Statist. A.*, **4**, 1–17.

[70] Weinert, H. L. and Kailath, T. (1974). *Ann. Statist.*, **2**, 787–794.

[71] Wendelberger, J. (1982). Smoothing Noisy Data with Multidimensional Splines and Generalized Cross-Validation. Ph.D. thesis, Statistics Dept., University of Wisconsin, Madison, WI.

[72] Winsberg, S. and Ramsay, J. O. (1983). University of Wisconsin, WI. *Psychometrika*, **48**, 575–595.

[73] Woltring, H. J. (1985). *Human Movement Sci.*, **4**, 229–245.

[79] Woltring, H. J. (1986). *Adv. Eng. Software*, **8**, 104–113.

(CURVE FITTING
GRADUATION

GRACE WAHBA

STATISTICAL MODELING

PREAMBLE

In applied probability* the primary objectives of modeling are the study, understanding, or prediction, by theory or simulation*, of the properties or behavior of a system. In statistical modeling the primary objective is inference*, with the aim of ascertaining the degrees to which observed data are compatible with alternative pictures of reality.

A statistical model has two main aspects: *structural*, concerning the *concrete* interpretation of the model in terms of facts and theories specific to the context, and *stochastic*, referring specifically to the *abstract* representation of sampling variability by well-defined probability distributions. These may coincide, especially in contexts involving deliberately randomized design or sampling (*see* RANDOMIZATION). Analysis based on descriptive statistics or exploratory data analysis* might make no stochastic assumptions (see also ref. 11).

This entry concentrates on establishing a general standpoint for statistics in the process of evolving new knowledge by the interplay between induction and deduction. The statistician assists this process by *criticism* and *estimation* [1]. The classical logic of the significance test is well adapted to criticism, while Bayesian methods offer the most power and flexibility for estimation. (Non-Bayesian methods will be called "classical" rather than "frequentist", since Bayesian probabilities bear a natural frequency interpretation in some contexts.)

Polarized opposition between Bayesian* and "classical" methodologies is misplaced in modeling. Bayesian posterior distributions* can be coherently manipulated and transformed by all the resources of probability calculus (in particular giving a straightforward approach to nuisance parameters*), and offer power and flexibility in the assessment of propositions involving parameter values, conditional on a structurally well-determined model. Classical methods, which can be logically based on the significance test* of a null hypothesis without explicit alternatives, are uniquely capable of indicating the inadequacy of a model and the need to formulate "something else" (a posterior probability* of "something else" cannot be computed [12, pp. 82–84]). Balanced accounts of these matters are rare: Box's [1] is outstanding.

STATISTICAL MODELS

In the "stochastic" sense, a *simple statistical model* is a unique probability distribution, asserted as a law of random variation in data. Since statistics envisages revision of such laws, it may also be called a *simple statistical hypothesis*. We generally represent a distribution by $P \equiv \{ p(x)|x \in \mathcal{X} \}$, where x is an instance ("outcome") of a random event X of any nature with outcomes in a sample space $\mathcal{X}$. A *compound statistical hypothesis* is a parametrized family $P \equiv \{ P_\theta | \theta \in \Theta \}$ of simple hypotheses. Abstractly, the θ-values are simply labels identifying the distributions, and their nature is irrelevant: Any one-to-one function $\phi(\theta)$ gives an equivalent parametrization, $P \equiv \{ P_\phi | \phi \in \Phi \}$, of P.

In the concrete or "structural" sense, a simple statistical hypothesis is any proposition sufficient to establish uniquely a probability distribution for the possible results of observing a system exhibiting random behavior. It is typically composed of elements referring to real features of the system and depends for its truth on several explicit physical conditions. The corresponding abstract model does not refer to these, and analysis based on it can only be interpreted when it has been reexpressed in concrete terms [7].

Structurally, a composite statistical hypothesis, or *statistical model*, is a family of simple ones, parametrized by variation of the elements or conditions; their nature is not relevant. If the same distribution should result from quite different conditions, the model is not *identifiable* in concrete terms.

IDENTIFIABILITY*

A model $P \equiv \{P_\theta | \theta \in \Theta\}$ is *identifiable* if whenever $\theta \neq \theta'$ the distributions P_θ, $P_{\theta'}$ are *distinct*, i.e., there is at least one event to which they assign different probabilities. The following straightforward result is important in some applications: Let $P = \{P_\theta\}$ be a model for a random event X, let the random event Y be a function of X, and let $Q = \{Q_\theta\}$ be the corresponding model for Y. Then if P is unidentifiable, so is Q.

CLASSICAL METHODOLOGY

The logical kernel of the classical approach is the significance test* of a null hypothesis, and its extension to hypothesis testing* within a family of alternative hypotheses.

Let $\Delta(D_0; H_0)$ be a *measure of discrepancy* between data D_0 and a hypothesis H_0: The larger Δ, the more remote H_0 as explanation of D_0. When H_0 holds, let $\Delta(D; H_0)$ have a definite distribution when D varies randomly under H_0, and let $\delta_0 = \Delta(D_0; H_0)$.

The specification of Δ induces a nested structure on the sample space, for given H_0, in terms of subsets such that each subset contains all observations D for which $\Delta(D; H_0) \geq \delta$ for some value of δ. Conversely, given a family $\{H\}$ of hypotheses and observed data D_0, a nested structure is induced on $\{H\}$ according to $\Delta(D_0; H) \geq \delta$ for different values of δ.

To each subset in the sample-space nesting can be attached its probability

$$\alpha = P_{H_0}[\Delta(D; H_0) \geq \delta]$$

under H_0. This is the basis for a significance test of a given hypothesis H_0, since when data D_0 are observed and also, for sufficiently small given α,

$$P_{H_0}[\Delta(D; H_0) \geq \delta_0]$$

$$= P_{H_0}[\Delta(D; H_0) \geq \Delta(D_0; H_0)] \leq \alpha,$$

then the observed datum belongs to an extreme (discrepant) class whose total probability is implausibly small, such hypotheses being "rejected at significance level α."

Conversely, given a family $\{H\}$ of hypotheses, each possible datum D_0 maps into the subset of $\{H\}$ not rejected at level α by the test based on $\Delta(D_0; H)$ when H is taken as null hypothesis. This set of hypotheses is a confidence set* at level $p = 1 - \alpha$, since if any $H_0 \in \{H\}$ is true, then the probability is at least p that the set so constructed contains H_0.

The above formulation is clearly very general and flexible, in that the choice of discrepancy function $\Delta(D; H)$ is open, and no fixed level of significance (α) or of confidence (p) is set. When "powerful" test procedures are available, these are embraced; but ad hoc or expedient procedures, nonparametric* or distribution-free* methods, and approaches to the testing of one dimension of a multiple parameter are also covered. The above dualism between hypothesis tests and confidence intervals is primary in the classical approach.

All discrepancy measures which are one-to-one monotonic functions of a given measure Δ are equivalent in that they will give rise to identical results for significance tests or confidence intervals. In applications, however, a given measure of discrepancy Δ will not arise arbitrarily: It will represent or summarize what the investigator perceives as being relevant and important features of the relationship between data and reality. In the course of any extended or complex investigation, the relevant features considered will vary kaleidoscopically as the problem is viewed under different aspects. The investigator will choose, among equivalent measures, one that immediately reflects his or her intuitive or reasoned perception of the current aspect.

The pure significance test, based on $\Delta(D; H_0)$, refers explicitly to only one hypothesis H_0. What constitutes departure from H_0 is subsumed in the form of Δ. It follows that significantly large values of $\Delta(D; H_0)$ evoke implicit *alternative* hypotheses H_1 for which $\Delta(D_0, H_1)$ should be more probable than $\Delta(D_0, H_0)$. For given Δ, only such alternatives are potentially "visible."

Formally, H_0 will typically be a logical conjunction $A_1 \wedge A_2 \wedge \cdots \wedge A_k$ of primary assumptions; if H_0 is "rejected" by a significance test*, the implication is that the logical disjunction (not A_1) $\vee$ (not A_2) $\vee \cdots \vee$ (not A_k) is true. The formal symbolic negation is a list of all the ways (not mutually exclusive) for the null hypothesis to fail. This ability to evoke alternatives is a valuable aid to model building. (See ref. 7 for further development of the above.)

BAYESIAN METHODOLOGY

Given a model $\{ p_\theta(x) | \theta \in \Theta \}$ for the data and a prior distribution $\pi(\theta)$ for the parameters, Bayes theorem* yields the posterior distribution

$$\pi(\theta | x) = \frac{p_\theta(x)\pi(\theta)}{\int p_\theta(x)\pi(\theta)\,d\theta}$$

on which all further inference is based. Joint specification of both $\{ p_\theta(x) | \theta \in \Theta \}$ and $\pi(\theta)$ gives a *Bayesian model*.

Subject to a justifiable specification of $\pi(\theta)$, it affords a uniquely powerful and flexible means of "estimation" within the parameter space Θ. The method of application is in principle straightforward: θ may be asserted to lie within a subset A of Θ with sufficiently high total posterior probability $\pi(A|x) = \int_A \pi(\theta|x)\,d\theta$. Indeterminacy in the choice of the subset A may be resolved by choosing it as the shortest, or so as to include all θ-values of sufficiently high likelihood* or posterior probability, and so on. Problems of decision (choice of action to

maximize expected gain) are also essentially straightforward (*see* DECISION THEORY). The weight of evidence in favor of a state of nature described by conditions satisfied within a region A of parameter space can be assessed, again, as the posterior probability $\pi(A|x)$ (though the integration may be technically formidable [39]).

Bayesian inference for a subparameter ϕ of a composite parameter $\theta = (\phi, \psi)$ depends on the marginal posterior distribution $\pi(\phi|x) = \int \pi(\phi, \psi|x)\,d\psi$; thus problems with nuisance parameters* are readily dealt with. When (as in computer diagnosis, etc.) inference must be made about a complex logical combination of propositions involving each of several parameters, the problems of combining the results of classical procedures (significance tests and confidence intervals) may prove insuperable; a Bayesian formulation may offer the only technically tractable approach.

Denoting by H all the primary assumptions underlying the choice of Bayesian model, i.e., of the parameter space Θ, the representations $\{ p_\theta(x) | \theta \in \Theta \}$ for the random variation of X given θ, and the prior distribution $\pi(\theta)$ of θ, the parameter and the data have a joint distribution

$$p(x, \theta | H) = p(x|\theta, H)p(\theta|H)$$
$$= p(\theta|x, H)p(x|H),$$

and these different factorizations can serve various purposes.

In particular, the factor $p(x|H) = \int p(x|\theta, H)p(\theta|H)\,d\theta$ can be regarded as a *Bayesian predictive distribution* for random data X, taking simultaneous account of the random variation X given θ and of the uncertainty in θ which are implied by H, and can be computed a priori. It can be used quite analogously to the significance test described under Classical Methodology: subject to establishment of a discrepancy function $\Delta(x; H)$, the data sample space can be structured into nested subsets corresponding to increasing discrepancy. If observed data x_0 fall into an extreme discrepant class of implausibly small total Bayesian predictive probability, the assumptions H underlying the Bayesian model may be "rejected" at the corresponding "significance level." [Box [1] proposes that the predictive distribution implies a discrepancy ranking: The smaller $p(x|H)$, the more discrepant is x from H. However, when $p(x|H)$ is a density for continuous X, this ranking is not invariant under transformation of x—an objection which does not apply to the use of the predictive distribution to evoke, in the manner described earlier, hypotheses H' for which $p(x|H') > p(x|H)$. In any case, as described above, interpretability of the modeling process goes with a discrepancy function interpretable in real terms.]

Although a Bayesian procedure has been used technically, the above inference scheme is not essentially Bayesian, since it does not yield a posterior probability for H; "rejection" is simply a message that "something else" (possibly not yet considered) must be sought. This is *global model criticism* [13].

In *local model criticism* [13] two or more basic hypotheses $H_1, \ldots, H_k$ will be explicitly present, with the specific Bayesian models they severally entail; it is in principle no embarrassment if the models involve completely disparate parametrizations. Given a prior distribution $\{w(H_i)\}$ of weights on the H_i, the posterior distribution is

$$w(H_i|x) \propto w(H_i) \int_{\Theta_i} p(x|\theta_i, H_i)\, d\theta_i,$$

$$\sum w(H_i|x) = 1,$$

where Θ_i is the parameter space for hypothesis H_i, θ_i the corresponding parameter. This clearly expresses which hypotheses remain plausible in the context of the given set $\{H_i\}$ (though, as Box [1] points out, the set as a whole is criticizable by the predictive "significance test").

CHOICE OF MODEL FOR THE DATA

Whether the context be Bayesian or not, the probability model $\{p_\theta(x|\theta)\}$ for the data x must be chosen. This may be done by mathematical deduction from a known or estab-lished random or sampling process underlying the observed system; by fitting frequency distributions to observed data; or even conventionally (as, no doubt, on most occasions when a normal distribution is assumed), simply to give the analysis "something to bite on." Whether the choice is appropriate or adequate may be investigated by goodness of fit* and other diagnostic* tests and by robustness* and sensitivity* analysis.

The data model $\{p_\theta(x|\theta)\}$ is common to Bayesian and non-Bayesian models [sometimes called the "agreed" or "public" element, as opposed to possible subjective specification of a Bayesian prior $\pi(\theta)$; but it may itself be subjective or arbitrary]. Both models are equally criticizable in this element. The Bayesian model is further criticizable in the choice of prior. While this is not usually directly verifiable, a difference between the results of a non-Bayesian significance test of H and of a Bayesian one (based on the predictive distribution) may give an indication of the appropriateness of choice of prior.

CHOICE OF BAYESIAN PRIOR DISTRIBUTION

Some contexts (e.g., games with randomized setup, certain genetic experiments) offer a well-defined objective prior distribution, deliberately created, or implied by established theory. In a wide class of other contexts, a process generating random parameter values is known to exist, and unknown parameters of the process can be estimated; this may be done by direct observation of θ-values (where possible), or indirect inference from a series of observed x-values (*see* EMPIRICAL BAYES THEORY, [10]).

Failing knowledge, or direct or indirect estimation, of the prior distribution, there are various resources for supplying one, more or less conventionally. The device of conjugate* priors allows a convenient tractability: the resulting posterior distribution is still in the same class as the prior; applying Bayes theorem amounts to updating a few parameter values, obviating problems of explicit integration.

It is common to use "vague" or "noninformative" priors. These are in effect nearly uniform over the range of parameter values for which the likelihood function is nonnegligible; the resulting posterior is nearly independent of the precise specification of the prior (note the essentially tautological nature of this remark). The procedure is an application of Savage's "*principle of precise measurement*" [12]. In effect, the prior is adopted without reference to knowledge of the context and the inference is determined by the likelihood function—the prior is merely fuel for Bayes theorem.

On a different plane are methods of prior specification which, while still context independent, have a theoretically interesting relationship with the data model $\{ p_\theta(x)|\theta \in \Theta \}$. These include Jeffreys' *invariant priors* [8, Sec. 3.10], with density function proportional to the square root of the determinant of the information matrix*, Bernardo's *reference priors* [23] and Jayne's *maximum-entropy* methods [34]. These proposals have spawned extensive literatures; they have at least the attraction of seeming to avoid capricious choice by appeal to some principle of universal character. All these approaches may give rise to *improper* priors, which should be used circumspectly, owing to the possibility of *marginalization* and other paradoxes (e.g., [4, 17]).

There remain the approaches characteristic of the *personalistic* interpretation of probability as rational degree of belief. These depend, in one way or another, on eliciting evidence about the form of the prior distribution from quantitative information about a person's appraisals of uncertainty or that person's reactions to possible outcomes, by methods which can range from the purest introspection to the use of very sophisticated questionnaires or situational experiments, nowadays often computer assisted.

Finally, even when a prior can be established it may in some contexts be inappropriate to use it. These are typically decision problems where the "rules of the game"

inhibit judging a particular case by reference to knowledge of related cases, exemplified by a rule of law that "the accused is deemed innocent until proved guilty beyond reasonable doubt." This amounts to a classical hypothesis test whose significance level limits the proportion of innocents convicted, a dominant consideration according to the rule. In consequence, the proportion of correct decisions is in general suboptimal.

By contrast, a Bayesian prior expressing the propensity to a crime within a cognate population allows an optimal overall proportion of correct decisions, at the possible expense of decreased safeguard for the innocent (thus partly judged by the behavior of his peers), and corresponds to a different rule of good judicial behavior. At bottom is an issue of *exchangeability**: An administrator may regard members of a population as permutable, while individuals are likely to take the contrary view.

References and Bibliography

As the list of cross-references at the end suggests, modeling ramifies into practically every area of statistical theory and practice. The following bibliography, grouped under the headings *General*, *Modeling and Analysis Techniques*, and *Case Studies*, is an eclectic and personal selection of literature; each item is worth reading for some kind of insight into the principles, techniques, or practice of modeling. The entries under *Case Studies* are chosen for good discussions of the general issues raised above, as they arise in practice.

General

[1] Box, G. E. P. (1980). Sampling and Bayes' inference in scientific modelling. *J. R. Statist. Soc. A*, **143**, 383–430 (with discussion). (Important analysis of frequentist and Bayesian issues in statistical modeling, enhanced by extensive discussion.)

[2] Box, G. E. P. and Hill, J. W. (1967). Discrimination among mechanistic models. *Technometrics*, **19**, 15–18.

[3] Box, G. E. P. and Tiao, G. C. (1973). *Bayesian Inference in Statistical Analysis*. Addison-Wesley, Reading, MA.

[4] Dawid, A. P., Stone, M., and Židek, J. V. (1973). Marginalization paradoxes in Bayesian and structural inference, *J. R. Statist. Soc. B*, **35**, 189–233 (with discussion).

[5] Dickey, J. M. (1973). Scientific reporting and personal probabilities: Student's hypothesis. *J. R. Statist. Soc. B*, **35**, 285–305.

[6] Diggle, P. J. and Gratton, R. J. (1984). Monte Carlo methods of inference for implicit statistical models. *J. R. Statist. Soc. B*, **46**, 193–227.

[7] Harding, E. F. (1986). Modelling: the classical approach. *The Statistician*, **35**, 103–114. (Develops classical methodology by several examples.)

[8] Jeffreys, H. (1939/1961). *Theory of Probability*, 3rd ed. Oxford University Press, London and New York. (Pioneering tract in "objective Bayes" theory; full of wisdom on modeling and analysis.)

[9] *J. R. Statist. Soc. A*, **147**, Part 2 (1984): 150th Anniversary Volume. (The "Personal Views" on present position and potential developments are excellent reading for modeling issues.)

[10] Morris, C. N. (1983). Parametric empirical Bayes inference. *J. Amer. Statist. Ass.*, **78**, 47–65 (with comments). (Readable comprehensive survey.)

[11] Reese, R. A. (1986). Data analysis: The need for models? *The Statistician*, **35**, 199–206. (A sceptical view.)

[12] Savage, L. J. and other contributors (1962). *The Foundations of Statistical Inference*, G. A. Barnard and D. R. Cox, eds. Methuen, London, England.

[13] Smith, A. F. M. (1986). Some Bayesian thoughts on modelling and model choice. *The Statistician*, **35**, 97–102. (Terse survey of up-to-date Bayes approaches.)

[14] Smith, A. F. M. and Spiegelhalter, D. J. (1980). Bayes factors and choice criteria for linear models. *J. R. Statist. Soc. B*, **42**, 213–220.

[15] Smith, T. M. F. (1983). On the validity of inferences from non-random samples. *J. R. Statist. Soc. A*, **146**, 394–403.

[16] *The Statistician*, **35**, (1986). Special Issue on Statistical Modelling.

[17] Stone, M. and Dawid, A. P. (1972). Un-Bayesian implications of improper Bayesian inference in routine statistical problems. *Biometrika*, **59**, 269–375.

Modeling and Analysis Techniques

[18] Aitchison, J. (1975). Goodness of prediction fit. *Biometrika*, **62**, 547–554.

[19] Aitchison, J. and Dunsmore, I. R. (1965). *Statistical Prediction Analysis*. Cambridge, University Press, Cambridge, England.

[20] Akaike, H. (1973). Information theory and an extension of the maximum likelihood principle. *2nd Int. Symp. Inf. Theory*, B. N. Petrov and F. Czaki, eds. Akademiai Kiado, Budapest, Hungary, pp. 267–281. (The "Akaike criterion.")

[21] Anderson, J. A. (1984). Regression and ordered categorical variables. *J. R. Statist. Soc. B*, **46**, 1–30 (with discussion).

[22] Anscombe, F. J. (1981). *Computing in Statistical Science through APL*. Springer, New York. (The many examples are very well discussed from the modeling aspect.)

[23] Bernardo, J. M. (1979). Reference posterior distributions for Bayesian inference. *J. R. Statist. Soc. B*, **41**, 113–147 (with discussion).

[24] Butler, R. W. (1986). Predictive likelihood inference with applications. *J. R. Statist. Soc. B*, **48**, 1–38 (with discussion).

[25] Chatfield, C. (1985). The initial examination of data. *J. R. Statist. Soc. A*, **148**, 214–253.

[26] Cook, R. D. (1986). Assessment of local influence. *J. R. Statist. Soc. B*, **48**, 133–169 (with discussion).

[27] Cook, R. D. and Weisberg, S. (1982). *Residuals and Influence in Linear Regression*. Chapman and Hall, New York and London.

[28] Cox, D. R. (1961). Tests of separate families of hypotheses. *Proc. 4th Berkeley Symp. Prob. Math. Statist.*, **1**, 105–123.

[29] Dempster, A. P., Laird, N. M., and Rubin, D. B. (1977). Maximum likelihood estimation from incomplete data via the EM algorithm. *J. R. Statist. Soc. B*, **39**, 1–38 (with discussion).

[30] Fuchs, C. (1982). Maximum likelihood estimation and model selection in contingency tables with missing data. *J. Amer. Statist. Ass.*, **77**, 270–278.

[31] Geisser, S. (1971). The inferential use of predictive distributions. In *Foundations of Statistical Inference*, V. P. Godambe and D. A. Sprott, eds. Holt, Rinehart, and Winston, Toronto, Canada, pp. 456–469.

[32] Geisser, S. and Eddy, W. F. (1979). A predictive approach to model selection. *J. Amer. Statist. Ass.*, **74**, 153–160.

[33] Green, P. J. (1984). Iteratively reweighted least squares for maximum likelihood estimation, and some robust and resistant alternatives. *J. R. Statist. Soc. B*, **46**, 149–192 (with discussion).

[34] Jaynes, E. T. (1968). Prior probabilities. *IEEE Trans. Systems Sci. Cybern.*, **SSC-4**, 227–291.

[35] Jeffreys, H. (1946). An invariant form for the prior probability in estimation problems. *Proc. R. Soc., London, A*, **186**, 453–461.

[36] Kalbfleisch, J. D. and Sprott, D. A. (1970). Application of likelihood methods to models involving large numbers of parameters. *J. R. Statist. Soc. B*, **32**, 175–208 (with discussion).

[37] Larson, M. G. and Dinse, G. E. (1985). A mixture model for the regression analysis of competing

risks data. *Appl. Statist.*, **34**, 201–211. (Mixtures, EM algorithm.)

[38] Lawrance, A. J. and Lewis, P. A. W. (1985). Modelling and residual analysis of non-linear autoregressive time series in exponential variables. *J. R. Statist. Soc. B*, **47**, 165–202 (with discussion).

[39] Naylor, J. C. and Smith, A. F. M. (1982). Applications of a method for the efficient computation of posterior distributions. *Appl. Statist.*, **31**, 214–225.

[40] Pettit, L. I. (1986). Diagnostics in Bayesian model choice. *The Statistician*, **35**, 183–190.

[41] Skene, A. M., Shaw, J. E. H., and Lee, T. D. (1986). Bayesian modelling and sensitivity analysis. *The Statistician*, **35**, 281–288.

[42] Spiegelhalter, D. J. and Smith, A. F. M. (1982). Bayes factors for linear and log-linear models with vague prior information. *J. R. Statist. Soc. B*, **44**, 377–387.

[43] Stewart, L. and Davis, W. W. (1986). Bayesian posterior distributions over sets of possible models computed by Monte Carlo simulation. *The Statistician*, **35**, 175–182.

[44] West, M. (1986). Bayesian model monitoring. *J. R. Statist. Soc. B*, **48**, 70–78.

[45] West, M., Harrison, P. J., and Migon, H. (1985). Dynamic generalized linear models and Bayesian forecasting. *J. Amer. Statist. Ass.*, **80**, 73–97 (with comments).

Case Studies

[46] Aitkin, M. and Longford, N. (1986). Statistical modelling issues in school effectiveness studies. *J. R. Statist. Soc. A*, **149**, 1–43 (with discussion).

[47] Aitkin, M., Anderson, D., and Hinde, J. (1981). Statistical modelling of data on teaching styles. *J. R. Statist. Soc. A*, **144**, 419–461 (with discussion).

[48] Andrews, D. F. and Herzberg, A. M. (1985). *Data —A Collection of Problems from Many Fields for the Student and Research Worker*. Springer, New York. (71 data sets to try modeling with.)

[49] Beck, M. B. (1986). The selection of structures in models of environmental systems. *The Statistician*, **35**, 151–162.

[50] Bennett, R. J. and Haining, R. P. (1985). Spatial structure and spatial interaction: Modelling approaches to the statistical analysis of geographical data. *J. R. Statist. Soc. A*, **148**, 1–27; Discussion, 27–36.

[51] Besag, J. E. (1986). On the statistical analysis of dirty pictures. *J. R. Statist. Soc. B*, **48**, 259–279. (Markov field models in image reconstruction.)

[52] Broadbent, S. (1980). Simulating the ley hunter. *J. R. Statist. Soc. A*, **143**, 109–140 (with discussion). (An archeological application.)

[53] Brooks, R. J., Dawid, A. P., Galbraith, J. I., Galbraith, R. F., Stone, M., and Smith, A. F. M. (1978). A note on forecasting vehicle ownership. *J. R. Statist. Soc. A*, **141**, 64–68. (See also ref. 64.)

[54] Chatfield, C. and Pepper, M. P. G. (1971). Time-series analysis: An example from geophysical data. *Appl. Statist.*, **20**, 217–237.

[55] Cruz-Orive, L. M., Hoppeler, H., Mathieu, O., and Weibel, E. R. (1985). Stereological analysis of anisotropic structures using directional statistics. *Appl. Statist.*, **34**, 14–32. (Anisotropic stereology of tissue capillaries.)

[56] Dinse, G. E. and Lagakos, S. W. (1983). Regression analysis of tumor prevalence data. *Appl. Statist.*, **32**, 236–248.

[57] Felsenstein, J. (1983). Statistical inference of phylogenies. *J. R. Statist. Soc. A*, **146**, 246–272 (with discussion).

[58] Hjorth, U. and Holmqvist, L. (1981). On model selection based on validation with applications to pressure and temperature prognosis. *Appl. Statist.*, **30**, 264–274. (Model selection, cross-validation, prediction, meteorology).

[59] Kempton, R. A. and Howes, C. W. (1981). The use of neighbouring plot values in the analysis of variety trials. *Appl. Statist.*, **30**, 59–70. (Nearest-neighbor models).

[60] Mountford, M. D. (1982). Estimation of population fluctuations with application to the common bird census. *Appl. Statist.*, **31**, 135–143.

[61] Oldham, P. D. (1985). The fluoridation of the Strathclyde Regional Council's water supply: Opinion of Lord Jauncey *in causa* Mrs. Catherine McColl against Strathclyde Regional Council: A review. *J. R. Statist. Soc. A*, **148**, 37–44.

[62] Racine, A., Grieve, A. P., Flühker, H., and Smith, A. F. M. (1986). Bayesian methods in practice: Experiences in the pharmaceutical industry. *Appl. Statist.*, **35**, 93–120; Discussion, 121–150.

[63] Smith, A. F. M. and West, M. (1983). Monitoring renal transplants: An application of the multi-process Kalman filter. *Biometrics*, **39**, 867–878.

[64] Tanner, J. C. (1978). Long-term forecasting of vehicle ownership and road traffic. *J. R. Statist. Soc. A*, **141**, 14–63 (with discussion). (See also ref. 53).

[65] Wachter, K. W. and Trussell, J. (1982). Estimating historical heights. *J. Amer. Statist. Ass.*, **77**, 279–303 (with comments).

[66] Wilkinson, G. N., Eckert, S. R., Hancock, T. W., and Moyo, O. (1983). Nearest neighbour (NN)

analysis of field experiments. *J. R. Statist. Soc. B*, **45**, 151–211 (with discussion).

[67] Yardi, Y., Shepp, L. A., and Kaufman, L. (1985). A statistical model for positron emission tomography. *J. Amer. Statist. Ass.*, **80**, 8–37 (with comments).

[68] Young, R. J. and Young, M. (1986). Use of prior information in specifying the dynamics of sales-advertising relationships. *The Statistician*, **35**, 263–269.

(Modeling plays a pervasive role in statistical work. The following list contains a representative selection of entries in which modeling is an important component. Although lengthy, it is by no means exhaustive.)

(ARCHAEOLOGY, STATISTICS IN
ASTRONOMY, STATISTICS IN
BIOSTATISTICS
BOX–JENKINS MODEL
BRADLEY–TERRY MODEL
CAPTURE–RECAPTURE METHODS
CATASTROPHE THEORY
COHERENT STRUCTURE THEORY
COHORT ANALYSIS
COMPARTMENT MODELS, STOCHASTIC
CONDITIONAL INFERENCE
CONSULTING, STATISTICAL
CORRELATION
CORRESPONDENCE ANALYSIS
COX'S REGRESSION MODEL
CUMULATIVE DAMAGE MODELS
DAMAGE MODELS
DAM THEORY
DENDROCHRONOLOGY
ECOLOGICAL STATISTICS
EXPLORATORY DATA ANALYSIS
FACTOR ANALYSIS
FALLACIES, STATISTICAL
FATIGUE MODELS
FISHERY RESEARCH, STATISTICS IN
FIXED-, RANDOM-, AND MIXED-EFFECTS
   MODELS
GENETICS, STATISTICS IN
GEOGRAPHY, STATISTICS IN
GEOLOGY, STATISTICS IN
GEOSTATISTICS
GROWTH CURVES
HUMAN GENETICS, STATISTICS IN
HYDROLOGY, STATISTICS IN
IDENTIFICATION PROBLEMS
INFERENCE, DESIGN BASED VS. MODEL
   BASED

INFERENCE, STATISTICAL, I AND II
LAG MODELS, DISTRIBUTED
LATENT STRUCTURE ANALYSIS
LAWLIKE RELATIONSHIPS
LEARNING MODELS
LINGUISTICS, STATISTICS IN
LOGIC OF STATISTICAL REASONING
LONGITUDINAL DATA ANALYSIS
METEOROLOGY, STATISTICS IN
MIXTURE DISTRIBUTIONS
MODEL CONSTRUCTION: SELECTION OF
   DISTRIBUTIONS
MODELS I, II, AND III
MULTIDIMENSIONAL SCALING
MULTIVARIATE COX REGRESSION
   MODEL
MULTIVARIATE PROBIT MODEL
NONPARAMETRIC CLUSTERING
   TECHNIQUES
PATTERN RECOGNITION
PLATEAU MODELS, LINEAR
POPULATION GROWTH MODELS
PREDICTIVE ANALYSIS
PSYCHOPHYSICS, STATISTICS IN
RANDOMNESS AND PROBABILITY
   —COMPLEXITY OF DESCRIPTION
RECURSIVE PARTITIONING
REGRESSION: CONFLUENCE ANALYSIS
REGRESSION DIAGNOSTICS
REGRESSION MODELS, TYPES OF
REGRESSION VARIABLES, SELECTION OF
REPRODUCTIVE MODELS
SCIENTIFIC METHOD AND STATISTICS
SHOCK MODELS
SIZE AND SHAPE ANALYSIS
SOCIAL NETWORK ANALYSIS
SPECIFICATION, PREDICTOR
STATISTICAL CATASTROPHE THEORY
STATISTICS, AN OVERVIEW
STATISTICS IN ANIMAL SCIENCE
STATISTICS IN CRYSTALLOGRAPHY
STATISTICS IN FORESTRY
STATISTICS IN PSYCHOLOGY
STRESS–STRENGTH MODELS
STRONG TRUE-SCORE THEORY
STRUCTURAL MODELS
SURVIVAL ANALYSIS
SYSTEMS ANALYSIS IN ECOLOGY
TEST FACTOR STRATIFICATION
TIME SERIES
TRACKING
TREND
ULTRASTRUCTURAL RELATIONSHIPS
UNFOLDING

E. F. HARDING

STATISTICS AND ARTIFICIAL INTELLIGENCE

In the context of statistical work, artificial intelligence (AI) consists of automated procedures for applying statistical procedures in the analysis of data. These will commonly include applications of exploratory data analysis, model construction, goodness of fit*, hypothesis testing*, and estimation*, and may include more specialized techniques appropriate to particular problems. As in all applications of AI, there is an implied authoritarianism, since certain analytical sequences are regarded as being "correct."

A key use of AI in statistics is in formalization of *strategies* of data analysis. Hand [13] defines "statistical strategy" as "a formal description of the choices, actions, decisions to be made while using statistical methods in the course of a study." Further discussion can be found in refs. 2, 9, and 18. The need for statistical strategies is especially acute at present (1988) to prevent, or reduce, misuse of statistical computer packages (*see* STATISTICAL SOFTWARE). Tools developed in AI may permit us to mechanize some statistical strategies; the success of "expert systems" suggests they may well be helpful in statistics. As Gale [8] points out: "AI techniques have been used in statistics for . . . demonstration that software can . . . translate a medical hypothesis into a statistical study, help a user to select an appropriate statistical technique and check automatically for the validity of the assumptions behind a statistical technique."

A typical example is the REX (Regression EXpert) system developed by Gale and Pregibon [9, 10]. This system encodes enough knowledge to apply simple linear regression*

"safely," and so provides a statistical strategy for a statistically naïve user. A similar system for MANOVA (multivariate analysis of variance*) has been constructed by Hand [14].

Other applications of AI are as "assistant" (software), more appropriate for use by professional statisticians. The concept was put forward by Huber [16]. It would perform automated record keeping and answer simple questions about correct data. A related development is the "student system" (Shafer [21]), based on selecting and working examples, and giving answers to questions.

Another issue is that of drawing a boundary between symbolically (algebraically) and numerically represented knowledge. An automated knowledge acquisition process, taking this into account, has been proposed by Ellman [4].

Some experts (e.g., Huber [16]) distrust the currently prevalent approach of using AI to enable "naïve users" to carry out sophisticated analysis in specialized fields, but emphasize the value of AI in extending human ability into areas in which they are presently unable to perform.

The collection of papers [7] includes thorough discussions of potentials of AI in statistics.

Statistical techniques, themselves, contribute to AI systems in the areas of (i) allowance for uncertainty and (ii) of learning concepts.

(i) Reasoning in AI systems must take into account the uncertainty of empirical relationships. In the domain of medical diagnosis, development of the MYCIN system [1] allowed for uncertainty as an essential ingredient in representing knowledge needed to use Bayesian methods. Shafer's [21] theory of evidence, and the earlier work of Dempster on upper and lower probability, have been applied in expert systems (see, e.g., Gordon and Shortliffe [11, 12] and a recent review by Spiegelhalter [22]). Zadeh [24] advocates the use of fuzzy sets* and "possibility theory" for the treatment of uncertainty in expert systems. However, Lindley [17] argues against the use of procedures which are not based on the calculus of probability*. Pearl [19] appears to support Lindley.

(ii) In the study of learning processes, the concept of *formation* (clustering in statistical terminology) is basic in AI. Although AI is primarily concerned with the clustering methods applicable to categorical types of data, while statisticians here have mainly been concerned with applications with variables measured on a "continuous" scale. (See Fisher and Langley [5] for a study of mutual interests in this area.)

Hora [15] investigated the use of statistical methods in assessing the performance—especially the learning capabilities—of intelligent machines.

Also, (iii) probabilistic methods for the combination of expert opinions can be used for the design of expert support systems and other decision-making techniques. Winkler [23] provides a comprehensive survey.

References

[1] Buchanan, B. G. and Shortliffe, E. H., eds. (1984). *Rule-Based Expert Systems*. Addison-Wesley, Reading, MA.

[2] Chambers, J. M. (1981). In *Computer Science and Statistics, Proc. 13th Symp. Interface Pittsburgh, PA*, W. F. Eddy, ed. Springer, New York, pp. 36–40.

[3] Charniak, E. (1983). *Proc. Natl. Conf. Artif. Intell.*, pp. 70–73.

[4] Ellman, T. (1986). In *Artificial Intelligence and Statistics*, W. A. Gale, ed. Addison-Wesley, Reading, MA, pp. 229–238.

[5] Fisher, D. and Langley, P. (1986). In *Artificial Intelligence and Statistics*, W. A. Gale, ed. Addison-Wesley, Reading, MA, pp. 77–116.

[6] Fox, J. (1986). In *Artificial Intelligence and Statistics*, W. A. Gale, ed. Addison-Wesley, Reading, MA, pp. 57–76.

[7] Gale, W. A., ed. (1986). *Artificial Intelligence and Statistics*. Addison-Wesley, Reading, MA.

[8] Gale, W. A. (1986). In *Artificial Intelligence and Statistics*, W. A. Gale, ed. Addison-Wesley, Reading, MA, pp. 1–16.

[9] Gale, W. A. and Pregibon, D. (1982). In *Comput. Sci. Statist. Proc. 14th Symp. Interface Washington, DC.*, K. W. Heines, R. S. Sacher, and J. W.

Williamson, eds. Springer, New York, pp. 110–117.

[10] Gale, W. A. and Pregibon, D. (1984). *AI Mag.*, **5**, 72–75.

[11] Gordon, J. and Shortliffe, E. H. (1984). In *Rule-Based Expert Systems*, B. G. Buchanan and E. H. Shortliffe, eds. Addison-Wesley, Reading, MA, pp. 272–292.

[12] Gordon, J. and Shortliffe, E. H. (1985). *Artif. Intell.*, **26**, 323–357.

[13] Hand, D. J. (1984). *The Statistician, London*, **33**, 351–369.

[14] Hand, D. J. (1986). In *Artificial Intelligence and Statistics*, W. A. Gale, ed. Addison-Wesley, Reading, MA, pp. 355–387.

[15] Hora, S. C. (1986). In *Artificial Intelligence and Statistics*, W. A. Gale, ed. Addison-Wesley, Reading, MA, pp. 117–131.

[16] Huber, P. J. (1986). In *Artificial Intelligence and Statistics*, W. A. Gale, ed. Addison-Wesley, Reading, MA, pp. 285–291.

[17] Lindley, D. V. (1986). The Calculus of Uncertainty in Artificial Intelligence and Expert Systems (Proc. of a Conference, Dec. 28–29, 1984). *Tech. Rep. GWU/SRRA/TR-86/2*, The George Washington University, Washington, D.C. pp. 112–128.

[18] Oldford, R. W. and Peters, S. C. (1984). In *Artificial Intelligence and Statistics*, W. A. Gale, ed. Addison-Wesley, Reading, MA, pp. 335–353.

[19] Pearl, J. (1986). *Artif. Intell.*, **28**, 9–15.

[20] Selig, S. M., ed. (1986). The Calculus of Uncertainty in Artificial Intelligence and Expert Systems (Proc. of a Conference, Dec. 28–29, 1984). *Tech. Rep. GWU/SRRA/TR-86/2*, The George Washington University, Washington, D.C.

[21] Shafer, G. (1976). *A Mathematical Theory of Evidence*. Princeton University Press, Princeton, NJ.

[22] Spiegelhalter, D. J. (1986). In *Artificial Intelligence and Statistics*, W. A. Gale, ed. Addison-Wesley, Reading, MA, pp. 17–55.

[23] Winkler, R. L. (1986). *Manag. Sci.*, **32**, 298–328.

[24] Zadeh, L. A. (1986). In The Calculus of Uncertainty in Artificial Intelligence and Expert Systems (Proc. of a Conference, Dec. 28–29, 1984). *Tech. Rep. GWU/SRRA/TR-86/2*, The George Washington University, Washington, D.C.

(BAYESIAN INFERENCE
CLASSIFICATION
DECISION THEORY
HIERARCHICAL CLUSTER ANALYSIS
INFERENCE, STATISTICAL—I, II
LEARNING MODELS)

STOCHASTICALLY CLOSED REFERENCE SETS

Randomization tests* are based on repeated permuting (dividing or rearranging) of data from randomized experiments to provide a reference set of data permutations and associated test statistic values. A reference set represents results for a set of alternative randomizations (random assignments), each member representing results for a particular randomization. The proportion of data permutations in the reference set that have test statistic values greater than or equal to the value for the experimental results is the *P* value*. (*See* RANDOMIZATION TESTS.)

Data-permuting procedures may be: (a) random, (b) nonrandom ("systematic"), or (c) partly random and partly systematic. Practical considerations may dictate which procedure to use. A single rationale can be employed for ensuring the valid application of data-permuting procedures. A data-permuting procedure is valid when a reference set it produces meets this criterion:

> Given reference set *R*, comprised of *n* members (data permutations), the conditional probability* that any particular member represents the experimental results is $1/n$, when the null hypothesis is true.

A reference set meeting this criterion is called a *stochastically closed reference set* [1, pp. 322–328].

The following example concerns a data-permuting procedure that is partly random and partly systematic and provides a stochastically closed reference set. To perform a one-tailed randomization test of the difference between treatments *A* and *B*, in a completely randomized equal-*n* design, let *A* be the treatment expected to provide the larger measurements, and let T_A, the total of the *A* measurements, be the test statistic. Compute T_A for the experimental results and for 999 *random* data permutations. Also, *systematically* permute the experimental results and each of the 999 random data per-

mutations by transposing the *A* and *B* measurements and compute T_A for each new data permutation... *or* perform the more economical but equivalent operation of subtracting T_A from the grand total of the *A* and *B* measurements to get a second test statistic value from each of those 1000 data permutations. The economical procedure only permutes the data implicitly and may appear invalid; however, when the systematic data-permuting component is made explicit, the reference set can readily be shown to be stochastically closed. Thus this method of increasing the sensitivity of a randomization test by doubling the number of test statistic values in a reference set is valid.

Reference

[1] Edgington, E. S. (1987). *Randomization Tests*, 2nd ed. Marcel Dekker, New York. (This revision includes additional randomization tests and computer programs, as well as a chapter on theory.)

(RANDOMIZATION TESTS
REFERENCE SET)

EUGENE S. EDGINGTON

STOCHASTIC COMPLEXITY

In MINIMUM-DESCRIPTION-LENGTH PRINCIPLE (MDL) a coding theoretic approach to statistical modeling was reviewed, in which the idea is to pick that model in a parametric class which allows encoding of the observed data with the fewest number of binary digits. Further developments have taken place, which we review here. The central concept is defined by a Bayesian* type of formula, of which the MDL criterion in the cited entry is just a computationally obtained upper bound. We can also interpret the approach equivalently as a maximum *unconditional* likelihood principle. The word "unconditional" means that the objective is to maximize a probability, or density, evaluated at the observations such that it only depends on the selected model class rather than on specific parameter values. The unconditional probability, or, alternatively, the code length, serves as a universal "utility function," which permits a fair comparison of model classes regardless of the number of parameters in them. Here, no "true" distribution assumption is needed, and on the whole an approach to statistical inquiry results which is free from arbitrary choices.

STOCHASTIC COMPLEXITY

Let $\{ f(\mathbf{x}|k, \boldsymbol{\theta}) : \boldsymbol{\theta} = (\theta_1, \ldots, \theta_k), \quad k = 1, 2, \ldots \}$ denote a parametric class of distributions, represented by densities, such that for each member the compatibility (marginality) conditions required for a random process are satisfied. Here, $\mathbf{x} = x_1, \ldots, x_n$ stands for a finite string of observations, also written as $\mathbf{x}^n$, to indicate its length. For each k, let $\pi(\boldsymbol{\theta}|k)$ be a strictly positive distribution in the k-dimensional parameter space. These "priors" need not be interpreted in any particular manner, because the end result will provide a justification for them. Relative to such a class of models, which by no means is meant to include any "true" distribution, we define the *stochastic complexity* of the data $\mathbf{x}$ to be

$$I(\mathbf{x}) = -\log \left[\frac{1}{n} \sum_{k=1}^{n} \int f(\mathbf{x}|k, \boldsymbol{\theta}) \, d\pi(\boldsymbol{\theta}|k) \right],$$
$$(1)$$

where the integration is over the k-dimensional space of the parameters. In the degenerate case without free parameters we define $I(\mathbf{x}) = -\log f(\mathbf{x})$, which is the *Shannon information**.

The density function $f(\mathbf{x}) = 2^{-I(\mathbf{x})}$ depends only on the chosen model class, and it satisfies the compatibility conditions because each member does it by assumption. This is why we may regard it as an unconditional likelihood function. On the other hand, the stochastic complexity $I(\mathbf{x})$, being the negative logarithm of a density rather than of a probability, differs from a code length only by an additive term, proportional to n and

the precision to which the observations are written. As to its minimality, one can show the following: Whenever the model class satisfies certain reasonable smoothness conditions (Rissanen [1]), and if $g(\mathbf{x})$ is any density function satisfying the marginality constraints required for a random process, then for every positive ϵ and essentially all values for $\boldsymbol{\theta}$ for each k,

$$E_{k, \boldsymbol{\theta}} \left[\log \frac{f(\mathbf{x})}{g(\mathbf{x})} \right] \geqslant -\epsilon \log n \qquad (2)$$

for all large enough values of n. This has the general implication that for increasingly long sequences it will be progressively more difficult to construct a density function which would assign a greater density to the data than $f(x) = 2^{-I(x)}$. Therefore, inasmuch as the fundamental problems of modeling and, in fact all of statistics, center on the construction of the most likely explanation of observations, we see that this goal is well achieved by the unconditional likelihood function $f(\mathbf{x})$ or its equivalent, the stochastic complexity.

Stochastic complexity is defined relative to a class of models consisting of $f(\mathbf{x}|k, \boldsymbol{\theta})$ and $\pi(\boldsymbol{\theta}|k)$. Hence it can be used to compare the goodness of any two model classes regardless of the number of parameters in them. So long as we have only a handful of candidate classes, we need not consider the code length required to describe them. If, however, the model classes themselves require a substantial amount of bits for their specification, then, of course, their description must be included in the total code length. In particular, the "priors" $\pi(\boldsymbol{\theta}|k)$ ought to be so found that the resulting complexity gets minimized, which differs from the aim with the usual priors, namely, to represent prior knowledge about some "true" parameter value. In our formalism, we see that a good prior will have the bulk of its probability mass near the maximum likelihood estimates. (More could be said about the difficult problem of formalizing prior knowledge.) Finally, since each model class may be held only until a better one is found, the purpose of statistical inquiry fundamentally becomes a search for a sequence of steadily improving model classes. No algorithm exists for finding the ultimate class, which leaves the task of suggesting the model classes to human intuition.

THREE-MODEL SELECTION CRITERIA

The stochastic complexity does not depend on an optimal model nor does it deliver one. Moreover, since the integral in (1) can be worked out in a closed form only for special model densities, it will be necessary to derive approximations, which act as model selection criteria. Here, we give three, the first two of which asymptotically approach the stochastic complexity, and the third in the Gaussian family of models with the conjugate* priors provides a close approximation to the stochastic complexity even for small samples.

By expanding the integrand in (1) in Taylor's series* about the maximum likelihood estimate $\hat{\theta}(n) = \hat{\theta}(\mathbf{x}^n)$ for each value of k, and by selecting a universal prior for the real numbers (Rissanen [1]), one can show that the MDL criterion

$$\min_{k,\theta} \left\{ -\log f(\mathbf{x}|k,\theta) \right.$$

$$\left. + \frac{k}{2} \log n + O(\log\log n) \right\} \quad (3)$$

provides an upper bound for $I(\mathbf{x})$. Let the data be ordered in some manner, and for each k let $\hat{\theta}(t)$ be the maximum likelihood estimate* with k components, computed from the partial data $\mathbf{x}^t = x_1, \ldots, x_t$. Then also the predictive criterion

$$\min_k \left(-\sum_{t=0}^{n-1} \log f\left(x_{t+1}|\mathbf{x}^t, k, \hat{\theta}(t)\right) \right) \quad (4)$$

provides an asymptotically accurate estimate of the stochastic complexity. The very first term requires a separate definition, and we evaluate it only for $k = 0$, after which k is incremented one by one until enough data have been gathered to admit a unique esti-

mate of the parameters. This criterion offers a third independent interpretation of the stochastic complexity as the accumulated prediction errors when the data sequence is predicted such that each observation is predicted only from the past. We see that such "honest" prediction automatically defines a proper density function $f(\mathbf{x})$. This contrasts with most of the usual criteria, some of which still are referred to quite inaccurately as "prediction" error criteria. As a special case this criterion gives a new predictive least-squares technique (Rissanen [2]).

The third criterion is

$$\min_k \left\{ \frac{n}{2} \log R_k(\mathbf{x}) + \tfrac{1}{2}\log|C_k(\mathbf{x})| \right\}, \quad (5)$$

where $R_k(\mathbf{x})$ denotes the sum of the least-squares deviations, obtained by fitting k parameters to the data, and $C_k(\mathbf{x})$ is the matrix defined by the double derivatives of the sum of the squared deviations, evaluated at the least-squares estimates. This criterion is independent of any ordering of the data; it is a refinement of the criterion (3) and also simpler to apply than the cited predictive least-squares criterion.

References

[1] Rissanen, J. (1986a). Stochastic complexity and modeling. *Ann. Statist.*, **14**, 1080–1100.

[2] Rissanen, J. (1986b). A predictive least squares principle. *IMA J. Math. Control Inform.*, **3**, 211–222.

[3] Rissanen, J. (1987). Stochastic complexity. *J. R. Statist. Soc. B*, **49**, 223–239.

(ALGORITHMIC INFORMATION THEORY
BAYESIAN INFERENCE
INFORMATION THEORY AND CODING
  THEORY
MAXIMUM LIKELIHOOD ESTIMATION
MINIMUM-DESCRIPTION-LENGTH
  PRINCIPLE
RANDOMNESS AND PROBABILITY
 —COMPLEXITY OF DESCRIPTION
STATISTICAL MODELING)

J. RISSANEN

STOCHASTIC REGRESSION MODELS

For many variants of the linear statistical model the regressors or instrumental variables are assumed to be fixed or nonstochastic. This assumption may be justified in many instances when the investigator can exercise some control over the data generation process. However, in many areas of science, for example economics, much of the data used for statistical analysis are passively generated, and the values the explanatory variables take on are determined within the system and are stochastic rather than deterministic. Under this type of sampling scheme we discuss, in the sections ahead, the statistical consequences of using alternative estimation rules.

In order of presentation we first consider the case where the regressors are stochastic but independent of the equation error term. In this case the maximum likelihood–least-squares* estimator is unbiased and consistent. Second, we consider the case when the regressors and the equation error term are not perfectly independent, in the sense that they are only independent of the contemporaneous (and sometimes succeeding) errors. In this case the least-squares estimator for the partially independent stochastic regressor model has the desirable asymptotic properties of being consistent and asymptotically efficient*. Next, we consider the case in which the regressors are not independent of the equation errors. In this case the least-squares estimator is not only biased but also inconsistent. In dealing with the general stochastic regressor models, all we can hope for are procedures that will produce estimators that have some desirable asymptotic properties. The instrumental variable* method, in the absence of other better alternatives, provides a consistent estimator, although it is not necessarily efficient. Finally, using a squared-error loss measure, we evaluate the performance of alternative prediction functions for the stochastic regression model.

THE STATISTICAL MODEL AND THE MAXIMUM LIKELIHOOD ESTIMATION RULE

Assume a sample of size T is randomly drawn from a $(K + 1)$ variate normal population where all parameters are unknown and $T > K$. Thus we observe independent random vectors, $z_1, z_2, \ldots, z_T$, each of dimension $(K + 1)$ and distributed as a multivariate normal* where

$$z_t = \begin{bmatrix} y_t \\ x_t \end{bmatrix}$$

for $t = 1, 2, \ldots, T$. The mean vector μ and nonsingular covariance matrix Σ of the distribution of the z_t are unknown and written as

$$\mu = \begin{bmatrix} \mu_y \\ \mu_x \end{bmatrix}$$

and

$$\Sigma = \begin{bmatrix} \sigma_y^2 & \Sigma'_{xy} \\ \Sigma_{xy} & \Sigma_{xx} \end{bmatrix}.$$

Therefore, μ_x is a $(K \times 1)$ vector and Σ_{xx} is a matrix of dimension $(K \times K)$. The conditional distribution of the random variable y_t given x_t, which may be specified as $(y_t|x_t) = \beta_0 + x'_t \beta + e_t$, is normally and independently distributed with mean $E[y_t|x_t] = \beta_0 + x'_t \beta$, and constant variance $\sigma^2 = \sigma_y^2 - \Sigma'_{xy} \Sigma_{xx}^{-1} \Sigma_{xy} = \sigma_y^2 (1 - \rho^2)$. The K-dimensional vector β is given by $\beta = \Sigma_{xx}^{-1} \Sigma_{xy}$, and $\beta_0 = \mu_y - \mu'_x \beta$. The corresponding population multiple correlation coefficient* ρ^2 is $\rho^2 = \Sigma'_{xy} \Sigma_{xx}^{-1} \Sigma_{xy} / \sigma_y^2$.

Further, let the corresponding mean and covariance sample statistics based on the T independent random vectors z_t be denoted by

$$\begin{vmatrix} \bar{y} \\ \bar{x} \end{vmatrix} \quad \text{and} \quad \begin{bmatrix} S_y^2 & S'_{xy} \\ S_{xy} & S_{xx} \end{bmatrix},$$

where $S_{xx} = \Sigma_t x_t x'_t - T\bar{x}\bar{x}'$, $S_{xy} = \Sigma_t x_t y_t - T\bar{x}\bar{y}$, $S_y^2 = \Sigma_t y_t^2 - T(\bar{y})^2$.

Given sample observations z_t, for $t = 1, 2, \ldots, T$, the maximum likelihood estimators of β and β_0 are $\tilde{\beta} = S_{xx}^{-1} S_{xy}$, and $\tilde{\beta}_0 =$

$\bar{y} - \bar{\mathbf{x}}'\tilde{\boldsymbol{\beta}}$. The sample multiple correlation coefficient is $R^2 = \mathbf{S}'_{xy}\mathbf{S}^{-1}_{xx}\mathbf{S}_{xy}/S^2_y$.

SAMPLING PROPERTIES OF THE ESTIMATOR

Given the underlying statistical model that describes the data generation process and the maximum likelihood rule for estimating the unknown parameters, let us consider various statistical models that arise from various assumptions regarding the sampling process underlying $\mathbf{x}_t$ and e_t.

$\mathbf{x}_t$ Distributed Independently of e_t

When the $\mathbf{x}_t$ are distributed independently of e_t then $E[e_t|\mathbf{x}'_t] = 0$ and $E[e^2_t|\mathbf{x}'_t] = \sigma^2$ and the maximum likelihood estimator of β_0 and $\boldsymbol{\beta}$ are unbiased. The estimator $\hat{\sigma}^2 = \tilde{\mathbf{e}}'\tilde{\mathbf{e}}/(T - K - 1)$, based on the residuals $\tilde{e}_t = y_t - \tilde{\beta}_0 - \mathbf{x}'_t\tilde{\boldsymbol{\beta}}$, is an unbiased estimator of σ^2 since $E[\hat{\sigma}^2|\mathbf{X}] = \sigma^2$ and $\hat{\sigma}^2 E[(\mathbf{X}'\mathbf{X})^{-1}]$ is an unbiased estimator of $\sigma^2 E[(\mathbf{X}'\mathbf{X})^{-1}]$. See Judge et al. [3] for a development of these results.

$\mathbf{x}_t$ Partially Independent of e_t

In many real-world data generation schemes the assumption of independence between $\mathbf{x}_t$ and e_t is untenable. A well-known example of this case is when one of the $\mathbf{x}_t$ is the lagged value of y_t, which is not independent of e_t. If we assume the e_t are serially uncorrelated, the least-squares estimator of β_0 and $\boldsymbol{\beta}$ is biased. However, the bias vanishes as the sample size approaches infinity and is thus consistent, that is plim$[\tilde{\beta}_0, \tilde{\boldsymbol{\beta}}] = \beta_0, \boldsymbol{\beta}$. The estimator $\hat{\sigma}^2$ is also consistent since the $\mathbf{x}_t$ and e_t are contemporaneously independent and thus plim $T^{-1}\mathbf{X}'\mathbf{e} = \mathbf{0}$. In this case the classical results hold and the limiting distribution of

$$\sqrt{T}\begin{pmatrix} \tilde{\beta}_0 - \beta_0 \\ \tilde{\boldsymbol{\beta}} - \boldsymbol{\beta} \end{pmatrix}$$

is normally distributed with mean vector zero and covariance σ^2plim$[\mathbf{X}'\mathbf{X}/T]^{-1}$. See Judge et al. [3] for the development.

$\mathbf{x}_t$ not Independent of e_t

If the e_t are *not* independently distributed random variables, then under the scenario of the preceding section the y_{t-1} are determined in part by e_{t-1} and are thus not independent of e_t. In this case the least-squares estimator is not consistent and is biased in finite samples. Alternatively, when one or more of the $\mathbf{x}_t$ are unobservable, the $\mathbf{x}_t$ are not contemporaneously uncorrelated with e_t and again the least-squares estimator is not consistent. Another case that is met in practice occurs when the $y_t, \mathbf{x}_t$ represent an instantaneous feedback system. In all of these cases the least-squares estimator of $\beta_0, \boldsymbol{\beta}$ is biased and inconsistent since plim $T^{-1}\mathbf{X}'\mathbf{e}$ does not vanish because $\mathbf{X}$ and $\mathbf{e}$ are contemporaneously correlated. See Judge et al. [3] for an analysis of these models.

A Consistent Estimator

The inconsistency of the least-squares estimator occurred in the previous section when plim$(\mathbf{X}'\mathbf{e}/T) \neq \mathbf{0}$. This suggests within the least-squares context that if we could find auxiliary variables $\mathbf{w}_t$ that are uncorrelated with e_t, so that plim$(\mathbf{W}'\mathbf{e}/T) = 0$, plim$(\mathbf{W}'\mathbf{X}/T) = \Sigma_{wx}$ exists and is nonsingular, and plim$(\mathbf{W}'\mathbf{y}/T) = \Sigma_{xy}$ exists, then we may, using the sample moments, use the instrumental variable estimator $\tilde{\boldsymbol{\beta}} = (\mathbf{W}'\mathbf{X})^{-1}\mathbf{W}'\mathbf{y}$ to estimate $\boldsymbol{\beta}$, where $w_1, w_2,\ldots, w_K$ are the instrumental variables. This estimator is consistent and if the asymptotic distribution of $\mathbf{W}'\mathbf{e}/\sqrt{T}$ is $N(0, \sigma^2$ plim$(\mathbf{W}'\mathbf{W}/T))$, then asymptotically $\sqrt{T}(\tilde{\boldsymbol{\beta}} - \boldsymbol{\beta})$ is

$$N\left(\mathbf{0}, \sigma^2 \text{ plim}\left[\frac{\mathbf{W}'\mathbf{X}}{T}\right]^{-1}\left[\frac{\mathbf{W}'\mathbf{W}}{T}\right]\left[\frac{\mathbf{X}'\mathbf{W}}{T}\right]^{-1}\right).$$

In this case the covariance of $\boldsymbol{\beta}$ which may be consistently estimated by the sample moments, is not necessarily a minimum since there may be many sets of instrumental variables. See Judge et al. [3].

A Prediction Function

Within the context of the first section consider the case where $\mathbf{x}_{(T+1)}$ is observed and $y_{(T+1)}$ is unobserved. The problem is that of predicting $y_{(T+1)}$ from $x_{(T+1)}$ using a prediction function based on the original sample. The maximum likelihood prediction function is $\tilde{y}_{(T+1)} = \mathbf{x}'_{(T+1)}\beta + \beta_0$. If we consider this problem using a squared-error loss measure as a basis for gauging predictor performance, Baranchik [1] has shown that if $K > 2$, a prediction estimator of the form

$$\tilde{\tilde{y}} = \tilde{y} - h\left(R^2/(1 - R^2)\right)\left(\mathbf{x}_{(T+1)} - \bar{\mathbf{x}}\right)\tilde{\beta}$$

is minimax and dominates $\tilde{y}$ if

(i) $0 \leqslant uh(u) \leqslant 2(K - 2)/(T - K - 1)$ for $u \geqslant 0$,

(ii) the derivative of $uh(u)$ is nonnegative for $u \geqslant 0$, and

(iii) $h(u) \leqslant (T - 3)/(K - 1)$ for $u \leqslant (K - 1)/(T - K - 2)$ when the derivative of $uh(u)$ is positive.

Alternative Stein-like prediction functions for the stochastic regressor case are considered by Judge and Bock [2]. If interest centers on parameter estimation rather than the prediction of $y_{(T+1)}$ the risk derivations of Judge and Bock [2, pp. 229–259] carry over directly for the stochastic regressor case.

SUMMARY

Whether or not the least-squares estimators of the regression parameters will be unbiased or consistent depends on whether the stochastic regressors are independent of, partially dependent on, or contemporaneously dependent on the errors e_t. Since the assumption of independent stochastic regressors is sometimes hard to justify we must seek alternative estimating procedures that will give consistent estimators. The anatomy

of the least-squares method suggests that the instrumental variable method produces consistent estimators provided that the instrumental variables are uncorrelated with the error disturbances but correlated with the stochastic regressors. An instrumental variable estimator may not be efficient, but in the absence of other alternatives at least provides a consistent estimator. Within the context of a squared-error loss measure Stein-like estimators offer a superior alternative to the conventional estimators usually applied in practice.

References

[1] Baranchik, A. J. (1973). *Ann. Statist.*, **1**, 312–321.

[2] Judge, G. G. and Bock, M. E. (1978). *The Statistical Implications of Pre-Test and Stein Rule Estimators in Econometrics*. North-Holland, Amsterdam, Netherlands. (Presents a range of Stein-like estimators for the regression model.)

[3] Judge, G. G., Hill, R. C., Griffiths, W. E., Lütkepohl, H., and Lee, T. C. (1988). *Introduction to the Theory and Practice of Econometrics*. Wiley, New York. (Specifies and analyzes a wide range of estimators for the regression model, and contains references to seminal papers in the area.)

(ECONOMETRICS
INSTRUMENTAL VARIABLE ESTIMATOR
JAMES–STEIN ESTIMATORS
LEAST SQUARES
REGRESSION (various entries)
STEIN EFFECT)

GEORGE G. JUDGE

STOCK MARKET PRICE INDEXES

Stock market price indexes are measures of the movements over time in the prices of stocks (or such other financial assets; for example, bonds) traded in particular stock markets such as the New York Stock Exchange (NYSE). Many of these indexes are calculated typically daily, or even several times a day, by stock brokerage firms, other financial institutions, and newspapers. They provide sensitive short-run indicators of the changing economic and political conditions affecting the market as reflected in changes in the prices of industrial, transport, and utility stocks.

Best-known examples of such indexes are the London *Financial Times* (FT) index, the NYSE Composite, the Standard and Poor's (S & P) Composite, and the Dow Jones Industrial Average (DJIA). The first three are index numbers proper in that they measure the *relative change* in the magnitude of a variable between two points (*see* INDEX NUMBERS), while the DJIA is a specialized version of the arithmetic mean of the prices of 30 stocks.

A stock price index is not designed as a "portfolio index," which is an indicator of the long-run performance of a real portfolio, or the actual holdings of securities by an individual or institution such as a pension fund. Yet the use of stock market price indexes is widespread for purposes of portfolio performance and risk measurement*, both among academics and commercially [2, 7, 11, 13].

BRIEF HISTORY

The origin of one of the world's earliest financial stock markets was in London in the late seventeenth century, where investors met together to purchase and trade issues of public debt. A similar need to raise capital for government led to the earliest U.S. "stock market" trading in the 1790s, following the War of Independence. The London Stock Exchange came formally into existence in 1802 and the New York Stock and Exchange Board in 1817. The trading of securities on these exchanges and subsequently on other stock markets throughout the noncommunist world brought a demand for the calculation of averages of the prices to establish overall "market" movements. These general movements in capital values could then be related to other economic aggregates, and general price changes separated from those specific to a particular industry or company. However, while the DJIA dates back to the 1880s [24], and the FT to 1930s [13], the prolifera-

tion of stock market price indexes is comparatively recent.

METHODS OF ESTIMATION

The estimation of stock price indexes generally involves three problems: sampling, weighting, and a method of aggregation* [12]. Stock price indexes are typically based on a purposive sample of leading stocks which are actively traded in the market. However, the samples of stocks such as those covered by S & P, DJIA, and FT indexes usually account for more than 50% of the value of all listed stocks on their respective stock exchanges. It is also thought that changes in all stock prices tend to follow a general pattern. The NYSE and the American Stock Exchange indexes are based on all stocks listed on their respective exchanges. The idea that stock market price indexes be based on random sampling was advocated in ref. 5, but henceforth not used.

The genesis of weighting in the specification of stock price indexes is essentially no different from that found in weighted least squares* when estimating the mean of a variable. However, stock price indexes are either equally weighted averages of the price variable, or weighted in proportion to the market value, or according to some other weighting system, reflecting the relative importance in the market of the stocks covered by the index. For example, the stock prices in the DJIA and FT index are equally weighted, while NYSE and S & P indexes are value weighted.

The methods of combining prices vary, but they are based on unweighted (equiweighted) or weighted averages of the price changes. When the index measures the change in prices in the current period relative to the base period, the base is usually set equal to 100 (FT index), but sometimes 50 (NYSE), or 10 (S & P). To explain the common methods of combining prices, let p_{it} be the price of the stock of the ith company (e.g., IBM or International Harvester, which are both included in DJIA; hence at a given point in time, the market value of the ith company and the price of its stock could be several times higher or lower than those of another company) for $i = 1, 2, \ldots, n$; observed in period $t = 0, 1$. The most common methods found in the literature are the following.

Equally Weighted Indexes

(i) Arithmetic index A_{01} is given by

$$A_{01} = \frac{1}{n} \sum_{i=1}^{n} \left(\frac{p_{i1}}{p_{i0}} \right). \tag{1}$$

(ii) Geometric index G_{01} is given by

$$G_{01} = \left[\prod_{i-1}^{n} \left(\frac{p_{i1}}{p_{i0}} \right) \right]^{1/n}. \tag{2}$$

For time periods, $t = 0, 1, \ldots, T$, the time series of price indexes, say, $P_{01}, P_{02}, \ldots, P_{0T}$, when P_{rs} is the price index in period s relative to period r, can be calculated in two ways. First, is to obtain the *direct comparison* index for the base period 0 and the current period $t = 1, 2, \ldots, T$, either from (1) or from (2). Second, is to obtain the *links* (binary indexes) $P_{01}, P_{12}, \ldots, P_{T-1,T}$, also either from (1) or from (2); and then to use the *chain* index method to form the time series $\bar{P}_{01}, \bar{P}_{02}, \ldots, \bar{P}_{0T}$, where

$$\bar{P}_{rs} = P_{r,r+1} \cdot P_{r+1,r+2} \cdots P_{s-1,s}, \tag{3}$$

for $0 < r < t$ and $r < s \leqslant T$. However, the direct comparison index and the corresponding chain index are identical if the links were obtained from (2), but they diverge if the links were obtained from (1). This divergence has been studied for U.K. stock price indexes for the period 1935–1970 in refs. 1 and 13.

The choice between the arithmetic index (1) and the geometric index (2) raises two other problems. First, as is well known, the geometric index is generally less than the corresponding arithmetic index. However, when the price changes are small, the Johnson approximation, often appealed to, especially in financial literature, provides a useful relation which shows that the ratio of (2) to (1) is approximately equal to $[1 - (\frac{1}{2})V^2]$, where V is the coefficient of variation of the

n price ratios [25]. (*See also* GEOMETRIC MEAN.) Second, the arithmetic index, unlike the geometric index, has a simple investment interpretation: It is the percentage change in the value of the portfolio (a linear combination of n different investment opportunities which has a positive market value) in the current period relative to the base period. Fortunately, if the change in stock prices between two consecutive periods is small, then it can be shown by a Taylor series* expansion [13, p. 319; 18, pp. 338–339] that the proportional change in the geometric index (2) is approximately equal to the percentage change in the arithmetic index (1). Further portfolio interpretations of (1) and (2) and implied investment strategies may be found in refs. 1, 7, 11, 13, and 18.

Weighted Indexes

Weighted versions of (1) and (2), but especially of (1), are also used. The weighted arithmetic index in period 1 relative to period 0, say P_{01}^*, is of the form

$$P_{01}^* = \frac{\sum_{i=1}^{n} w_i (P_{i1}/P_{i0})}{\sum_{i=1}^{n} w_i}, \tag{4}$$

where w_i is the weight of the stock of the ith company. Particular versions of (4) are used, for example, in the construction of NYSE Composite, Financial Times Actuaries, S & P Composite, Moody's Averages [3, 9, 12, 22], among many others. Indexes so computed are sometimes referred to as value weighted, Laspeyres*, Paasche*, value ratio, or by some other label according to the choice of the weights w_i, which in general might be the number of stocks outstanding, average earnings, or the average number of stocks sold over a past period [9, 12, 14, 17, 23].

Dow Jones Industrial Average

The DJIA is the arithmetic mean of the prices themselves of 30 stocks rather than of the price ratios as in (1). However, the denominator (or what is commonly called, the "divisor"), of this average is adjusted every time there is a stock "split" or stock "replacement." Reference 22 contains an arith-

Table 1 Stock Price Indexes in 1983

	Dow Jones (30 Stocks)		Standard & Poor's	
At close of	Average	Divisor	S & P Composite (500 stocks) (1941–1943 = 10)	Transportation (20 stocks) (1982 = 100)
March 31	1130.03	1.292	153.00	135.90
June 30	1221.96	1.248	168.10	153.90
Sept. 30	1233.13	1.230	166.00	153.80
Dec. 30	1258.94	1.230	164.90	158.80

Source: References 17 and 23.

metic example of how the DJIA divisor is adjusted, and ref. 17 lists the values of the DJIA divisor, and the reasons for its changes, from November 1928, when it was 16.02, to August 1983, when it became 1.23. A recent comprehensive study of the history, compilation, and interpretation of the DJIA is given in ref. 24.

STOCK PRICE INDEXES AND FINANCIAL ECONOMICS

Stock market indexes have played a twofold part in the theory and applications of portfolio management:

(i) *Empirical*: The work of King [10] and others expressed the return on industrial securities as a function largely of one or more indexes. To evaluate minimum variance portfolios of n securities required input of a covariance matrix with $n(n + 1)/2$ elements. A single-index model reduces the number of parameters to $2n + 1$ (see ref. 8 for details). For computational purposes this development is now less important. More important is the implication of the single-index or "market" model that it is only that proportion of a security's variance explained by the market index which is relevant for a mean variance efficient portfolio. The nonmarket or specific risk can be reduced or even eliminated by diversification of the portfolio.

(ii) *Theoretical*: Here the index has played the role of an investment portfolio. The capital asset pricing model (CAPM) of Sharpe, Lintner, Mossin, and Black (see, for example, refs. 2 and 15) shows, given certain

restrictive assumptions, that the riskiness of an asset for an investor is measured by its covariance with the "market" portfolio of all investors. An alternative theory, the arbitrage pricing model [20] gives a similar risk measure but is based upon the factor structure of returns (and hence indexes). The empirical merits of the two theories are a matter of current controversy [4, 15, 19].

Similarly, market indexes considered as portfolios have played the role of "benchmark" or "randomly selected" for the purposes of portfolio performance comparisons. Measures of portfolio performance have been developed within the CAPM framework, using indexes adjusted for risk. These measures, like the tests of the CAPM, are subject to criticisms as discussed in ref. 19.

While the use of stock price indexes is widespread for the purposes of risk measurement and portfolio performance, the controversy among the academic community concerns the justification for their use, rather than whether or not they should be used [15, 19–21].

References

[1] Allen, R. G. D. (1975). *Index Numbers in Theory and Practice*. Macmillan, London, England. (Elementary; British indexes.)

[2] Copeland, T. and Weston, J. F. (1983). *Financial Theory and Corporate Policy*, 2nd ed. Addison-Wesley, Reading, MA.

[3] Crowe, W. R. (1965). *Index Numbers*. Macdonald and Evans, London, England. (Elementary textbook.)

[4] Dhrymes, P. J., Friend, I., and Gultekin, N. B. (1984). *J. Finance*, **39**, 323–346.

[5] Drakatos, C. (1962). *The Banker's Magazine* (*London*), **193**, 465–473.

[6] Fama, E. R. (1976). *Foundations of Finance*. Basic Books, New York.

[7] Fisher, L. (1966). *J. Bus.*, **39**, 191–225.

[8] Francis, J. C. and Archer, S. H. (1972). *Portfolio Analysis*. Prentice-Hall, New York.

[9] Kekish, B. J. (1967). *Financial Analysts J.*, **23**, 65–69. (Explains Moody's averages.)

[10] King, B. F. (1966). *J. Bus.*, **39**, 139–190.

[11] Latane, H. A., Tuttle, D. L., and Yong, W. E. (1971). *Financial Analysts J.*, **27**, 75–85. (Elementary; discusses indexes for portfolio management.)

[12] Lorie, J. H. and Hamilton, M. T. (1973). *The Stock Market*. Irwin, Homewood, IL. (Elementary; contains a chapter on stock market indexes.)

[13] Marks, P. and Stuart, A. (1971). *J. Inst. Actuaries*, **97**, 297–324. (In-depth analysis of the FT index and its arithmetized version.)

[14] McIntyre, F. (1938). *J. Amer. Statist. Ass.*, **33**, 557–563. (Of historical interest.)

[15] Merton, R. C. (1982). In *Handbook of Mathematical Economics*, Vol. 2, K. J. Arrow and M. D. Intrilligator, eds. North-Holland, Amsterdam, Netherlands, pp. 601–669. (Technical; theory of investment under uncertainty.)

[16] Mitchell, W. C. (1916). *J. Polit. Econ.*, **24**, 625–693. (Of historical interest.)

[17] Pierce, P. S. (1984). *The Dow Jones Investor's Handbook*. Dow Jones–Irwin, Homewood, IL.

[18] Rich, C. D. (1948). *J. Inst. Actuaries*, **74**, 338–339.

[19] Roll, R. (1978). *J. Financial Econ.*, **33**, 1051–1069.

[20] Ross, S. A. (1976). *J. Econ. Theory*, **13**, 341–360.

[21] Shiller, R. J. (1984). *Brookings Papers on Economic Activity*, **2**, 457–498.

[22] Smith, G. (1985). *Statistical Reasoning*. Allyn and Bacon, Boston. (Elementary textbook.)

[23] Standard & Poor's Statistical Service (1984). *Security Price Index Record*. Standard & Poor's Corporation Publishers, New York.

[24] Stillman, R. J. (1986). *Dow Jones Industrial Average*. Dow Jones–Irwin, Homewood, IL. (Elementary; comprehensive study and history.)

[25] Young, W. E. and Trent, R. H. (1969). *J. Financial Quant. Anal.*, **4**, 179–199. (Study various approximations of the geometric mean.)

(INDEX NUMBERS
RISK MANAGEMENT, STATISTICS OF
RISK MEASUREMENT, FOUNDATIONS OF)

NURI T. JAZAIRI
JOHN MATATKO

SYNTHETIC ESTIMATION

Synthetic estimation is a simple method of small-area estimation that makes use of information collected at high levels of geographic aggregation* and applies this information without change at lower levels of aggregation.

Assume that we have a finite population residing in a large area, divided into I disjoint subareas and also divided into J disjoint subgroups through the use of some covariates. Let N_{ij} be the population of the jth subgroup in the ith subarea and let $P_{ij} = N_{ij}/N_i$. be the proportion of the population of subarea i which belongs to subgroup j.

Now assume that there is a characteristic which is observed at the level of the large area for each subgroup, denoted $F_{\cdot j}$. What is desired is an estimate at the subarea level across subgroups, denoted $\hat{F}_{i\cdot}$. If we had been fortunate enough to observe an F_{ij} for each subarea–subgroup combination, a reasonable estimate would be

$$\sum_{j=1}^{J} P_{ij} F_{ij},$$

weighting each F_{ij} according to its frequency in subarea i. The *synthetic assumption* is that F_{ij} is much more varied by covariate than by geography and so one can replace F_{ij} by $F_{\cdot j}$, in the above, to get

$$\hat{F}_{i\cdot} = \sum_{j=1}^{J} P_{ij} F_{\cdot j}.$$

This estimate has been used for a variety of purposes, some described in Gonzalez [4]. Hill [6] has argued that this methodology could be used to estimate census undercoverage for small areas. This would proceed as follows. Let $F_{ij} = (M_{ij} - N_{ij})/N_{ij}$, where M_{ij} represents an unobserved "superior" count of the population in that subarea for that subgroup. Let N_{ij} here specifically represent the corresponding census count. Therefore F_{ij} is the rate of actual census undercoverage in the ijth subarea–subgroup. Assume that over subareas, a superior count for the jth subgroup is observable and let this be denoted $M_{\cdot j}$. Such superior counts may be obtained through the use of demographic analysis, for example. Using the synthetic assumption, an estimate of the rate of undercoverage for the ith subarea is

$$\hat{F}_{i\cdot} = \sum_{j=1}^{J} P_{ij} F_{\cdot j} = \sum_{j=1}^{J} \frac{N_{ij}}{N_{i\cdot}} \left[\frac{M_{\cdot j}}{N_{\cdot j}} - 1 \right]. \quad (1)$$

This particular application of synthetic estimation has been investigated by Schirm and Preston [9].

Some theory has been developed for synthetic estimation, e.g., Gonzalez and Waksberg [5] and Holt et al. [7], where a testing framework is presented so that one can determine the appropriateness of the synthetic assumption.

Now $N_{i\cdot}(\hat{F}_{i\cdot} + 1)$ estimates the population count for the ith subarea, say $\hat{M}_i$. Utilizing (1), we see that

$$\hat{M}_{i\cdot} = N_{i\cdot}(\hat{F}_{i\cdot} + 1) = \sum_{j=1}^{J} N_{ij} \left[\frac{M_{\cdot j}}{N_{\cdot j}} \right].$$

Simplifying this to the case of $J = 1$, we get

$$\hat{M}_i = N_i \left[\frac{M_\cdot}{N_\cdot} \right].$$

This estimator was originally derived by Deming [2] (see also Deming and Stephan [3]) as an answer to the problem of estimating several means M_i when it is known that the distribution of N_i is normal with parameters (M_i, N_i) $i = 1, \ldots, I$, and

$$\sum_{i=1}^{I} M_i = M_\cdot,$$

which is assumed to be known. (Note that in this model the variance is proportional to the observed value.) This estimator was suggested for use in surveying when the estimates of the three interior angles of a triangle do not sum to 180°. Cohen and Zhang [1] extended this result to the case where

$$(N_1, \ldots, N_I) \sim \mathbf{N}(\mathbf{M}, \mathbf{\Sigma})$$

and, in particular, when $\mathbf{\Sigma}$ is diagonal with ith diagonal element V_i. In this case the estimate is

$$\hat{M}_i = N_i + (M_\cdot - N_\cdot) \left[V_i \middle/ \sum_{j=1}^{I} V_j \right].$$

We note, finally, that the estimate

$$\hat{M}_i = N_i \left(\frac{M_\cdot}{N_\cdot} \right)$$

is the analog to iterative proportional fitting applied to a *one-way* table. For more details, *see* ITERATIVE PROPORTIONAL FITTING, Deming and Stephan [3], and Ireland and Kullback [8].

References

[1] Cohen, M. L. and Zhang, X. D. (1988). The Difficulty of Improving Statistical Synthetic Estimation. *Statist. Res. Div. Rep. No. Census/SRD/RR-88/12*, Bureau of the Census, U.S. Department of Commerce, Washington, D.C.

[2] Deming, W. E. (1943). *Statistical Adjustment of Data*. Wiley, New York.

[3] Deming, W. E. and Stephan, F. F. (1940). On a least squares adjustment of a sampled frequency table when the expected marginal totals are known. *Ann. Math. Statist.*, **11**, 427–444.

[4] Gonzalez, M. E. (1973). Use and evaluation of synthetic estimates. *Proc. Amer. Statist. Ass. Soc. Statist. Sect.*, 33–36.

[5] Gonzalez, M. E. and Waksberg, J. L. (1975). Estimation of the Error of Synthetic Estimates. Unpublished paper presented at the first meeting of the International Association of Survey Statisticians, Vienna, Austria.

[6] Hill, R. B. The synthetic method: Its feasibility for deriving the census undercount for states and local areas. *Proc. 1980 Conference on Census Undercount*. Bureau of the Census, U.S. Department of Commerce, Washington, D.C.

[7] Holt, D., Smith, T. M. F., and Tomberlin, T. J. (1979). A model-based approach to estimation for small subgroups of a population. *J. Amer. Statist. Ass.*, **74**, 405–410.

[8] Ireland, C. T. and Kullback, S. (1968). Contingency tables with given marginals. *Biometrika*, **55**, 179–188.

[9] Schirm, A. L. and Preston, S. H. (1987). Census undercount adjustment and the quality of geographic population distributions. *J. Amer. Statist. Ass.*, **82**, 965–990.

(DEMOGRAPHY
GEOGRAPHY, STATISTICS IN
ITERATIVE PROPORTIONAL FITTING
STOCHASTIC DEMOGRAPHY
UNDERCOUNT IN THE U.S. DECENNIAL
  CENSUS)

MICHAEL L. COHEN

T

TIED-DOWN BROWNIAN MOTION
See BROWNIAN BRIDGE

U

UNDERCOUNT IN THE U.S. DECENNIAL CENSUS

INTRODUCTION

A problem that afflicts virtually all census-like activities around the world is actually counting all of the people eligible to be counted. Even in small countries with relatively homogeneous populations and population registers, such as Norway and Sweden, there is some undercount [7]. This undercount problem has been of special concern in the United States, and currently is the focus of a major political controversy. This entry discusses the census* undercount in the U.S. context, but the statistical techniques described are potentially applicable in all countries.

Concerns about the accuracy of the census counts in the United States have existed almost as long as the census itself. For example, in 1791, following the first census, Thomas Jefferson [12] wrote:

> Nearly the whole of the states have now returned their census. I send you the result, which as far as founded on actual returns is written in black ink, and the numbers not actually returned, yet pretty well known, are written in red ink. Making a very small allowance for omissions, we are upwards of four millions; and we know that omissions have been very great.

Almost 100 years later, in Vol. 2 of the *Journal of the American Statistical Association*, General Francis A. Walker [21], Superintendent of the U.S. Censuses of 1870 and 1880, writing about the undercount of blacks in the 1870 census, elicited one of the earliest statistical proposals for adjustment for the undercount from H. A. Newton [15] and H. S. Pritchett [17], both of whom used the method of least squares* to fit a third-degree polynomial to census data for 1790 to 1880 and then measured the undercount for 1870 as a residual from the fitted curve. (See the description in Stigler [20].)

Beginning with the 1940 census, the Bureau of the Census estimated the size of the undercount by race, using a technique known as demographic analysis (described in greater detail below). The estimated differential undercount between blacks and whites has remained between 5% and 6% up through the 1980 census. Details are provided in Table 1. While the explanations for the undercount have changed over the decades, as techniques for taking the census have changed, the differences in undercount among population groups have been a major concern for demographers and statisticians.

181

Table 1   Estimated Net Census Undercount from 1940 to 1980, as Measured by Demographic Analysis

Year	Black	White	Difference	Overall Undercount
1940	10.3%	5.1%	5.2%	5.6%
1950	9.6%	3.8%	5.8%	4.4%
1960	8.3%	2.7%	5.6%	3.3%
1970	8.0%	2.2%	5.8%	2.9%
1980[a]	5.9%	0.7%	5.2%	1.4%

Source: Fay et al. [10].
[a] The figures for 1980 are based on the assumption that 3 million undocumented aliens were living in the United States at the time of the census. The registration of approximately 2.2 million previously undocumented aliens in 1987–1988, suggests that the estimated undercount percentage for whites may be substantially higher.

STATISTICAL TECHNIQUES FOR ESTIMATING THE UNDERCOUNT

Prior to the 1980 census, there was extensive discussion in the statistical community regarding the advisability of adjusting the census counts to correct for the undercount (for example, see Committee on National Statistics [7]), and following the census there were a number of articles published on the topic (for example, see Ericksen and Kadane [9], Freedman and Navidi [11], Savage [18], and Schirm and Preston [19], each of which is followed by extensive commentary).

There are basically two techniques that have been used to estimate the undercount: demographic analysis (see Coale [6]) and the dual-system or capture–recapture* technique (e.g., see Bishop et al. [3]). Demographic analysis combines birth, death, immigration, and emigration records with other administrative records to carry forward the population from one census to the next, deriving an estimate of the overall population size, and thus the undercount (see Fay et al. [10]). The methodology can be used to provide population and undercount figures by age, race, and sex, but only at a national level. Thus demographic analysis cannot be used to provide reliable state, regional, and local estimates, principally because of the absence of accurate data on migration*.

In the dual-system estimation approach, those included in the census are matched with a second source (e.g., a random sample of the population or a list based on administrative records). Suppose there are x_{1+} individuals counted in the census and the second source contains x_{+1} individuals, x_{11} of whom match with individuals in the census. Then the traditional capture–recapture estimate for the overall population size, N, is

$$\hat{N} = x_{1+}x_{+1}/x_{11} \qquad (1)$$

(Petersen [16]). This estimate is easily modified to deal with the use of a sample from the census results rather than the census itself, and with complex sample surveys used as the second source (for details see Wolter [24] and Cowan and Malec [8]). In essence, (1) is replaced by

$$\hat{N} = x_{1+}^* x_{+1}^* / x_{11}^*, \qquad (2)$$

where $x_{1+}^* = x_{1+} - II - \widehat{EE}$, II is the weighted number of people in census enumeration with insufficient information to be matched, $\widehat{EE}$ is the estimate of number of erroneous enumerations in census enumeration, x_{+1}^* is the weighted number of people selected for the sample used as second source, and x_{11}^* is the weighted number of people in both the census and the sample.

As with the demographic analysis method, the dual-system estimation method is based on a set of assumptions. The three assumptions most widely discussed are:

(i) *Perfect matching.* The individuals in the second list can be matched with those in the second list, without error.

(ii) *Independence of lists.* The probability of an individual being included in the census does not depend on whether the individual was included in the second list.

(iii) *Homogeneity.* The probabilities of inclusion do not vary from individual to individual.

The failure of assumption (ii) is referred to as *correlation bias* and alternative estimators are available in such circumstances using multiple lists and the techniques developed for estimation in multiple-capture problems (e.g., see Bishop, Fienberg, and Holland [3]). In the presence of positive correlation bias (being missed in the census is positively correlated with being missed in the second list), however, the traditional estimator tends to underestimate the actual population size but yields an improvement over the unadjusted value (see Childers et al. [4]).

The dual-system estimation approach was used in conjunction with the censuses of 1970 and 1980 to evaluate population coverage as part of what was called the postenumeration survey (PES) program. In the 1980 PES program, a sample of 100,000 records from the census was matched with data from households in the April and August *Current Population Survey*, each containing approximately 70,000 households and about 185,000 individuals, and estimates of the undercount were produced for the United States as a whole as well as for all 50 states and several large local areas. Perhaps the greatest problem with the dual-systems approach in 1980 was the rate of matching errors [the failure of assumption (i) above]. While it is widely believed that there was positive correlation bias in the 1980 PES estimates (e.g., see Freedman and Navidi [11]), there is little solid empirical evidence on the issue.

DISPUTE OVER THE 1980 CENSUS RESULTS

For the 1980 census, a decision was made shortly before the reporting deadline, in December 1980, not to adjust the results for the anticipated differential undercount (see Mitroff, Mason, and Barabba [14]), although Kadane [13] and others contend that this decision was in fact made prior to the taking of the census. A lawsuit was filed on the census day by the city of Detroit requesting that the 1980 census be adjusted for the undercount, and this action was followed by 52 others, 36 of which requested adjustment. One of these cases, brought by the state and city of New York gained considerable attention, with a large number of statisticians testifying for and against adjustment (see the editorial prologue to Ericksen and Kadane [9]).

The New York lawsuit, known as *Cuomo v. Baldrige*, ultimately went to trial in January 1984, but the judicial opinion was not issued until December 1987. The judge ruled that no adjustment be made. He argued that, because statisticians and demographers can and do disagree on the reliability of an adjustment of the 1980 census, it would be inappropriate for the court to substitute its judgment for that of the experts at the Census Bureau. The articles by Ericksen and Kadane [9] and by Freedman and Navidi [11] reflect some of the statistical arguments presented in court in this case.

ADJUSTMENT FOR UNDERCOUNT IN THE 1990 CENSUS

Simultaneously with these activities, the Census Bureau* launched a major research program to improve the methodology used for census adjustment and it commissioned the Committee on National Statistics* (at the National Academy of Sciences) to establish a Panel on Decennial Census Methodology, whose charge included the review of the census undercount research program. The panel's 1985 report (Citro and Cohen [5]) outlined the basic issues that needed to be addressed in the adjustment research program. In two subsequent letter reports, issued in 1986 and 1987, the panel reviewed the proposed methodology for adjustment in 1990 and made a positive assessment of its technical feasibility. Childers et al. [4] report on this methodology for adjustment in 1990 and its technical justification.

In the summer of 1987, the Subcommittee on Population and Census of the U.S. House of Representatives held two hearings on the topic of census adjustment, at which several statisticians testified (for a detailed report, see Wallman [22]). In her presidential address to the American Statistical Association* in August of 1987, Barbara Bailar [2] noted:

> A sizeable group of eminent statisticians now believe that adjustment of the 1990 census is feasible, that it has been successfully demonstrated in test censuses and that it would substantially improve the accuracy of the 1990 census. Those who press for adjustment say that an undercount in 1990 is inevitable, a view that the Census Bureau largely shares. They argue that even an imperfect adjustment will be a move in the right direction and will increase the accuracy of census data for its many uses.

Then in October 1987, the Department of Commerce, in which the Bureau of the Census is located, announced that the 1990 census would not be adjusted for the differential undercount. This decision has been widely criticized in the statistical community, and yet another congressional hearing was held in March 1988, focussing, yet again, on the technical feasibility and advisability of adjustment (Wallman [23]).

While considerable uncertainty and controversy surrounds the use of adjustment methodology in the 1990 census, virtually all of those familiar with the census procedures expect that the differential undercount in 1990 will be at least as large as that in previous census years. At the time of preparation of this entry, legislation mandating the adjustment of the 1990 census had been introduced into the U.S. Congress.

Acknowledgment

The preparation of this entry was supported in part by the National Science Foundation under Grant SES-8701606 to Carnegie Mellon University.

References

[1] American Statistical Association (1982). Report of the ASA Technical Panel on the Census Undercount (with discussion). *Amer. Statistician*, **38**, 252–260.

[2] Bailar, B. A. (1988). Statistical practice and research: The essential interactions. *J. Amer. Statist. Ass.*, **83**, 1–8.

[3] Bishop, Y. M. M., Fienberg, S. E., and Holland, P. W. (1975). *Discrete Multivariate Analysis: Theory and Practice*. M.I.T. Press, Cambridge, MA, Chap. 6.

[4] Childers, D., Diffendal, G., Hogan, H., Schenker, N., and Wolter, K. (1987). The Technical Feasibility of Correcting the 1990 Census. Paper delivered at the Annual Meetings of the American Statistical Association.

[5] Citro, C. F. and Cohen, M. L., eds. (1985). *The Bicentennial Census. New Directions for Methodology in 1990*. Report of the Panel on Decennial Census Methodology of the Committee on National Statistics. National Academy Press, Washington, D.C.

[6] Coale, A. J. (1955). The population of the United States in 1950 by age, sex, and color—A revision of census figures. *J. Amer. Statist. Ass.*, **50**, 16–54.

[7] Committee on National Statistics (1978). *Counting the People in 1980: An Appraisal of Census Plans*. Report of the Panel of Decennial Census Plans. National Academy of Sciences, Washington, D.C.

[8] Cowan, C. D. and Malec, D. (1986). Capture–recapture models when both sources have clustered observations. *J. Amer. Statist. Ass.*, **81**, 347–353.

[9] Ericksen, E. P. and Kadane, J. B. (1985). Estimating the population in a census year: 1980 and beyond (with discussion). *J. Amer. Statist. Ass.*, **80**, 98–131.

[10] Fay, R. E., Passel, J. S., Robinson, J. G., and Cowan, C. C. (1988). *The Coverage of Population in the 1980 Census*. Bureau of the Census, U.S. Department of Commerce, Washington, D.C.

[11] Freedman, D. and Navidi, W. C. (1986). Regression models and adjusting the 1980 census (with discussion). *Statist. Sci.*, **1**, 3–39.

[12] Jefferson, T. (1986). Letter to David Humphreys. In *The Papers of Thomas Jefferson*, Charles T.

Cullen, ed. Princeton University Press, Princeton, NJ, Vol. 22, p. 62.

[13] Kadane, J. B. (1984). Book review of Mittroff, Mason and Barabba. *J. Amer. Statist. Ass.*, **79**, 467–469.

[14] Mittroff, I. I., Mason, R. O., and Barabba, V. P. (1983). *The 1980 Census: Policymaking and Turbulence.* D. C. Heath, Lexington, MA.

[15] Newton, H. A. (1891). Note on President Walker's article on statistics of the colored race. *Publications* [later *Journal*] *of the American Statistical Association*, **2**, 221–223.

[16] Petersen, C. G. J. (1896). The yearly immigration of young plaice into the Limfjord from the German Sea. *Rep. Dan. Bio. Stn.* (1895) **6**, 5–84.

[17] Pritchett, H. S. (1891). A formula for predicting the population of the United States. *Publications* [later *Journal*] *of the American Statistical Association*, **2**, 278–286. [Reprinted from *Transactions of the Academy of Science, St. Louis* (1891).]

[18] Savage, I. R. (1982). Who counts? (with discussion). *Amer. Statist.*, **36**, 195–216.

[19] Schirm, A. L. and Preston, S. H. (1987). Census undercount adjustment and the quality of geographic population distributions (with discussion). *J. Amer. Statist. Ass.*, **82**, 965–990.

[20] Stigler, S. M. (1988). The Centenary of *JASA*. *J. Amer. Statist. Ass.*, **83**, 583–587.

[21] Walker, F. A. (1890). Statistics of the colored race in the United States. *Publications* [later *Journal*] *of the American Statistical Association*, **2**, 91–106.

[22] Wallman, K. K. (1988a). Adjusting the census: A tale of two cities. *Chance*, **1** (No. 2), 48–52.

[23] Wallman, K. K. (1988b). A tale of two cities, Act III. *Chance*, **1** (No. 3), 55–57.

[24] Wolter, K. (1986). Some coverage error models for census data. *J. Amer. Statist. Ass.*, **81**, 338–346.

(BUREAU OF LABOR STATISTICS
BUREAU OF THE CENSUS, U.S.
CAPTURE–RECAPTURE METHODS
CENSUS
DEMOGRAPHY
JOURNAL OF THE AMERICAN
   STATISTICAL ASSOCIATION)
NATIONAL STATISTICS, COMMITTEE ON

STEPHEN E. FIENBERG

WEATHER MODIFICATION—II

Weather modification studies have been carried out all over the world since about 1950, and quite a number have incorporated randomized allocation of treatments. The common feature of all these studies is that on some occasions clouds were *seeded* by introducing an agent (silver iodide or dry ice) intended to increase precipitation on a well-defined *target* area, while on other occasions clouds were left *unseeded*. The effect of seeding was assessed by statistical comparison of target precipitation on seeded occasions with that on unseeded occasions. Most studies concentrated on testing the null hypothesis of zero effect, though many also tried to provide effect estimates and confidence bounds.

Design problems were mostly concerned with definition of the "occasions," or experimental units, and their allocation to be seeded or unseeded. Meteorology* does not provide well-defined "natural" units, since entities such as storms are very elusive to unequivocal a priori definition that allows randomized allocation of seeding. Time units were commonly used because they are well defined. Other problems concerned precipitation measurements, which varied from averages of a few target rain gauge readings to elaborate integrations of radar observations of rain water reflectivity. Since the variability of all such measurements was very high —coefficients of variation of about 1/3 even when the units were entire years—sample sizes needed to be large (usually five years or more) to yield reasonable power. Attempts to reduce variability centered on introducing concomitant variables*, but the only effective ones were precipitation data in nearby *control* areas which seemed out of the range of likely seeding effects.

Analyses consisted basically of comparing the seeded and unseeded samples, possibly after adjusting for concomitants and stratifying the units into some meaningful classifications. Some effort went into the choice of test statistics, which required assumptions about precipitation distributions and type of effect. The issue of multiplicity arose in the wide selection of strata, in the variety of measurements and statistics, and in the choice of covariates. It is an issue of paramount importance in experiments that last a number of years, and which inevitably generate novel ideas and techniques as the study progresses. Valid probabilistic evaluation, on the other hand, demands rigid adherence to a predetermined protocol and analysis. It is no mean matter to reconcile these claims.

A few of the better known rainfall stimulation studies are described here, chosen because they illustrate the statistical issues of design and analysis, and not because they were thought to be "representative." They are written by workers involved in the original analyses of these studies, and tend to sound rather sanguine. That is one of the problems encountered by statisticians involved in such studies.

For overall appraisals of weather modification studies by meteorologists, see refs. 1, 4, and 15. Collections of papers on designs, analyses, and methodological issues appear in refs. 1 (discussion), 14, 22, and 23, which include, respectively, experimental results for Whitetop in Missouri, several Australian experiments, Israeli I and the Swiss Grossver-

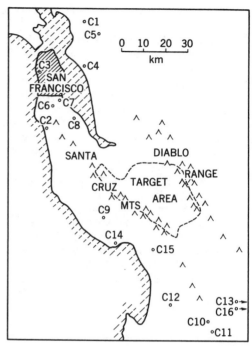

Figure 1 Map of Santa Clara cloud seeding project. Twenty-five rain gauges in target area and 18 control stations were selected by mutual agreement of operator and sponsor in 1955 for evaluation purposes. Two of the control stations were eventually dropped, leaving the 16 shown here (C1–C16). From A. S. Dennis (1980). *Weather Modification by Cloud Seeding*. Academic, New York.

such III, and Santa Barbara II, and the Tasmanian experiment.

SANTA CLARA COUNTY SEEDING OPERATIONS [5]

Commercial seeding operations were directed at a target area in California's Santa Clara county during 10 winters. It was agreed at the outset to evaluate success by comparing rainfall at 25 target stations and 18 control stations located around the target but not downwind of it (for prevailing winds on rainy days); Fig. 1 shows the stations, two of the control stations having been omitted for the actual analysis.

The operational winters were compared statistically with historical data on earlier

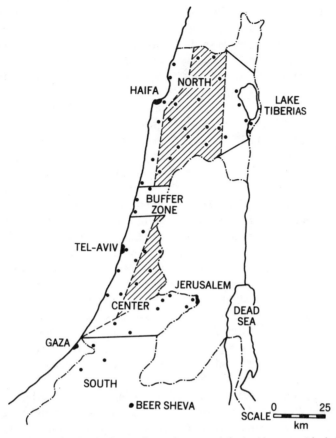

Figure 2 Map of Israel showing both experimental areas and the interior areas (shaded). Dots indicate rain gauges used in analysis of 1964–1965. From ref. 14.

winters. The analysis compared each winter's precipitation on an average target station with that on an average control station. All winters were defined as December–March, irrespective of exact seeding dates. Cloud seeding had actually been carried out on a more flexible schedule, but analyses used that objectively defined unit to avoid obvious biases and excluded years where special operations could have introduced bias. The statistical analysis compared the (target, control) scatter for the operational winters with the corresponding scatter for the historical winters. Linear regressions (without transformations) estimated precipitation on target in terms of precipitation on control during the unseeded winters, and also during the seeded winters. For further details see ref. 5.

Applying normal theory would have shown significance at roughly 10%, as would have rerandomization tests and nonparametric tests. The test statistic

$$\text{Double ratio} = \cfrac{\cfrac{(\text{Target seeded total})}{(\text{Control seeded total})}}{\cfrac{(\text{Target unseeded total})}{(\text{Control unseeded total})}},$$

on the other hand, would have given a one-sided P-value of 0.018. Distributional properties of this statistic are discussed in ref. 9.

Because of concern with validity, the original paper [5] did not publish significance or confidence statements. Indeed, this method of historical regression is fraught with risks [2], because the weather on successive sets of seasons is unlikely to be distributed like two independent random samples. (The bias introduced by such assumptions was studied later [11].) Other possible sources of bias are related to the initiation of commercial seeding operations as a result of preceding droughts and to the influences that determine the termination of such operations [3].

THE RANDOMIZED EXPERIMENTS IN ISRAEL [9, 10, 13]

The two experiments, Israeli I (1961–1966/ 1967) and Israeli II (1969/ 1970–1974/ 1975), randomly assigned seeding to 24-hour experimental units in the fall–winter seasons.

A changeover design* was chosen for Israeli I with alternate target areas—Fig. 2 — in the north and center of Israel. On each day, seeding was assigned to one of the targets, the other serving as that day's control. This was a highly efficient design because (i) each day's target area could be compared with its control area with which it was highly correlated and (ii) seeding was expected to affect the target–control difference on each day, whereas in an ordinary design there would only have been half the effect since half the days would have remained unseeded (see Gabriel in ref. 14). A buffer zone was introduced between the two areas to minimize the contamination of either target by seeding in the other.

The analysis was restricted to 391 rainy days and excluded the other two-thirds of the winter days as those were days with very little rain on the targets (only about 2% of all

rain), and they provided essentially no seeding opportunities in either area. The definition of "rainy days" was in terms of any rain in a nearby area that was never seeded. The analysis included all rainy days, even those which were not actually seeded (except for a small number of days when a pilot's strike grounded the aircraft); in order to avoid possible biases, days without seeding opportunities on the target were not excluded from the analysis.

The definition of the 24-hour day was altered during the experiment (from a period commencing at 8 P.M. to one commencing at 8 A.M.). This decision caused concern in certain quarters [20], even though the analysis was always consistent with the randomization and seeding of each day, so that no bias was possible [8, 10].

The target areas were defined in the original design and adhered to in the analysis. Analyses of interior subareas of the target were later considered preferable, but were not used for the critical analysis because that might have introduced bias.

The Mann–Whitney–Wilcoxon statistic* was chosen early on for the crucial test of significance*. No changes were allowed in the test—even though the results suggested that a nonparametric test was not optimal—because data-directed choice of the test statistic would have invalidated any significance statement.

The a priori chosen test statistic suggested a positive effect of seeding. Various other analyses, albeit data driven, supported such a finding even more strongly. Detailed fol-

Table 1   Various Analyses of the Israeli Experiments: One-Sided P-Values* against the Alternative of a Positive Effect

Target	Mann–Whitney–Wilcoxon	Normal Two-Sample	Double Ratio Rerandomized
Israeli I			
Entire areas	0.054[a]	0.009	0.025
Interior subareas	0.013	0.002	0.002
Israeli II			
Catchment area			0.017[a]

[a] The a priori chosen test statistic.

low-up analyses within a variety of strata gave even stronger support for such effects on days when the clouds were of intermediate temperatures.

Israeli II was designed principally to replicate Israeli I, but the analyses published so far relate to a single target—the Lake of Galilee catchment area—and use an upwind control area near Haifa on the coast [13]. The target was highly correlated with the control, so that target–control comparisons were considerably less noisy than the target data itself. The potential benefits of a changeover design could not be obtained with this control because it was upwind of the target and therefore could not be seeded without contaminating the catchment area. An alternative target area farther south was also not suitable for such a design because the catchment area was too far away and too poorly correlated with it.

The analyses of Israeli II were focused on a narrower target than those of Israeli I and on days with intermediate cloud-top temperatures. These choices resulted from the findings of Israeli I, which also made sense in terms of cloud physics. The double ratio statistic was chosen for the analysis, again because it had been found sensitive in Israeli I. It attained a rerandomization P-value* of 0.017, confirming the positive results of Israeli I.

The estimates of increase of precipitation due to seeding were $22(\pm 7)\%$ for the interior areas in Israeli I and $18(\pm 8)\%$ for the target of Israeli II. Later exploratory analyses suggest the effect is strongest at cloud-top temperatures of about -12 to $-21°C$ and consists mostly of an increase in the duration of precipitation [12].

A variety of detailed analyses of the data from both experiments by regression methods, poststratifications, and multivariate analyses, fleshed out the picture of the apparently successful experiments. However, these were described as exploratory analyses. The pitfalls of multiplicity were largely avoided, and the indications from Israeli I were verified by means of the replication in Israeli II.

CLIMAX I AND II EXPERIMENTS IN COLORADO [17–19]

Climax I and Climax II were replicated wintertime orographic cloud seeding experiments, intended to examine if seeding clouds with silver iodide could increase snowfall amounts over what is naturally expected. Climax I ran during 1960–1965 and Climax II during 1965–1970. The target area of Climaxes I and II was the summit of Fremont pass, a few miles north-northeast of Leadville, Colorado. For complete experimental design descriptions, see ref. 18 and Grant and Mielke in ref. 14.

The experimental unit for both experiments was a 24-hour period and the treatment involved seeding clouds with silver iodide from ground generators positioned at different locations to account for various wind directions. A specified 24-hour period was declared an experimental unit if certain wind conditions were satisfied and the U.S. Weather Bureau duty forecaster at Denver, Colorado, forecast at least 0.01 in. of precipitation at Leadville, Colorado. The decision to seed or not seed each experimental unit was randomized.

Snowboard measurement sites were placed at specified intervals over Fremont pass and also over the neighboring Vail and Hoosier passes. Two precipitation measurement sites termed CRG (climax recording gauge) and HAO (high altitude observatory snowboard) were independently operated by the Weather Bureau (i.e., both the CRG and HAO measurements were collected by the same operator) and were located within 1 m of each other. A recently noted blunder in the recording of the data by Colorado State University personnel was that the CRG and HAO measurements were accidentally reversed in four of the five years the Climax II experiment occurred. A pooled measurement termed TGM (target mean) that involved the average of the nonmissing measurements from CRG, HAO, and seven other target area measurement sites near the summit of Fremont pass, was not affected by this blunder.

Table 2   Climax Experiments: One-Sided *P*-Values of the Target Group Mean Measurements for Two Strata and Three Successive Analyses

Stratum	Analysis		
	(A1)	(A2)	(A3)
500-mb temperature (-20 to $-11°C$)			
Climax I	0.1314	0.0244	0.0339
Climax II	0.1271	0.0110	0.0048
700-mb wind direction (190–250)			
Climax I	0.0084	0.0344	0.0471
Climax II	0.0367	0.0075	0.0064

The results of the Climax experiments were analyzed several times. (A1) The initial joint analysis, reported in 1971 [19], was based primarily on two-sample Mann–Whitney–Wilcoxon tests for the target area measurements, and did not take into account the measurements from eight gauges at sites in the control area. (A2) Reanalyses reported in 1981 [18], had been initiated because of a concern of Mielke (in the discussion of ref. 1) that the 1971 results may have been the consequence of a type I statistical error, which accounted for the measurements of the eight previously defined gauges in the control area. (A3) Further reanalyses involving metric-based statistical methods carried out in 1982 [17], were prompted by concern about the nonmetric nature of the two-sample Mann–Whitney–Wilcoxon tests* of (A1) and (A2) and the least-squares method used in (A2). Comparisons of the three analyses for two important meteorological strata are given in Table 2 for the TGM measurements [17, 18]. These results imply that the 1979 concern of Mielke that questioned the credibility of analysis (A1) was unjustified. Any doubts involving the Climax I and II experimental results must be attributed to the inadequacy of (A1). That is surely the wrong reason for doubting the conclusions of a 10-year project involving two carefully replicated 5-year experiments. Raising concerns regarding any complicated scientific experiment is natural and demands attention, but the present situation demonstrates how an unjustified minor concern could have falsely

discredited the findings of a major scientific study.

THE FLORIDA AREA CUMULUS EXPERIMENTS [24–26]

The Florida Area Cumulus Experiments (FACE) comprised a two-stage program for investigating the potential of "dynamic seeding" to enhance summertime convective rainfall over a sizable target area (1.3×10^4 km^2) in south Florida. In its time, it was the only program ever conducted in the United States whose stated objective was to increase areal precipitation by altering cloud dynamics. The first stage of the program was an exploratory experiment (FACE-1, 1970–1976); the second stage was a fully planned and implemented confirmatory experiment (FACE-2, 1978–1980).

FACE-1

The FACE-1 experiment utilized the following: (1) A design document, (2) a single fixed target area (i.e., a quadrilateral extending from Fort Myers and Naples on the west coast of Florida to West Palm Beach and Fort Lauderdale on the east coast), (3) some "screening" criteria (e.g., a one-dimensional numerical model for estimating seedability, airborne assessment of cloud fields, etc.) for selection of operational days, (4) a randomized assignment of treatment (e.g., either pyrotechnic AgI flares or placebo sand flares) to each selected operational day, and (5) hourly adjusted radar estimated rainfall volumes in the target area (e.g., the hour before treatment, the six hours after treatment initiation, etc.).

The design and implementation of FACE-1 changed markedly during the period of the experiment [25]. Its implementation during the summers of 1970, 1971, 1973, 1975, and 1976, resulted in operational changes (e.g., reformulation of the AgI treatment flares, changes in the gauge network used to adjust the radar estimation of rainfall, yearly changes in the randomization plan, the distinction of A and B operational days, where B days were those in which the cloud field received 60 or more flares). A number of these changes were due to the lack of consistent yearly support and funds. The term "exploratory" was therefore well deserved [6]. The five summers of operations produced a total of 104 operational days with 75 B days (39 seeded and 36 unseeded).

The exploratory analyses, back-to-back stem-and-leaf displays* of treated vs. nontreated daily rainfall volumes for the B days, produced indications of a positive treatment effect in radar estimated rainfall for both the floating target (FT) and total target (TT), and also showed that relatively few days were responsible for the indicated difference. Comparisons of means, medians, and interquartile ranges suggested a treatment effect of about 45% for the FT and about 10–20% for the TT with supporting Mann–Whitney–Wilcoxon one-tailed P-values* less than 0.05 and 0.18, respectively [24, 25]. Similar results appear to hold for the A and B days combined. However, due to their highly exploratory nature, all calculated P-values were viewed as only suggestive (if they should have been presented at all).

Finally, an analysis of covariance* attempted to account for a few potentially interfering meteorological factors (e.g., prewetness, wind speed, etc.). A further linear "sweepout" analysis (more consistent with the data analysis ideas of Tukey) used the same four covariates and explored the residuals for evidence of treatment. Point

estimates of treatment in both analyses increased to about 60 and 30% for the FT and TT, respectively. Additional analyses suggested that the greater rainfall on treated days was due to larger rain areas and higher rain rates [25].

FACE-2

Given the encouraging results of FACE-1 and its exploratory nature, the NOAA eventually undertook a confirmatory experiment. A detailed design was prepared with the basic characteristics unchanged from FACE-1. The confirmatory analyses were further clarified prior to the disclosure of the treatment decisions and the commencement of analysis [24, 25].

The implementation was carefully conducted in each of three summers (1978, 1979, 1980) according to the design and operations documents, and no deviations or changes were allowed. This resulted in a total of 75 operational days with 25 treated and 26 nontreated B days.

The exact criteria for confirmatory analysis of FACE-2 were subject to some debate by scientists and statisticians both within and outside the program. The majority opted for testing a "nested" set of three hypotheses by means of six rerandomization [9] analyses of single and double ratios [24, 25] and "adjusted" P-values. A minority preferred two linear models (i.e., FT and TT variables), which adjusted for the same four covariates that FACE-1 analyses had used.

In each of these analyses, none of the three null hypotheses could be rejected at any reasonable level. Thus the FACE-2 results did not confirm the FACE-1 indications of a positive treatment effect [26].

Subsequently, a linear model cross-validation* analysis, with covariates guided by meteorological considerations, indicated that FACE-1 had a treatment effect of about 45–50% and FACE-2 of about 15% [7]. However, although the 95% confidence interval for the FACE-1 result lay entirely above the null value, the FACE-2 interval clearly straddled it. This analysis did sharpen the estimate of treatment effects in both FACE-1 and -2, but did not provide generally acceptable evidence of a treatment effect in FACE-2.

The FACE scientists and statisticians concluded that the failure to confirm a seeding effect was probably due to three factors: "1) an unknown and possibly intermittent seeding effect, 2) inadequate predictor equations, and 3) a limited sample size" [26]. This highlights the need for careful thought (e.g., conceptual models), planning, and exploratory analyses before undertaking confirmatory experiments.

HIPLEX-1 EXPERIMENT [16, 21]

The purpose of HIPLEX-1, the first in a sequence of high plains weather modification experiments, was to investigate the effect of dry ice cloud seeding on a chain of physical events associated with small cumulus clouds. In that, HIPLEX-1 differed from earlier experiments that had largely considered the events in the clouds as a "black box" and concentrated on measuring the clouds' precipitation "output."

The choice of dry ice was to avoid residual effects of seeding agents such as silver iodide. This randomized cloud seeding experiment was conducted by the U.S. Bureau of Reclamation in the vicinity of Miles City, Montana. The experimental units were individual clouds belonging to three classes, and randomized separately within each class. A double-blind protocol governed the randomized cloud-seeding operation from a jet aircraft.

Since this experiment was concerned with a chain of physical events represented (as a rational approximation) by a sequence of random variables, multivariate analyses for 12 primary response variables and 11 secondary response variables were essential. In order to avoid intractable multivariate distributional problems, multiresponse permutation procedures* (MRPP) were used to analyze the HIPLEX-1 data [16].

HIPLEX-1 was conducted only during the summers of 1979 and 1980, after which it was terminated by federal budget cuts. By that time only 20 experimental units had been obtained. While substantial changes were noted with the response variables measured during the first 5 min after the seeding treatment, the more subtle changes associated with response variables measured more than 5 min after seeding were not obvious. Descriptions of other aspects of HIPLEX-1 are presented in refs. 16 and 21.

References

[1] Braham, R. R., Jr. (1979). *J. Amer. Statist. Ass.*, **74**, 57–68; discussion, pp. 68–104. (A meteorologist's carefully reasoned account of weather modification experimentation and of cooperation, or otherwise, with statisticians. Comments by some of the latter.)

[2] Brier, G. W. and Enger, I. (1952). *Bull. Amer. Meteor. Soc.*, **33**, 208. (On biases of historical comparisons.)

[3] Court, A. (1960). *J. Amer. Soc. Civil Eng., Irrig. Drain. Div.*, **86**, 121–126.

[4] Dennis, A. S. (1980). *Weather Modification by Cloud Seeding*. Academic, New York. (Currently the only text covering the entire topic written by a meteorologist.)

[5] Dennis, A. S. and Kriege, D. F. (1966). *J. Appl. Meteor.*, **5**, 684–691. (Analysis of the Santa Clara operations.)

[6] Flueck, J. A. (1986). In *Precipitation Enhancement —A Scientific Challenge*, Vol. 21, No. 43, R. R. Braham, Jr., ed. Amer. Meteor. Society, Boston, MA, Chap. 16.

[7] Flueck, J. A., Woodley, W. L., Barnston, A., and Brown, T. (1986). *J. Climate Appl. Meteor.*, **25**, 546–564.

[8] Gabriel, K. R. (1980). *Commun. Statist. A*, **9**, 1963–1973; 1997.

[9] Gabriel, K. R. and Feder, P. (1969). *Technometrics*, **11**, 149–160. (Rerandomization of the double ratio statistic.)

[10] Gabriel, K. R. and Neumann, J. (1978). *J. Appl. Meteor.*, **17**, 552–554. (Comments on Israeli I.)

[11] Gabriel, K. R. and Petrondas, D. (1983). *J. Climate Appl. Meteor.*, **22**, 626–631. (On correcting for bias of historical comparisons.)

[12] Gagin, A. and Gabriel, K. R. (1986). *J. Appl. Meteor.*, **28**, 913–921.

[13] Gagin, A. and Neumann, J. (1981). *J. Appl. Meteor.*, **20**, 1301–1311. (Analysis of Israeli II.)

[14] Le Cam, L. and Neyman, J., eds. (1967). *Proc. Berkeley Symp. Math. Statist. Prob.: Weather Modification*, **5**, University of California Press, Berkeley, CA. (Papers on experiments in Colorado, Israel, Australia, Switzerland, and other places, as well as comments on methodology, especially by Neyman and Scott.)

[15] Mason, J. (1980). *The Meteorological Magazine*, **109**, 335–344. (A meteorologist's assessment of the evidence for cloud seeding effects in Florida, Israel, and Tasmania.)

[16] Mielke, P. W., Berry, K. J., Dennis, A. S., Smith, P. L., Miller, J. R., and Silverman, B. A. (1984). *J. Climate Appl. Meteor.*, **23**, 513–522. (Final analysis of HIPLEX-1.)

[17] Mielke, P. W., Berry, K. J., and Medina, J. G. (1982). *J. Appl. Meteor.*, **21**, 788–792. (Geometrically consistent analysis of Climax.)

[18] Mielke, P. W., Brier, G. W., Grant, L. O., Mulvey, G. J., and Rosenzweig, P. N. (1981). *J. Appl. Meteor.*, **20**, 643–659. (Reanalysis of Climax.)

[19] Mielke, P. W., Grant, L. O., and Chappell, C. F. (1971). *J. Appl. Meteor.*, **10**, 1198–1212; Corrigendum, **15**, 801. (Analysis of Climax.)

[20] Neyman, J. (1979). In *Applications of Statistics*, P. R. Krishnaiah, ed. North-Holland, Amsterdam, Netherlands, pp. 1–25.

[21] Smith, P. L., Dennis, A. S., Silverman, B. A., Super, A. B., Holroyd, E. W., Cooper, W. A., Mielke, P. W., Berry, K. J., Orville, H. D., and Miller, J. R. (1984). *J. Climate Appl. Meteor.*, **23**, 497–512. (Discussion of HIPLEX-1.)

[22] Wegman, E. J. and DePriest, D. J., eds. (1980). *Statistical Analysis of Weather Modification Experiments*. Dekker, New York. (Mostly on statistical methodology and logic of experimentation.)

[23] Wierkowski, J. J. and Odell, P. L., eds. (1979). *Commun. Statist. Theor. Meth.*, **8**, Nos. 10, 11. (Special Issues on statistical analysis of weather modification experiments. Papers on analysis and methodology, mostly by statisticians.)

[24] Woodley, W. L., Jordan, J., Simpson, J., Biondini, R., Flueck, J. A., and Barnston, A. (1982). *J. Appl. Meteor.*, **21**, 139–164. (This and the next two references provide detailed analyses of FACE.)

[25] Woodley, W. L., Flueck, J. A., Biondini, R., Sax, R. I., Simpson, J., and Gagin, A. (1982). *Bull. Amer. Meteor. Soc.*, **63**, 263–276.

[26] Woodley, W. L., Barnston, A., Flueck, J. A., and Biondini, R. (1983). *J. Climate Appl. Meteor.*, **22**, 1529–1540.

(CHANGE-OVER DESIGNS
CONCOMITANT VARIABLES
DESIGN OF EXPERIMENTS
GEOSTATISTICS

HYDROLOGY, STOCHASTIC
METEOROLOGY, STATISTICS IN
RANDOMIZATION
SPATIAL DATA ANALYSIS
SPATIAL SAMPLING
TIME SERIES
WEATHER MODIFICATION—I)

K. R. GABRIEL
A. S. DENNIS
J. A. FLUECK
P. W. MIELKE

WEIERSTRASS APPROXIMATION THEOREM

If $f(x)$ is a continuous real-valued function of x for $0 \leqslant x \leqslant 1$, then, as $n \to \infty$,

$$P_n(x) = \sum_{j=0}^{n} f\left(\frac{j}{n}\right)\binom{n}{j} x^j (1-x)^{n-j} \to f(x)$$

uniformly over $0 \leqslant x \leqslant 1$. The polynomials $P_n(x)$ are Bernstein polynomials*.

(INTERPOLATION
MATHEMATICAL FUNCTIONS,
    APPROXIMATIONS TO)

WIENER–KOLMOGOROV PREDICTION THEORY

A principal question examined by Wiener [17] and Kolmogorov [10] is how to predict the unknown future values of a time series* on the basis of the known past. Suppose that the observations X_t, $t = 1, \ldots, n$, on a single time series are made at consecutive time periods, and it is desired to estimate a future observation X_{n+h}, $h \geqslant 1$. The observed time series may be viewed as a (part) realization of a stochastic process*, $\{x_t\}$, $t = 0, \pm 1, \ldots$. Wiener and Kolmogorov assumed that $\{x_t\}$ is weakly stationary* with mean 0, the infinite past $\{x_t, t \leqslant n\}$ has been observed and considered only the linear least-squares* predictors. Thus x_{n+h} is estimated by a linear function of known values

$$\hat{x}_n(h) = \sum_{j=0}^{\infty} \delta_h(j) x_{n-j},$$

where the $\delta_h(j)$ are chosen so that the mean square error of prediction

$$V(h) = E\left[\{\hat{x}_n(h) - x_{n+h}\}^2\right]$$

is minimized. In this case all one need know of the $\{x_t\}$ process to determine the prediction constants, $\delta_h(j)$, from the least-squares principle is its covariance function, $R(s) = E[x_t x_{t+s}]$, or equivalently its spectral density function

$$f(\mu) = (2\pi)^{-1} \sum_{s=-\infty}^{\infty} R(s) \exp(-is\mu).$$

Note that if $\{x_t\}$ is a Gaussian process*, then the linear least-squares predictor reduces to the unrestricted, and so possibly nonlinear, least-squares predictor. Also, although unrealistic, the assumption that the infinite past has been observed is made for convenience.

A full mathematical derivation of the Wiener–Kolmogorov prediction theory, in a Hilbert space setting, has been given by Doob [8]; see also PREDICTION AND FILTERING, LINEAR. A simplified derivation is given below by assuming that

$$\sum_{s=-\infty}^{\infty} |R(s)| < \infty, \quad f(\mu) > 0, \quad \text{all } \mu.$$

$$(1)$$

These assumptions are more restrictive than is necessary for deriving the linear least-squares predictor. They however ensure that

$$\int_{-\pi}^{\pi} \log f(\mu) \, d\mu > -\infty \quad (2)$$

so that the process is purely nondeterministic and free from deterministic components which could be predicted exactly from the infinite past. If (2) does not hold, then x_t admits the Wold decomposition and it may be written as $x_t = w_t + v_t$, where v_t is purely nondeterministic while w_t, uncorrelated with v_t, is deterministic.

When assumptions (1) hold, x_t admits the infinite moving-average* representation

$$x_t = \sum_{j=0}^{\infty} b(j) \epsilon_{t-j}, \quad b(0) = 1, \quad (3)$$

and an infinite autoregressive representation

$$\sum_{j=0}^{\infty} a(j) x_{t-j} = \epsilon_t, \quad a(0) = 1, \quad (4)$$

in which $\{\epsilon_t\}$ is a sequence of uncorrelated random variables with mean 0 and variance σ^2, say, and the $a(j)$ and $b(j)$ are absolutely summable.

If $f(\mu)$ is known exactly, then σ^2 and the $a(j)$ and $b(j)$ may be determined by the Wiener–Hopf factorization of $f(\mu)$. One sets

$$\sigma^2 = 2\pi \exp\{c(0)\},$$

$$b(j) = (2\pi)^{-1} \int_{-\pi}^{\pi} B(\mu) \exp(ij\mu) \, d\mu, \quad (5)$$

$$a(j) = (2\pi)^{-1} \int_{-\pi}^{\pi} A(\mu) \exp(ij\mu) \, d\mu,$$

where

$$B(\mu) = \exp\left\{\sum_{v=1}^{\infty} c(v) \exp(-iv\mu)\right\}$$

and

$$A(\mu) = \{B(\mu)\}^{-1}$$

give the transfer functions of the $b(j)$ and $a(j)$, respectively, and

$$c(v) = (2\pi)^{-1} \int_{-\pi}^{\pi} \log f(\mu) \exp(iv\mu) \, d\mu.$$

The $c(v)$ introduced above are known as *cepstral correlations* (Parzen [12]). They are also the parameters of an exponential model proposed by Bloomfield [6] for $f(\mu)$, with $c(v) = 0$, $|v| > p$, say. This model provides an alternative to the standard autoregressive–moving-average process*; see BOX–JENKINS MODEL, for which $f(\mu)$ is a rational function.

As in Whittle [16], let

$$\hat{x}_n(h) = \sum_{j=0}^{\infty} \delta_h(j) x_{n-j} = \sum_{j=0}^{\infty} \varphi_h(j) \epsilon_{n-j}.$$

The generating functions, $D_h(z)$ and $\Phi_h(z)$

of the $\delta_h(j)$ and $\varphi_h(j)$, respectively, are connected by the relation

$$D_h(z) = \sum_{j=0}^{\infty} \delta_h(j)z^j = B(z)\Phi_h(z), \quad (6)$$

where $B(z)$ denotes the corresponding generating function of the $b(j)$. Now,

$$
\begin{aligned}
V(h) &= E\left[\{x_{n+h} - \hat{x}_n(h)\}^2\right] \\
&= E\left[\left\{\sum_{j=0}^{h} b(j)\epsilon_{n+h-j} \right.\right. \\
&\quad \left.\left. + \sum_{j=0}^{\infty}\left(b(j+h) - \varphi_h(j)\right)\epsilon_{n-j}\right\}^2\right] \\
&= \sigma^2 \sum_{j=0}^{h} b^2(j) \\
&\quad + \sum_{j=0}^{\infty}\left\{b(j+h) - \varphi_h(j)\right\}^2 \quad (7)
\end{aligned}
$$

and attains its smallest value when $\varphi_h(j) = b(j+h)$. Thus

$$D_h(z) = \{B(z)\}^{-1}\sum_{j=0}^{\infty} b(j+h)z^j, \quad (8)$$

$$V(h) = \sigma^2 \sum_{j=0}^{h} b^2(j). \quad (9)$$

Equations (8) and (9) completely determine the linear least-squares predictor when $f(\mu)$ is known exactly. Moreover, in view of (4), we may write

$$\hat{x}_n(1) = -\sum_{j=1}^{\infty} a(j)x_{n-j},$$

$$\hat{x}_n(h) = -\sum_{j=1}^{h-1} a(j)\hat{x}_n(h-j) - \sum_{j=h}^{\infty} a(j)x_{n+h-j}. \quad (10)$$

In words, the rule for determining $\hat{x}_n(h)$ is to replace the ϵ_{n+j}, $j = 1, \ldots, h-1$, which have not yet happened by their mean value of 0, to replace the x_{n+j} $j = 1, \ldots, h-1$, which have not been observed by their optimal linear predictor $\hat{x}_n(j)$, and the x_{n-j}, $j = 0, 1, \ldots$, which have already been observed are left unchanged.

As an example, suppose that x_t follows a moving-average process of order 1,

$$x_t = \epsilon_t - \beta\epsilon_{t-1}, \quad |\beta| < 1.$$

Then

$$\hat{x}_n(1) = -\sum_{j=1}^{\infty} \beta^j x_{n+1-j},$$

$$\hat{x}_n(h) = 0, \quad h > 1,$$

and for $h > 1$ the observed past conveys no information on x_{n+h}. In general, for a moving-average process of order q, $\hat{x}_n(h) = 0$, $h > q$. Similarly, if x_t is an autoregressive process of order m, then $\hat{x}_n(h)$ depends only on the last m observed x's, alone.

Note that once an additional observation, x_{n+1}, becomes available, the $\hat{x}_n(h)$ may be updated by the formula (Box and Jenkins [7, p. 134])

$$\hat{x}_{n+1}(h) = \hat{x}_n(h+1) + b(h)\{x_{n+1} - \hat{x}_n(1)\}.$$

The basic linear least-squares prediction theory as described above has been extended in several directions. First, to the problem of linear interpolation. Suppose that the entire realization of $\{x_t\}$ has been observed except for a single value, x_0, say. The linear least-squares interpolator of x_0 can be determined by an analogous argument; the mean squared interpolation error is the harmonic mean of $f(\mu)$ whereas that of one-step prediction is the geometric mean of $f(\mu)$. Applications include the problems of estimating missing values and outlier detection in time series, and analysis of one-dimensional spatial data (Battaglia and Bhansali [2]).

Second, to the problem of predicting one time series y_t from only the past of x_t. This is the problem of Wiener filtering and includes as its special case that of estimating a signal in the presence of noise, the transfer function modeling, and estimation of a distributed lag relationship; see Bhansali and Karavellas [4].

Finally, to a continuous time process, *see* PREDICTION AND FILTERING, LINEAR.

In practice, $f(\mu)$ is invariably unknown and the prediction constants need to be estimated from an observed realization of $\{x_t\}$. A "parametric" and a "nonparametric" approach is available for estimating the Wiener–Kolmogorov predictor. In the parametric approach, $f(\mu)$ is postulated to be a known function of a finite number of unknown parameters, which are then estimated from the observed time series. A popular model is the autoregressive–moving-average process; *see* PREDICTION AND FORECASTING. In the nonparametric approach, $f(\mu)$ is explicitly recognized to be a function of an infinite number of parameters, but since only n observations are available, a model with k parameters is fitted such that $k \rightarrow \infty$ as $n \rightarrow \infty$. This approach may be implemented by fitting an autoregressive model of order k; *see* Parzen [12]. An alternative is to factorize a "window" spectral estimate by numerically adapting the procedure described earlier; *see* Bhansali [3].

When an estimated model is used for prediction, expression (9) for the mean squared error of prediction needs to be modified to allow for the effect of parameter estimation. Bloomfield [5] and Yamamoto [18], amongst others, show that the effect is to increase the mean square error of prediction by a term of order n^{-1}, and develop expressions for evaluating the increase. Another problem of considerable interest is that of model selection, i.e., which model to use for prediction. In this context, the use of a model selection criterion has received considerable attention. It is pertinent to note that a derivation of Akaike's [1] final prediction criterion, which is the forerunner of his information criterion*, is essentially based on the idea of selecting the model so as to minimize the expected one-step mean square error prediction when the parameters of each candidate model have also been estimated. For an autoregressive model, expressions for evaluating the mean square error of prediction when the order as well as the parameters have been estimated are derived by Shibata [14, 15] and when the order is unknown by Bhansali [3a]. Hannan and Nicholls [9], amongst others, consider the question of directly estimating σ^2 from (5) by replacing $f(\mu)$ by the raw, or slightly smoothed, periodogram*; this question is of interest in assessing the adequacy of a fitted model and for model selection.

The stationarity assumption made so far is not realistic from the point of view of practical applications. A standard approach is to try and transform the observed nonstationary series to one that is approximately stationary, e.g., by differencing or by estimating and removing a "trend" by linear regression, *see* PREDICTION AND FORECASTING and BOX–JENKINS MODEL. A number of widely used ad hoc forecasting procedures provide optimal linear forecasts for special cases of the former procedure, see Box and Jenkins [7].

Priestley [13] considers a class of nonstationary processes, called oscillatory processes, with time-varying spectra and develops a theory analogous to that already described for their linear least-squares prediction.

For simplicity, the discussion has so far been confined to univariate processes. An exposition of the multivariate prediction theory may be found in Whittle [16]; *see also* PREDICTION AND FILTERING, LINEAR.

References

[1] Akaike, H. (1969). *Ann. Inst. Statist. Math.*, **21**, 243–247.

[2] Battaglia, F. and Bhansali, R. J. (1987). *Biometrika*, **74**, 771–779.

[3] Bhansali, R. J. (1974). *J. R. Statist. Soc. B*, **36**, 61–73.

[3a] Bhansali, R. J. (1981). *J. Amer. Statist. Ass.*, **76**, 588–597.

[4] Bhansali, R. J. and Karavellas, D. (1983). In *Time Series in the Frequency Domain, Handbook of Statistics*, Vol. 3, D. R. Brillinger and P. R. Krishnaiah, eds. North-Holland, Amsterdam, Netherlands, pp. 1–19.

[5] Bloomfield, P. (1972). *Biometrika*, **59**, 501–508.

[6] Bloomfield, P. (1973). *Biometrika*, **60**, 217–226.

[7] Box, G. E. P. and Jenkins, G. M. (1970). *Time Series Analysis: Forecasting and Control*. Holden Day, San Francisco.

[8] Doob, J. L. (1953). *Stochastic Processes*. Wiley, New York.

[9] Hannan, E. J. and Nicholls, D. F. (1977). *J. Amer. Statist. Ass.*, **72**, 834–840.

[10] Kolmogorov, A. N. (1939). *C. R. Acad. Sci. Paris*, **208**, 2043–2045.

[11] Parzen, E. (1969). In *Multivariate Analysis—II*, P. R. Krishaniah, ed. Academic, New York, pp. 389–409.

[12] Parzen, E. (1983). In *Time Series in the Frequency Domain, Handbook of Statistics*, D. R. Brillinger and P. R. Krishnaiah, eds. North-Holland, Amsterdam, Netherlands, pp. 221–247.

[13] Priestley, M. B. (1981). *Spectral Analysis and Time Series*, Vol. 2. Academic, New York.

[14] Shibata, R. (1976). *Biometrika*, **63**, 117–126.

[15] Shibata, R. (1980). *Ann. Statist.*, **8**, 147–164.

[16] Whittle, P. (1963). *Prediction and Regulation by Linear Least Squares Methods*. English Universities Press, London, England.

[17] Wiener, N. (1949). *Extrapolation, Interpolation and Smoothing of Stationary Time Series*. Wiley, New York.

[18] Yamamoto, T. (1976). *Appl. Statist.*, **25**, 123–127.

(AKAIKE'S INFORMATION CRITERION
AUTOREGRESSIVE–INTEGRATED
   MOVING-AVERAGE (ARIMA) MODELS
AUTOREGRESSIVE–MOVING-AVERAGE
   (ARMA) MODELS
BOX–JENKINS MODEL
FORECASTING
KOLMOGOROV, ANDREI NIKOLAYEVICH
PREDICTION AND FORECASTING
STATIONARY PROCESSES
STATISTICAL MODELING
STOCHASTIC PROCESSES
WIENER, NORBERT)

R. J. BHANSALI

WOODROOFE'S INVERSION FORMULA

Let X and Y be independent, positive random variables with distribution functions F and G. Assume that only observation of $(X, Y | Y \leqslant X)$ is available. When and how may F and G be recovered from the marginal distributions F_* of $X | Y \leqslant X$ resp. G_* of $Y | Y \leqslant X$? This problem was solved by Woodroofe [3], who was concerned with the *truncated data* problem: from n independent and identically distributed (i.i.d.) replications from the conditional distribution of

(X, Y) given $Y \leqslant X$, estimate the distribution functions F and G nonparametrically.

Define $\alpha = P\{Y \leqslant X\}$ and (assuming $\alpha > 0$) the joint conditional distribution function $H_* = T(F, G)$ by

$$H_*(x, y) = P\{X \leqslant x, Y \leqslant y | Y \leqslant X\}$$

$$= \alpha^{-1} \int_0^x G(y \wedge z)\, dF(z),$$

where $y \wedge z = \min(y, z)$; then $F_*(x) = H_*(x, \infty)$ and $G_*(y) = H_*(\infty, y)$. Let F and G have supports $[a_F, b_F]$ and $[a_G, b_G]$. Assuming that there exist F and G with $\alpha > 0$ such that $H_* = T(F, G)$, F and G are unique among distributions with supports satisfying $a_G \leqslant a_F$, $b_G \leqslant b_F$, $a_G < b_F$, and given by

$$\Lambda(x) = \int_0^x dF(z)/[1 - F(z-)]$$

$$= \int_0^x dF_*(z)/C(z),$$

$$\mathrm{M}(y) = \int_y^\infty dG(z)/G(z)$$

$$= \int_y^\infty dG_*(z)/C(z),$$

with $C(z) = G_*(z) - F_*(z-)$.

Here $C(z) = P\{Y \leqslant z \leqslant X | Y \leqslant X\}$, $\Lambda(x)$ is the cumulative hazard and $\mathrm{M}(y)$ is a "backwards cumulative hazard":

$$\mathrm{M}(y) = \int_y^\infty P\{Y \leqslant z - dz | Y \leqslant z\}\, dz.$$

From the cumulative hazards, the distribution functions are recovered via the usual product–integral representation, e.g.,

$$(1 - F(x))\prod_0^x [1 - \Lambda(dx)]$$

$$\propto \prod_{\substack{z \in D \\ z \leqslant x}} [1 - \lambda(z)]\exp[-\Lambda_c(x)],$$

where D is the set of discontinuity points of F, $\lambda(z) = \Lambda(z) - \Lambda(z-)$, and Λ_c is the continuous part of Λ. For product integrals, see, e.g., Johansen [2] or Gill and Johansen [1].

References

[1] Gill, R. D. and Johansen, S. (1987). *Tech. Rep.*, Dept. of Mathematical Statistics, Center for Mathematics and Computer Science, Amsterdam, Netherlands, and Institute of Mathematical Statistics, University of Copenhagen, Denmark. (To appear in *Ann. Statist.*)

[2] Johansen, S. (1987). *CWI Newsletter*, **12**, 3–13.

[3] Woodroofe, M. (1985). *Ann. Statist.*, **13**, 163–177.

(HAZARD RATE AND OTHER CLASSIFICATIONS OF DISTRIBUTIONS SURVIVAL ANALYSIS)

NIELS KEIDING

WORSLEY-HUNTER BOUNDS

This is an improved Bonferroni *upper bound** proposed by Worsley [8] in 1982. A similar result was obtained independently by Hunter [3]. A somewhat different approach for obtaining Bonferroni-type bounds was developed by Galambos [1]. In its general formulation, the Worsley bound is given by the following result:

Let events $A_1, \ldots, A_n$ be represented as vertices $v_1, \ldots, v_n$ of a graph* G, where v_i and v_j are joined by an edge e_{ij} if and only if the events A_i and A_j are not disjoint.

Let T be a spanning tree of G. Then

$$\Pr\left[\bigcup_{i=1}^{n} A_i\right] \leqslant \sum_{i=1}^{n} \Pr(A_i)$$
$$- \sum_{(i, j): \, e_{ij} \in T} \Pr(A_i \cap A_j). \quad (1)$$

The sharpest bound is obtained by finding the tree for which the second term on the right hand side of (1) is maximum.

[Note that the upper Bonferroni bound simply states that

$$\Pr\left[\bigcup_{i=1}^{n} A_i\right] \leqslant \sum_{i=1}^{n} \Pr(A_i).]$$

If T is a tree it is always possible to find a permutation $p_1, \ldots, p_n$ of $\{1, \ldots, n\}$ so that

v_{p_i} is joined to some $v_{p_j} = v_{p_{i'}}$, say, for some $j < i$, $i = 2, \ldots, n$, and $\bigcup_{i=1}^{n} A_i$ can be written as

$$\bigcup_{i=1}^{n} A_i = A_{p_1} \cup \left(A_{p_2} \setminus A_{p_{2'}}\right) \cup \cdots$$
$$\cup \left(A_{p_n} \setminus A_{p_{n'}}\right). \quad (2)$$

Applying the Bonferroni bound to (2) we obtain (1). If $T \equiv G$, then the events on the right-hand side of (2) are disjoint and (1) becomes equality (by the well-known properties of the union operator $\cup$).

As a corollary, letting

$$T = (e_{12}, e_{23}, \ldots, e_{n-1, n})$$

we have

$$\Pr\left[\bigcup_{i=1}^{n} A_i\right] \leqslant \sum_{i=1}^{n} \Pr(A_i)$$
$$- \sum_{i=1}^{n-1} \Pr(A_i \cap A_{i+1}).$$

Kwerel's [3] refinement of the Bonferroni bound given by

$$\Pr\left[\bigcup_{i=1}^{n} A_i\right] \leqslant \sum_{i=1}^{n} \Pr(A_i)$$
$$- (2/n) \sum_{i<j} \Pr(A_i \cap A_j)$$

is a particular case of the Worsley bound, as is the more general Kounias bound [4]:

$$\Pr\left[\bigcup_{i=1}^{n} A_i\right] \leqslant \sum_{i=1}^{n} \Pr(A_i)$$
$$- \sum_{k=1}^{m} \sum_{i \in J_k \setminus j_k} \Pr(A_i \cap A_{j_k}),$$

where $(J_1, \ldots, J_m)$ is a partition of $\{1, \ldots, n\}$ and $j_k \in J_k$, $k = 1, \ldots, m$. See Worsley [8] for more details and applications. Also compare with the Galambos [1] approach.

There are enlightening discussions of the Worsley-Hunter bound in [2], [6] and, more recently [7]. In the latter paper, Seneta presents a sharpening of the bounds, and an illuminating numerical example.

References

[1] Galambos, J. (1977). *Ann. Prob.*, **5**, 577–581.

[2] Hochberg, Y. and Tamhane, A. C. (1987). *Multiple Comparison Procedures*, Wiley, New York.

[3] Hunter, D. (1976). *J. Appl. Prob.*, **13**, 597–603.

[4] Kounias, E. G. (1968). *Ann. Math. Statist.*, **39**, 2154–2158.

[5] Kwerel, S. M. (1975). *J. Amer. Statist. Ass.*, **70**, 472–479.

[6] Mărgăritescu, E. (1986). *Biom. J.*, **28**, 937–943.

[7] Seneta, E. (1988). *Austral. J. Statist.*, **30A**, 27–38.

[8] Worsley, K. J. (1982). *Biometrika*, **69**, 297–302.

(BONFERRONI INEQUALITIES AND INTERVALS
GRAPH THEORY)

Y

YULE DISTRIBUTIONS

Yule* [4] described the frequencies X of biological species in families by the distribution with probability mass function

$$f(x) = K_\rho \frac{x!\Gamma(\rho + 1)}{\Gamma(r + \rho + 2)}$$

$$\propto B(x, \rho + 1),$$

$$x = 1, 2, \ldots; \rho > 0, \qquad (1)$$

where $B(\cdot, \cdot)$ is the beta function and K_ρ is a constant, equal to ρ if ρ is a positive integer. Yule derived (1) as a compound geometric distribution* [2, pp. 245–246]. If $\rho = 1$, (1) takes the form

$$f_1(x) = [x(x + 1)]^{-1}, \qquad x = 1, 2, \ldots. \qquad (2)$$

For large values of x, $f_1(x) \cong x^{-2}$, and is thus nearly equivalent to a zeta distribution* with $\rho = 1$, an approximation that has been applied to word frequency distributions [3]. The expected value of the distribution (2) is infinite. More generally, the rth ascending factorial moment* of (1) when ρ is a positive integer is

$$\mu[s] = E[X(X + 1)(X + 2)$$

$$\ldots (X + s - 1)] = E[X[s]]$$

$$= \begin{cases} \dfrac{\rho\Gamma(\rho + 1)}{(\rho - s)(s + 1)^{[\rho - s]}}, & s < \rho - 1, \\ \infty, & s \geqslant \rho - 1. \end{cases}$$

The distribution (1) is a special case of generalized Waring distributions (*see* FACTORIAL SERIES DISTRIBUTIONS and Irwin [1]). It is a discrete analog of the Pareto distribution*. See Xekalaki and Panaretos [7] for a discussion of this interrelation. Xekalaki [5], [6] describes applications of Yule and modified Yule distributions.

References

[1] Irwin, J. O. (1975). *J. R. Statist. Soc. A*, **138**, 18–31. (See also *ibid.*, pp. 204–227 and 374–384, for a full discussion of generalized Waring distributions.)

[2] Johnson, N. L. and Kotz, S. (1969). *Distributions in Statistics: Discrete Distributions*. Wiley, New York, pp. 244–247.

[3] Simon, H. A. (1954). *Biometrika*, **42**, 425–440.

[4] Yule, G. U. (1924). *Philos. Trans. R. Soc. Lond. B*, **213**, 21–87.

[5] Xekalaki, E. (1983). *Commun. Statist. A*, **12**, 1181–1189.

[6] Xekalaki, E. (1984). *J. Econ.*, **24**, 397–403.

[7] Xekalaki, E. and Panaretos, J. (1988). *Teor. Veroyat. Primer.*, **33**, 206–210.

(FACTORIAL SERIES DISTRIBUTIONS
ZETA DISTRIBUTIONS
ZIPF'S LAW)

Z

ZELEN'S INEQUALITIES

An explicit form of the Chebyshev–Markov inequalities [1] for the case when the first four moments $(\mu_1', \mu_2, \mu_3, \mu_4)$ of the distribution of a random variable X are known, given by Zelen [2]. For a distribution with support* on $(-\infty, +\infty)$ the inequalities are (with $\sigma = \sqrt{\mu_2}$, $\sqrt{\beta_1} = \mu_3/\mu_2^{3/2}$, $\beta_2 = \mu_4/\mu_2^2$):

For $t < -\frac{1}{2}\{|\sqrt{(\beta_1 + 4)}| - \sqrt{\beta_1}\}$,

$$\Pr[X \leqslant \mu_1' + t\sigma] \leqslant A(t);$$

for $-\frac{1}{2}\{|\sqrt{(\beta_1 + 4)}| - \sqrt{\beta_1}\} \leqslant t \leqslant \frac{1}{2}\{|\sqrt{(\beta_1 + 4)}| + \sqrt{\beta_1}\}$,

$$B(t) \leqslant Pr[X \leqslant \mu_1' + t\sigma] \leqslant A(t) + B(t);$$

for $t > \frac{1}{2}\{|\sqrt{(\beta_1 + 4)}| + \sqrt{\beta_1}\}$,

$$\Pr[X \leqslant \mu_1' + t\sigma] \geqslant 1 - A(t);$$

where

$A(t)$

$$= \frac{\beta_2 - \beta_1 - 1}{(1 + t^2)(\beta_2 - \beta_1 - 1) + (t^2 - t\sqrt{\beta_1} - 1)^2},$$

$B(t)$

$$= \frac{1 + \frac{1}{2}t\{|\sqrt{(\beta_1 + 4)}| + \sqrt{\beta_1}\}}{|\sqrt{(\beta_1 + 4)}|[\frac{1}{2}\{|\sqrt{(\beta_1 + 4)}| + \sqrt{\beta_1}\} - t]}.$$

Note that when $t^2 - t\sqrt{\beta_1} - 1 = 0$ [i.e., $t = \frac{1}{2}\{-\sqrt{\beta_1} \pm |\sqrt{(\beta_1 + 4)}|\}$], $A(t) = (1 + t^2)^{-1}$.

When the range of variation (support) is not unlimited, there are more complex formulas; see ref. 2 for details.

References

[1] Shohat, J. A. and Tamarkin, J. D. (1943). *The Problem of Moments*. American Mathematical Society, New York.

[2] Zelen, M. (1954). *J. Res. Natl. Bur. Stand.*, **53**, 377–381.

(CAMP–MEIDELL INEQUALITY
CANTELLI'S INEQUALITY)
CHEBYSHEV'S INEQUALITY
PROBABILITY INEQUALITIES FOR
   RANDOM VARIABLES
WINCKLER–GAUSS INEQUALITIES)

ZERO-ORDER CORRELATION

In behavioral sciences the simple correlation between two variables is sometimes called the *zero-order correlation* (as opposed to partial correlations).

(CORRELATION)

ZETA DISTRIBUTIONS

The *zeta* or *discrete Pareto distribution** has probability mass function (PMF)

$$\Pr[X = x] = f(x) = [\zeta(\rho + 1)]^{-1} r^{-(\rho+1)},$$

$$x = 1, 2, \ldots; \rho > 0, \quad (1)$$

where $\zeta(\cdot)$ is the Riemann zeta function, defined by $\zeta(s) = \sum_{r=1}^{\infty} r^{-s}$, and tabulated in Abramowitz and Stegun [1, p. 811] for $s = 2(1)42$.

The kth moment,

$$\mu_k' = \zeta(\rho - k + 1)/\zeta(\rho + 1), \quad (2)$$

is finite only for $k < \rho$, and is infinite for $k \geqslant \rho$.

If $(X_1, \ldots, X_n)$ is an independent random sample based on (1), the maximum likelihood* estimator $\hat{\rho}$ of ρ is a solution of the equation [5]

$$n^{-1} \sum_{i=1}^{n} \log X_i$$

$$= -\frac{\partial}{\partial \rho} \{\zeta(\rho + 1)\} / \zeta(\rho + 1). \quad (3)$$

Values of the right-hand side of (3) for $0.1 \le \rho \le 4$ are tabulated in Johnson and Kotz [3, p. 242]. For $\rho > 4$,

$$\frac{\partial}{\partial \rho} \{\zeta(\rho + 1)\} / \zeta(\rho + 1)$$

$$\cong (1 + 2^{\rho+1})^{-1} \log_e 2.$$

The zeta distribution appears as the size–frequency form of Zipf's law*, and, in particular, has been used to describe word frequency in sentence construction. Nanopoulos [4] establishes a weak law of large numbers under (1). For further discussion of (1) see ref. 3.

The distribution with PMF

$$\Pr[X = x] = g(x)$$

$$= \frac{1}{(2x - 1)^\alpha} - \frac{1}{(2x + 1)^\alpha},$$

$$x = 1, 2, \ldots ; \alpha > 0, \quad (4)$$

and having mean value $(1 - 2^{-1/\alpha})\zeta(1/\alpha)$, was called a "zeta distribution" by Haight [2], but it differs from (1), and was derived as the limit of a generalized harmonic distribution* (as $Z \to \infty$) with PMF

$$\Pr[X = x] = h(x)$$

$$= (2Z)^{1/\alpha} \left\{ \left[\frac{2Z}{2x - 1} \right]^{1/\alpha} - \left[\frac{2Z}{2x + 1} \right]^{1/\alpha} \right\},$$

$$Z > 0, \quad (5)$$

where [ ] denotes the greatest integer value. Here α is usually close to 1; when $\alpha = 1$, $g(x) = 2(4x^2 - 1)^{-1}$, and $h(x)$ is proportional to the number of quantities Z/j (j being a positive integer) for which the nearest integer is x. Equation (5) was also derived to describe word frequencies (see ref. 3, p. 247), and is related to the Yule distribution*.

References

[1] Abramowitz, M. and Stegun, I. A., eds. (1964). *Handbook of Mathematical Functions*. *Appl. Math. Series No. 55*, National Bureau of Standards, Washington, D.C.

[2] Haight, F. A. (1966). *J. Math. Psychol.*, **3**, 217–233.

[3] Johnson, N. L. and Kotz, S. (1969). *Distributions in Statistics: Discrete Distributions*. Wiley, New York.

[4] Nanopoulos, P. (1977). *C. R. Acad. Sci., Paris*, A & B, **285**, 875–876.

[5] Seal, H. L. (1952). *J. Inst. Actuar.*, **78**, 115–121.

(PARETO DISTRIBUTION
YULE DISTRIBUTIONS
ZIPF'S LAW)

CUMULATIVE INDEX